The Maillard Reaction
Chemistry, Biochemistry and Implications

The Maillard Reaction
Chemistry, Biochemistry and Implications

Harry Nursten
The University of Reading, Reading, UK

advancing the chemical sciences

QZ
40
N974m
2005

Cover photograph: Micrograph of sucrose crystals in polarised light.

ISBN 0-85404-964-9

A catalogue record for this book is available from the British Library

© The Royal Society of Chemistry 2005

All rights reserved

Apart from fair dealing for the purposes of research for non-commercial purposes or for private study, criticism or review, as permitted under the Copyright, Designs and Patents Act 1988 and the Copyright and Related Rights Regulations 2003, this publication may not be reproduced, stored or transmitted, in any form or by any means, without the prior permission in writing of The Royal Society of Chemistry, or in the case of reproduction in accordance with the terms of licences issued by the Copyright Licensing Agency in the UK, or in accordance with the terms of the licences issued by the appropriate Reproduction Rights Organization outside the UK. Enquiries concerning reproduction outside the terms stated here should be sent to The Royal Society of Chemistry at the address printed on this page.

Published by The Royal Society of Chemistry,
Thomas Graham House, Science Park, Milton Road,
Cambridge CB4 0WF, UK

Registered Charity Number 207890

For further information see our web site at www.rsc.org

Typeset by Macmillan India Ltd, London, UK
Printed by Athenaeum Press Ltd, Gateshead, Tyne and Wear, UK

Preface

The Maillard reaction was given its name in honour of Louis-Camille Maillard, who first described it in 1912. It is the reaction between an amino compound, often an amino acid, peptide, or protein, and a carbonyl compound, usually a reducing sugar, such as glucose, fructose, or lactose. Since such compounds are present in virtually every cell, the ramifications of the Maillard reaction are almost boundless and understanding it is of fundamental importance to food science and to the functioning of living cells. However, it has significance also in many other areas, such as soil science, aspects of textiles, and pharmaceuticals.

Its importance is such that well attended international symposia have been devoted to it on a regular four-yearly basis since 1979. Each of these symposia has led to a book of collected papers and poster presentations. There have also been many more local symposia, particularly in Japan, but, up to the present, there is no single-author volume, summarising the Maillard reaction and inter-relating its many facets. This is what is attempted here, with the intention of ensuring that the advantages of more disciplined and uniform treatment outweigh the inevitable shortcomings.

Personally, I became aware of the Maillard reaction as early as 1947, as part of the final year course in Colour Chemistry given by E.J. Cross at the University of Leeds and my interest was reinforced by the lectures of Marc Karel at the Massachusetts Institute of Technology in 1961 on its role in food science and technology. The latest chromatographic techniques were involved at each stage of my other research work and it seemed that the separation capabilities had become such as to be able to achieve progress with unravelling the almost prohibitively complex Maillard systems. Research into aspects of the reaction therefore followed, largely through collaboration with a series of postgraduate students and postgraduate fellows, encompassing both aroma compounds and coloured compounds originating in the reaction. I am greatly indebted to each of these colleagues and to the many fellow scientists, who became my friends, over the years all over the world.

Contents

Preface			v
Abbreviations			x
Chapter	**1**	**Introduction**	**1**
	1	Categorisation	1
	2	Historical: Louis-Camille Maillard	1
	3	The Maillard Reaction	2
	4	The Literature	4
Chapter	**2**	**The Chemistry of Nonenzymic Browning**	**5**
	1	Reaction A: Sugar–Amine Condensation	5
	2	Reaction B: Amadori Rearrangement	5
	3	Reaction C: Sugar Dehydration	13
	4	Reaction D: Sugar Fragmentation	15
	5	Reaction E: Strecker Degradation	18
	6	Reaction F: Aldol Condensation	19
	7	Reaction G: Aldehyde–Amine Condensation	20
	8	Standard Melanoidin	24
	9	Melanoidins: Properties	25
	10	Reaction H: Free-Radical Reactions	26
Chapter	**3**	**Recent Advances**	**31**
	1	Introduction	31
	2	Effect of pH	31
	3	Effect of High Pressure	33
	4	Fluorescence	35
	5	Determination of α-Dicarbonyl Intermediates	36
	6	Control of Aldol/Retroaldol Reactions	36

		7	Kinetics of the Maillard Reaction	37

       7 Kinetics of the Maillard Reaction   37
       8 Effect of Glass Transition Temperature T_g 42
       9 Amines other than Amino Acids as Reactants   43
      10 Effect of Lipids   46
      11 Sites of Protein Glycation   47
      12 Effect of Oligo- and Polysaccharides   50

Chapter 4 Colour Formation in Nonenzymic Browning 52

      1 General   52
      2 Coloured Polymeric Maillard Products   57

Chapter 5 Flavour and Off-Flavour Formation in Nonenzymic Browning 62

      1 Flavour   62
      2 Volatile Compounds   62
      3 Sapid Compounds   86
      4 Summary   89

Chapter 6 Toxicological and Protective Aspects 90

      1 Toxic Products   90
      2 Carcinogenic Products   90
      3 Mutagenic Products   93
      4 Allergenicity   98
      5 Other Toxicological Aspects   98
      6 Protective Effects   99

Chapter 7 Nutritional Aspects 101

      1 Effects on Availability of Essential Amino Acids   101
      2 Effects on Ascorbic Acid and Related Compounds   101
      3 Effects on Enzyme Activity   101
      4 Interactions with Metals   102
      5 Absorption/Elimination of Amadori Compounds   103
      6 Digestion of Melanoidins   103

Chapter 8 Other Physiological Aspects 105

      1 Introduction   105
      2 Formation of Amadori Compounds *in Vivo*   105

	3	Formation of 'Advanced Glycation Endproducts' *in Vivo*	112
	4	Glycation of Phospholipids	121
	5	Glycation of Nucleic Acids and their Components	121
	6	The Role of the Maillard Reaction in the Lens	121
	7	Role of the Maillard Reaction in Nephropathy	122
	8	Role of the Maillard Reaction in Cancer	123

Chapter 9 Other Consequences of Technological Significance — **124**

1	Introduction	124
2	Effect on a_w	124
3	Effect on pH	124
4	Effect on Redox Potential	125
5	Effect on Solubility	140
6	Effect on Texture	140
7	Effect on Foamability and Foam Stability	141
8	Effect on Emulsifying Power	141
9	Volatile Formation on Storage	141
10	Binding of Volatiles	142
11	Other Losses of Functionality	142

Chapter 10 Implications for other Fields — **143**

1	Soil Science: Humic Substances	143
2	Textiles	144
3	Pharmacology	144

Chapter 11 Nonenzymic Browning Mainly Due to Ascorbic Acid — **146**

Chapter 12 Caramelisation — **150**

Chapter 13 Inhibition of Nonenzymic Browning in Foods — **152**

1	Introduction	152
2	Six Main Ways to Inhibit Nonenzymic Browning	152
3	Chemistry of the Inhibition of Nonenzymic Browning by Sulfite	154
4	Analysis	155
5	Effect of a_w	158

Chapter 14 Inhibition of the Maillard Reaction in Vivo — **161**

    1 Introduction — 161
    2 Trapping Agents — 162
    3 Intervention through Enzymes — 168
    4 Hypoglycaemic Agents — 170
    5 Summary — 171

Bibliography — **172**

    1 Books of the International Maillard Symposia — 172
    2 References — 172

Subject Index — **208**

Abbreviations

ABAP	AAPH, 2,2'-azobis(2-amidinopropane) dihydrochloride
aBTS	2,2'-Azinobis-3-ethylthiazoline-6-sulfonate
aDIBA	α,α'-Azodiisobutyramidine dihydrochloride
AGE	advanced (intermediate) glycation endproduct
ALE	advanced lipoxidation endproduct
AMP	AMRP, advanced Maillard reaction product
AOXP	antioxidative potential
AP	amadori product
AU	absorbance unit
a_w	water activity
BHA	butylated hydroxyanisole
BHT	butylated hydroxytoluene
BSA	bovine serum albumin
CAV	colour activity value
CML	N^ε-carboxymethyllysine
CP-MAS	cross-polarization-magic angle spinning
cTDA	comparative taste dilution analysis
CCR	cytochrome c oxidoreductase
DFG	N-(1-deoxyfructos-1-yl)glycine
3-DG	3-deoxyglucosone, 3-deoxyglucosulose
DH	deoxyhexosone, deoxyhexosulose
DMPD	N,N-dimethyl-p-phenylenediamine
DPC	degree of phosphate catalysis
dPPH	α,α-Diphenyl-β-picrylhydrazyl
DTPA	diethylenetriaminepentaacetic acid
EAGLE	either an advanced (intermediate) glycation or an advanced lipoxidation endproduct
EDTA	ethylenediaminetetraacetic acid
ESI	electrospray ionisation
FAB	fast atom bombardment
FL	fructosyllysine
FRAP	ferric reducing ability of plasma
GIM	glucosylisomaltol
GST	glutathione S-transferase
HAA	heterocyclic aromatic amines
HDMF	furaneol™, 4-hydroxy-2,5-dimethylfuran-3-one

HEPES	4-(2-hydroxyethyl)piperazine-1-ethylsulfonic acid
HMF	hydroxymethylfurfural
HNE	4-hydroxynonenal
IARC	International Agency for Research on Cancer
LPP	limit-peptide pigment
MALDI-TOF	matrix-assisted laser desorption/ionisation time-of-flight
MDA	malondialdehyde
ORAC	oxygen radical antioxidant capacity
PBN	phenyl N-t-butylnitrone, N-t-butyl-α-phenylnitrone
PIPES	piperazine-1,4-bis(2-ethanesulfonic acid)
SMP	skim milk powder
STZ	streptozotocin
TEAC	trolox equivalent antioxidant capacity
TBA	thiobarbituric acid
TES	N-tris(hydroxymethyl)methyl-2-aminoethanesulfonic acid
TFA	trifluoroacetic acid
TPTZ	2,4,6-tris(2-pyridyl)-s-triazine
TRAP	total radical trapping activity of plasma
UHT	ultra-high temperature
VCEAC	vitamin C equivalent antioxidant capacity

CHAPTER 1

Introduction

1 Categorisation

There are two main types of mechanism by which browning of food occurs, depending on whether the process is mediated by enzymes or not. The division is not precise and, in a specific case, it is usually difficult to rule out one or the other mechanism, unless conditions are such, for instance, during heat processing, that enzymes would have been inactivated.

Under such conditions, only nonenzymic browning can occur. Nonenzymic browning itself is subdivided roughly (again because there is an overlap) into three types of reactions.

The first, called the Maillard reaction,[1] occurs between a carbonyl compound, which here is usually a reducing sugar, and an amine, which here is usually an amino acid, a peptide, or a protein. The second is caramelisation, a reaction where the sugars react on their own, but normally requires more drastic conditions. (Some discuss this under the heading of 'active' aldehydes.) The third is ascorbic acid oxidation. The last, although it need not involve any enzyme at all, is nearest to enzymic browning, since it often does involve ascorbic acid oxidase, which, however, does not affect the phenols, which are the normal substrate in enzymic browning, but may involve other enzymes, *e.g.,* laccase or peroxidase.

Here, much attention will be given to the Maillard reaction, since one can consider caramelisation and ascorbic acid oxidation as special cases of it. Also, the Maillard reaction is the one of physiological significance.

2 Historical: Louis-Camille Maillard

Maillard (1878–1936) qualified in medicine from the University of Nancy in 1903, where he joined the Chemical Division of the School of Medicine.[2] In 1914, he became the head of a biological group in the Chemical Laboratory, University of Paris, and, in 1919, he was appointed as the Professor of Biological and Medical Chemistry at the University of Algiers.

Maillard was interested in Emil Fischer's synthesis of peptides, which he thought, correctly as it turned out, could be achieved under milder conditions by the use of glycerol.[3] This logically led on to the use of sugars as another type of polyhydroxy compound to bring about the formation of peptides and to the discovery that reducing sugars showed extra reactivity.[1] He published seven more papers on the sugar–amino acid reaction (see ref. 2). Further light has recently been shone on Maillard's career.[639]

Robert Ling (1861–1937), Lecturer in Brewing and Malting at the Sir John Cass Institute, London, had earlier noted that kilning led to amino compounds being produced from proteins, and that these reacted at 120–140 °C with sugars, such as glucose and maltose, produced simultaneously, giving what he thought were probably glucosamine-like compounds.[4]

3 The Maillard Reaction

The Maillard reaction is incredibly complex. For instance, a simple example such as the reaction of glucose with ammonia gives evidence, using simple methods, of the formation of more than 15 compounds and the reaction of glucose with glycine gives more than 24. Using HPLC and TLC on solvent-soluble material only [0.1% (w/w) of reactants], about 100 components are detectable as reaction products of xylose and glycine.[5]

In order to understand something so complex, it is necessary to draw up a simplified scheme of the reactions involved. This has been done most successfully by Hodge[6] (see Scheme 1.1). The discussion here is based on this.

Hodge subdivides the Maillard reaction as follows:

  I *Initial stage:* products colourless, without absorption in the ultraviolet (about 280 nm).
     Reaction A: Sugar–amine condensation
     Reaction B: Amadori rearrangement
 II *Intermediate stage:* products colourless or yellow, with strong absorption in the ultraviolet.
     Reaction C: Sugar dehydration
     Reaction D: Sugar fragmentation
     Reaction E: Amino acid degradation (Strecker degradation)
III *Final stage:* products highly coloured.
     Reaction F: Aldol condensation
     Reaction G: Aldehyde–amine condensation and formation of heterocyclic nitrogen compounds

It is worth noting that Mauron[7] calls the three stages Early, Advanced, and Final Maillard reactions, respectively. The way these reactions fit together is outlined in Scheme 1.1. The final products of nonenzymic browning are called melanoidins to distinguish them from the melanins produced by enzymic browning. Theoretically, the distinction is clear; however, in practice, it is very difficult to classify the dark-brown products formed in foods, since they tend to be very complex mixtures and are chemically relatively intractable.

Reaction H has been inserted into Scheme 1.1. It represents the much more recently discovered free-radical breakdown of Maillard intermediates (see Chapter 2).

Oxygen plays an essential role in enzymic browning, but is not essential for nonenzymic browning. It may help in fact, for example, in the formation of

Introduction

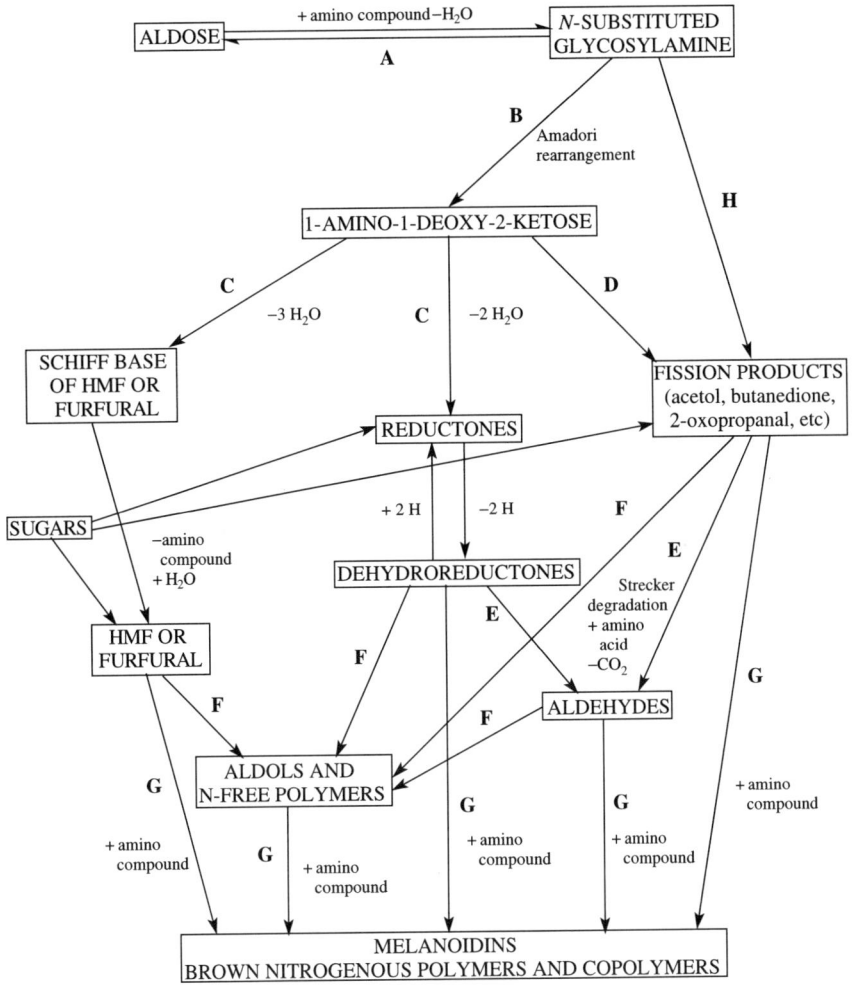

Scheme 1.1 *Nonenzymic browning (based on Hodge[6])*

reductones, such as dehydroascorbic acid, but it may also hinder the progress of the reaction, for example, in oxidising 2-oxopropanal to 2-oxopropanoic acid.

Table 1.1 lists 12 symptoms of nonenzymic browning and shows how these develop in relation to the three stages of it. Note in particular that, as far as browning itself is concerned and also off-flavour production, there is an induction period.

The Maillard reaction is exceptionally widespread. It occurs virtually ubiquitously in foodstuffs, particularly during processing at elevated temperatures (roasting, baking, extruding) or during storage for prolonged periods. It is important in the manufacture of reaction flavours, coffee, and chocolate. It occurs also in textiles. It plays a complex role in humic substances in the soil and the sea. The changes it brings about in food have both nutritional and toxicological effects. It has important medical implications, since it occurs in the body wherever there is contact between

Table 1.1 *Nonenzymic browning*

No.	Symptom	Initial	Stage Intermediate	Final
1.	Production of colour of discoloration	−	+	+++
2.	Production of flavour or off-flavour	−	+	++
3.	Production of water	+	+	+
4.	Production of carbon dioxide	?	+	?
5.	Lowering of pH	?	?	?
6.	Increasing reducing power (antioxidant activity)	+	+	+
7.	Decreasing solubility	−	−	+
8.	Loss of vitamin C activity	+	−	−
9.	Loss of biological value of protein	+	+	+
10.	Chelation of metals	−	?	+
11.	Development of toxicity	−	?	?
12.	Production of fluorescence	−	+	+

After Karel[8] and Nursten.[9]

amino compounds and reducing sugars, particularly for prolonged periods (*e.g.*, aging, cataract, diabetes, and dialysis fluid).

The reactions contributing to the Maillard reaction will be discussed further individually in the next chapter.

4 The Literature

The most important set of volumes on the Maillard reaction are the books recording the papers presented at the international Maillard symposia, held about every four years, since the first, in Uddevalla, Sweden, in 1979.[A–G] Also important are the books resulting from the European cooperation in the field of science and technical research, COST Action 919.[10–13] There are many other important books that deal with aspects of the Maillard reaction, and reference will be made to these, when papers in them are quoted. Three books specifically on the Maillard reaction are those by Baynes and Monnier,[14] Ikan,[15] and Fayle and Gerrard.[16] Important review articles are those by Reynolds,[17,18] Namiki,[19] and Ledl and Schleicher.[20]

CHAPTER 2

The Chemistry of Nonenzymic Browning

Hodge's scheme was outlined in Chapter 1, where it was shown to be composed of eight types of reactions, A to H. These will now be considered further in turn:

1 Reaction A: Sugar–Amine Condensation

This reaction can be formulated as in Scheme 2.1.

It should be noted that each of the steps in Scheme 2.1 is reversible. The amine can be a protein, and it has been shown that insulin will react with glucose at a significant rate even at room temperature.

There is no fundamental reason why the glycosylamine should not act as the amine for a further molecule of aldose, thus giving a diglycosylamine.

N-Substituted glycosylamines on mild heating give fluorescent nitrogenous compounds, which can rapidly react with glycine to give melanoidins. It seems therefore that a reaction of type G can, under certain circumstances, short-circuit Reaction B, *etc.* (*cf.* Reaction H).

Lysine locked up as ε-glycosylamine appears to be nutritionally available.[7]

2 Reaction B: Amadori Rearrangement

This reaction, which is thought to be acid-catalysed, can be depicted as in Scheme 2.2.

It is important to note that overall the Amadori rearrangement is not reversible (but see below). The reaction takes place spontaneously even at 25 °C. The mechanism depicted by Scheme 2.2 is supported by the fact that, if the hydroxy group at C-2 is blocked by, *e.g.*, a methyl group, rearrangement becomes impossible. Further support comes from the long-established facts that 11 fructosylamino acids and two difructosylamino acids have been found in stored, freeze-dried apricots and peaches and some have also been detected in dehydrated carrots, cabbage, spray-dried tomato powder, soy sauce, tea (derived from glutamic acid, theanine), beet molasses, liquorice, sterilized and condensed milk, dried skim milk and whey, infant food, roasted meat, cartilage collagen, and calf and hog liver extracts.

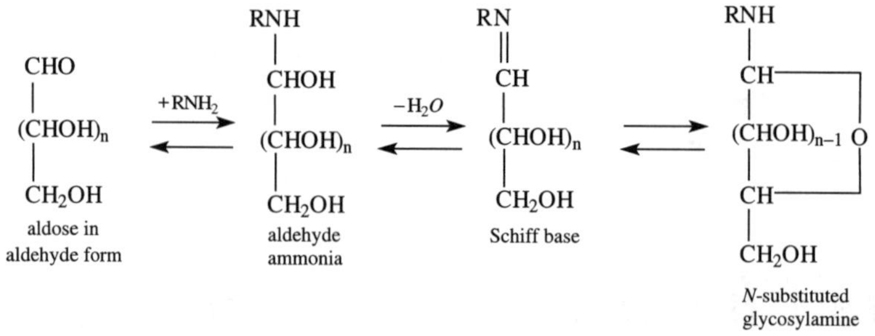

Scheme 2.1 Sugar-amine condensation to form N-substituted glycosylamine

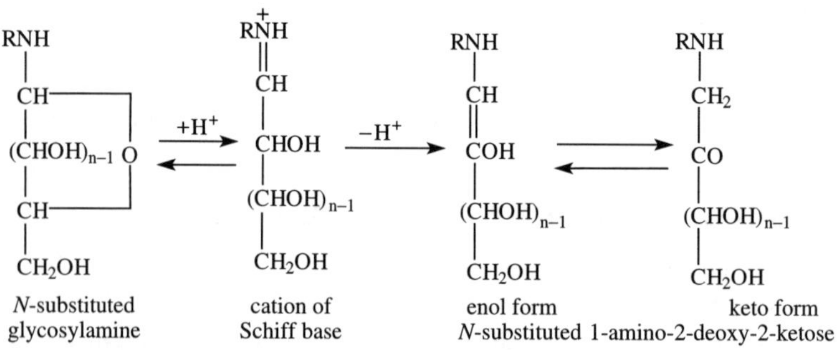

Scheme 2.2 Amadori rearrangement leading to the Amadori compound, the N-substituted 1-amino-2-deoxy-2-ketose

More recently, Eichner et al.[22] have determined up to 12 Amadori compounds in tomato powder and in dried bell and red peppers, asparagus, cauliflower, carrot, and celery, accounting for up to more than 9% dry weight (tomato powder). Fructosylpyrrolidonecarboxylic acid was among the Amadori compounds detected. It can be formed from the glutamic acid or glutamine compound, its production indicating more extensive exposure to heat. In the case of malts, dark malts contain more Amadori compounds than light ones, but none survive into very dark malts, exposed to 200 °C. Eichner et al. found eight Amadori compounds in cocoa beans (almost 100 mg total per 100 g fat-free) before roasting, implying their formation under the relatively mild conditions of fermentation and drying of the beans.

Eichner et al.[23] also showed that fructosylglycine decomposes progressively at 90 °C in a citrate buffer (pH 3.0), 3-deoxyglucosone (3-DG) reaching a maximum concentration in about 15 h, whereas hydroxymethylfurfural (HMF) increases continuously, but more slowly after 24 h. The difference between the glycine liberated and the fructosylglycine lost widens progressively, implying that glycine is involved in further reactions. Whereas the Amadori compound is decomposed to >90% in 3 days at pH 3, at pH 7, it only takes 8 h. At pH 7, HMF is not detected and 3-DG is only formed in small amounts, but more browning is evident.

Fresh garlic contains no N^α-fructosylarginine; however, in the manufacture of aged garlic extract (over 10 months or so), it increases progressively,[24] providing potent antioxidant activity. The two pairs of tetrahydro-β-carbolines, isomeric at position 1, the 1-methyl-3-carboxy and the 1-methyl-1,3-dicarboxy, are also not present in fresh garlic, but increase progressively in aged garlic extract.[25] They too are potent antioxidants.

The analysis for Amadori compounds has been reviewed by Yaylayan and Huyghues-Despointes.[26] Recently, high-performance ion-exchange chromatography coupled with pulsed amperometric detection has been advocated as a method that gives excellent resolution, yet is highly selective and sensitive for the detection of sugar without derivatisation at picomole levels with minimal clean-up.[21] For glucose and for fructosylglycine, a linear response was obtained up to 100 nmol mL^{-1}, with a detection limit of about 200 pmol mL^{-1}. Good separation of Amadori compounds from Pro, Gly, Val, Ile, and Met was demonstrated.

More recently, acid hydrolysates have been analysed by HPLC.[27] Furosine (see below) was the main product observed for dried figs and apricots, whereas furosine and the γ-aminobutyric acid derivative, in about equal amounts, were the main products for prunes and dates. 2-Furoylmethyl-γ-aminobutyric acid and -arginine were the most abundant products observed for commercial raisins, ranging from about 10 to 75 mg per 100 g of sample each. Most of the Amadori compounds present in raisins seem to have been formed during storage rather than processing.

Using their method, Blank et al.[21] examined the stability of fructosylglycine in solutions at different pH values. At 100 mM, 90 °C, 7 h, in water at pH 5, 6, 7, and 8, about 70, 31, 3, and 0% remained undegraded, whereas in 0.1 M phosphate the corresponding values were about 35, 0, 0, and 0%. After 1 h, the values had been about 98, 90, 60, and 34% and 95, 63, 30, and 24%, respectively. The degradation was clearly favoured both by higher pH and by the presence of phosphate.

Compared with the N-substituted glycosylamines, the 1-amino-1-deoxy-2-ketoses are more stable to moist acid atmospheres, but are still heat-labile and decompose rapidly in mild alkali. They exert greater reducing power, although less than reductones. They brown more easily with amino acids. Acid hydrolysis gives much compared with little HMF, but *no* hexose is recovered, in keeping with the reaction's being irreversible (however, see below).

Ketoses undergo a similar series of reactions, leading to 2-amino-2-deoxyaldoses (Heyns rearrangement). However, browning reactions of fructose differ from those of glucose, *e.g.*, loss of amino acid or of free amino groups (casein) is much lower.[28]

Thermal degradation of Amadori compounds, *e.g.*, at 250 °C for 15 min, has been investigated by Birch et al.[29] They used glucose, ^{14}C-labelled at C-1, C-2, or C-6, and β-alanine. C-2 was retained, but most of C-1 and/or C-6 was lost, depending on the product.

The pathways by which Amadori compounds lead to other products are given in more detail in Scheme 2.3.

The following should be noted:

1. The relatively stable Amadori compounds can react essentially by two routes:
   - 1,2-enolisation via 3-deoxy-1,2-dicarbonyls,
   - 2,3-enolisation via 1-deoxy-2,3-dicarbonyls,

   the choice being effected mainly through pH, a low pH favouring 1,2-enolisation and *vice versa*.

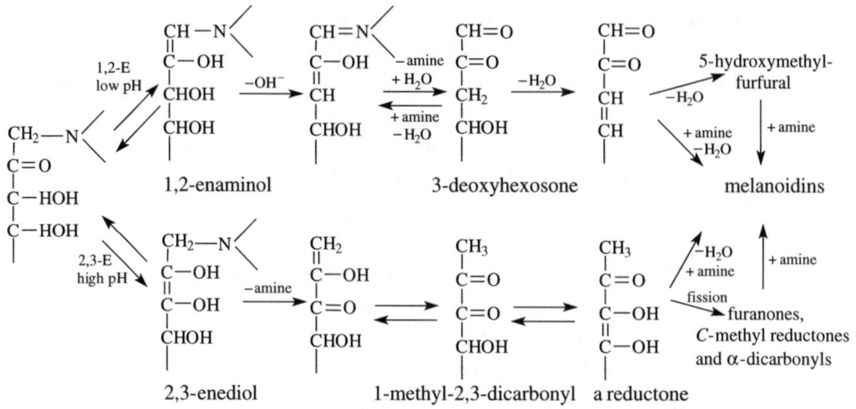

Scheme 2.3 *Maillard reactions: the two major pathways from Amadori compounds to melanoidins (based on Hodge[273])*

2. Certain key steps appear to be irreversible (but see below):
   - *N*-substituted glycosylamine → 1,2-enaminol (Scheme 2.2),
   - 1,2-enaminol → 3-deoxy-1,2-dicarbonyl,
   - loss of amine from 2,3-enediol.
3. The relation between the two competing routes within the 1,2-enolisation pathway (pyrraline:β-dicarbonyl) has been found to be 70 : 30 by means of mass spectrometry applied to [1-^{13}C and 5-^{13}C]pentose systems.[30]

Lysine locked up in ε-Amadori compounds becomes nutritionally not available; this poses an analytical problem.

2.1 Determination of Nutritionally Blocked Lysine

Since lysine is an essential amino acid, its ε-amino group has special significance. The reaction with glucose gives ε-*N*-deoxyfructosyllysine (see Scheme 2.4).

Because of the partial recovery of lysine, the interpretation of the results of the analysis is complicated. It is important to bear in mind that lysine locked up in the Amadori compound, although partially recoverable by amino acid analysis, is no longer nutritionally available.

When lactose reacts with protein in dairy products, lactulosyllysine residues are formed. On hydrolysis, these give yields of 40% recovered lysine and 32% furosine. The lysine residues nutritionally blocked can then be calculated as follows:

$$\% \text{ Lys nutritionally blocked} = \frac{3.1 \text{ furosine}}{\text{total-Lys} + 1.87 \text{ furosine}} \times 100$$

The factors are derived as follows: 100/32 = 3.1 and 60/32 = 1.87. Total-Lys represents the total lysine recovered in the analysis, *i.e.*, unreacted lysine plus lysine recovered from lactulosyllysine residues. The results obtained are presented in Table 2.1.

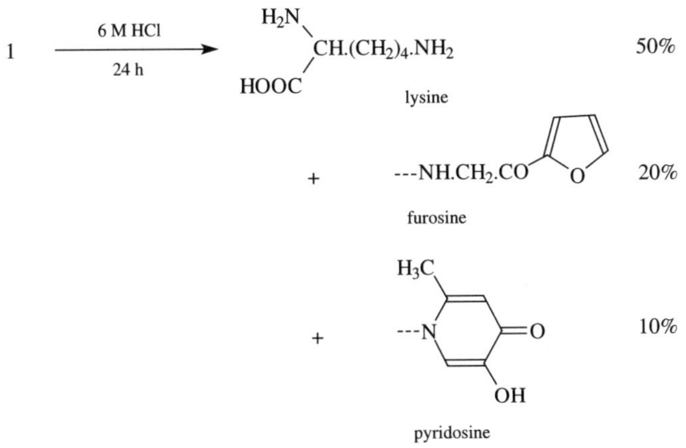

Scheme 2.4 *The products of hydrolysis of ε-N-deoxyfructosyllysine under protein-analysis conditions*

Table 2.1 *Lysine damage in good manufacturing practice*[7,31]

	%
Raw or freeze-dried milk	0
Pasteurised (74 °C, 40 s)	0–2
HTST pasteurised (135–150 °C, a few seconds)	0–3
HTST sterilized	5–10
UHT	0–2
Spray-dried powder	0–3
Sweetened condensed	0–3
Sterilised fluid	8–15
Roller-dried (without precondensation)	10–15
Evaporated	15–20
Roller-dried (conventional)	20–50

The requirement for lysine is high in infants (103 as compared with 12 mg kg^{-1} d^{-1} for adults) and, consequently, formula is recommended to contain not less than 6.7 g per 100 g protein. Cow's milk fortunately contains at least 20% more lysine than human milk,[31] but, in any case, roller drying is no longer used in Western countries to produce formula.

That the available lysine in milk-based infant formulae falls on storage and that the furosine obtained increases has been confirmed by Ferrer *et al.*[32] They showed that for an adapted infant formula, the former decreased from 9.78 to 7.85 and 7.45 g kg^{-1} sample in 24 months at 20 and 37 °C, respectively. Furosine increased correspondingly from 187 to 750 and 1001 µg kg^{-1} formula, respectively. For a follow-up infant formula, the corresponding figures were: available lysine from 12.63 to 6.62 and 6.48 g kg^{-1} and furosine from 225 to 758 and 1121 µg kg^{-1}, respectively. It appeared that the furosine increase correlated well with lysine loss over the first 12 months of storage, but less well thereafter.

Henle et al.[33] have developed an alternative method for assessing the degree of lysine modification. It is based on complete enzymic hydrolysis by means of pepsin, followed by pronase E and then by aminopeptidase M and prolidase, the resultant lactulosyllysine being determined by ion-exchange chromatography. Results with skim milk and skim milk powder, exposed to different degrees of heat, showed levels of modified lysine 2.5–3.6 times as high as indicated by furosine determination. For infant formula from a range of manufacturers, the ratios were 3.2–5.6. These results have serious, even legal, implications, which need to be addressed.

The kinetics of formation for furosine, as well as those for HMF and lactulose, was investigated as time–temperature integrators for thermal processing of milk in relation to its fat content.[34] From previous experiments, under isothermal and non-isothermal conditions, formation of the three compounds could be described as following pseudo-zero-order kinetics. Knowing the kinetic model allowed the experimental design to be simplified. The kinetics of formation of HMF and lactulose was found not to be affected by fat content; however, for furosine, significant differences were observed between the kinetic parameters for whole, semi-skimmed, and skimmed milk. Nevertheless, these differences were negligible in the context of process impact.

Furosine has also been explored as a quality indicator for foods other than dairy products.[35] Jams gave values of 72.6–629.3 mg per 100 g protein, but reduced-sugar jams (40–55% sugar, rather than 60% or more) gave lower values, 15.1–335.4 mg per 100 g protein. Higher values may thus be due to extra sugar, but also due to lower a_w, higher pH, and more severe processing. Fruit-based infant foods contained 44.0–178 mg per 100 g protein, the presence of citrus juice leading to the highest values.

Earlier, Sanz et al.[36] had found tomato pulp to give a little furosine (7.3 mg per 100 g), with only a trace of 2-furoylmethyl-γ-aminobutyric acid, but, on storage for 4 d at 50 °C and an a_w of 0.44, the corresponding amounts were 70.9 and 245.4 mg per 100 g and the amounts for the 2-furoylmethyl derivatives of Ala, Ser + Thr + Glu, and Asp + Asn had come up from 0 to 60.9, 103.5, and 300.5 mg per 100 g, respectively. 2-Furoylmethyl-γ-aminobutyric acid was also the most abundant furoylmethyl derivative obtained from commercial tomato products, ranging from 0 for whole peeled tomatoes to 87.6 mg per 100 g dry matter for double-concentrated tomato paste. Furosine was usually the second most abundant derivative obtained, ranging up to 42.8 mg per 100 g, but its amount exceeded that for 2-furoylmethyl-γ-aminobutyric acid in the two samples of tomato pulp analysed.

Furosine content can also serve as an indicator of the freshness of hen's eggs.[37] It is present in the albumen of fresh eggs at about 10 mg per 100 mg protein, but increases on storage, depending on the temperature, 40 d at 20 °C leading to a value 10 times as high, whereas the concentration in the yolk hardly changes. Maxima of 60 and 90 mg per 100 mg protein in albumen were suggested for EC Grade A-extra and Grade A eggs, respectively. Whole egg was surprisingly found to contain less furosine than expected on the basis of the separate results obtained for albumen and yolk.

Furosine did not prove very useful for the evaluation of dried pasta,[38] but 2-acetyl-3-D-glucopyranosylfuran ranged from < 1 with spaghetti dried under mild conditions to up to 20 ppm with some commercial samples.

Similarly, lysylpyrrolealdehyde increases when spaghetti are dried close to 80 °C or higher with moisture at 15% or lower ($a_w = 0.8$–0.7), commercial samples giving values of 0–40 ppm.

The Amadori compounds of alanine, lysine, and arginine, and γ-aminobutyric, pyrrolidonecarboxylic, and aspartic acid were shown to be present in freeze-dried orange juice and tomato pulp, stored at 50 °C for 14 and 11 d, respectively, through their 2-furoylmethyl derivatives.[39] In addition, the Amadori compound of proline was present in the orange juice and that of serine in the tomato pulp.

The concentration of furosine in some enteral formulae has been determined, ranging from 245 to 441 mg per 100 g protein.[40] It seems to decrease on further processing and does not appear to be a useful indicator of quality. Furosine content seems to be stable in enteral formulae on storage for >36 weeks at 4 °C, 24 weeks at 20 °C, and >12 weeks at 30 °C.[41] Stability at 55 °C was down to 3–8 weeks, depending on protein content, whereafter furosine content declined. Reactivity of lysine residues with o-phthaldialdehyde declined roughly in parallel with furosine formation.

On the other hand, furosine is suitable to assess the quality and freshness of royal jelly, the essential food for the larvae of queen bees, secreted by nurse worker bees.[42] Commercial royal jelly gave contents of 37–113 mg per 100 g, whereas samples of known origin averaged 42 mg per 100 g. A sample with 72 mg per 100 g when stored at 4 °C or room temperature for 10 months reached 100 and 501 mg per 100 g, respectively. The latter represents blockage of only 12% of lysine residues.

During investigations of heat treatment of milk-resembling models or of milk, formation of advanced (intermediate) Maillard products (AMPs) seems to take place in parallel with the ability to generate furosine,[43] but, after some time, furosine reaches a steady-state concentration whereas AMP fluorescence continues to increase, remaining correlated with lysine blockage. AMP fluorescence thus offers a rapid alternative to furosine quantification. The FAST (Fluorescence of AMP and Soluble Tryptophan) index [100 (AMP fluorescence/Trp fluorescence)], determined on the transparent soluble fraction of milk at pH 4.6, using $\lambda_{ex} = 290$ nm and $\lambda_{em} = 340$ nm for tryptophan and $\lambda_{ex} = 330$ nm and $\lambda_{em} = 420$ nm for AMP, gave a correlation with furosine concentration with $r = 0.962$.

It is possible to assess lysine availability using an *Escherichia coli* lysine auxotroph,[44] but further development is required before such a test can compete with furosine determination.

Baby cereals, containing a high proportion of maltose and glutamine, may in consequence contain glucosylisomaltol (GIM). Guerra-Hernández et al.[45] used the sample containing the least GIM (0.48 mg kg^{-1}) for spiking and determining the recoveries, which were 95–100% (mean 96.9%). The detection limit was 0.14 mg kg^{-1}. For bread, GIM correlated with baking time ($r^2 = 0.682$) and HMF ($r^2 = 0.999$), but baby cereals often contain dehydrated fruits and caramel, which give rise to HMF themselves, interfering with the use of HMF as an indicator of browning. The furosine content is very low, but increases greatly in the drying process, for which furosine is therefore a sensitive indicator. However, during storage, conditions are not favourable for furosine formation, making GIM the more useful indicator, increasing in 4 weeks at 55 °C from 0.48 to 7.69 mg kg^{-1}. GIM in baked bread, heated at 190 °C for 30 min, increased from not detected to 20.9 mg kg^{-1}.

Furosine formation (on acid treatment) was used as a marker for the presence of Amadori compounds in model systems of lysine (0.1 M) or BSA (10 mg L^{-1}) with glucose (0.2 M) for 7 d under physiological conditions of pH and temperature.[46] Addition of Cu^{2+} and Fe^{3+} significantly decreased furosine formation, but promoted browning (A_{420}), whereas EDTA and DTPA (diethylenetriaminepentaacetic acid) almost completely reversed the effects. The results suggest that metal catalysis converts Amadori compounds into AGEs, whereas oxygen does not affect furosine formation, but markedly suppresses browning. The formation of furosine, chromophores, and fluorophores ($\lambda_{ex} = 370$ nm/$\lambda_{em} = 440$ nm) increased with increasing concentrations of phosphate and carbonate buffers, but not with HEPES, TES, and PIPES buffers. Incubation of Amadori compounds in phosphate and carbonate buffers did not show any differences in browning and fluorescence intensity from incubation in the other buffers. This suggests that phosphate and carbonate accelerate the formation of Amadori compounds, i.e., the early stages of the Maillard reaction.

In whey protein isolate, some of the β-lactoglobulin is glycated, some even triply, as shown by electrospray ionisation (ESI) mass spectroscopy.[47] Attachment of lactose to purified β-lactoglobulin, which still has 35% of the singly glycated protein as the sole contaminant, was investigated under different conditions: (a) 'solubilised' by reaction in an aqueous system, pH 7.2, (b) 'dry-way' simply at 65 °C, and (c) 'combined' with 'dry-way' at 50 °C for 96 h, followed by 'solubilised' at 50 °C for up to 4 d. Maximum efficiency of attachment was found with 'dry-way' at 65 °C, the highest temperature used, for 3 h, when 4–11 lactose molecules had been attached, 22.6% (the largest proportion) of the ESI signal being contributed by the form with seven lactose units.

Skim milk has necessarily to be exposed to heat during the manufacture of skim milk powder (SMP). Even standard low-heat SMP on capillary zone electrophoresis[48] demonstrates that α-lactalbumin and α_{s1}-casein give rise to three peaks each, each variant of β-casein A to 2–3 peaks, κ-casein to two peaks, and β-lactoglobulin to seven peaks, the additional peaks being attributed to glycation. The concentration step (45 °C under vacuum) produced some lactosylation only occasionally and it was the spray-drying step which was critical. The inlet temperature (T_{in}) had a positive linear effect, whereas the effect of the outlet temperature (T_{out}) was positive, but quadratic, $T_{in}(T_{out})^2$, explaining about 70% of the variation in the response. This led to setting a low T_{out} with T_{in} at the best compromise between low for minimising lactosylation, yet high enough to give a rapid drying rate. With T_{in} at 173 °C and T_{out} at 75 °C, the resulting e-gram remained close to that for the skim milk. However, such SMP still had a water content > 5 % (w/w), whereas 4 % or below is required to prevent bacterial growth and quality loss, such as from lactose changes. Freeze-drying lowered moisture < 2.5%, without change in the e-gram.

Such SMP stored at 52 °C developed a brown colour and caramel odour in only 3.5 weeks, no peaks being evident on the e-gram. Even with storage at 37 °C, both β-lactoglobulin and β-casein suffered more modification than at lower temperatures, starting at week 2.

Caseinoglycomacropeptide is split off from κ-casein during renneting and thus is present in cheese whey, where it is subject to reaction with lactose. Moreno et al.[49] have examined ovine macropeptide and found, by LC-ESIMS, that it was about 60%

lactosylated on reaction with lactose for 11 d at 40 °C under 44% r.h., with 29% mono-, 18% di-, 8% tri-, and 6% tetra-lactosylated, there being three Lys residues plus Met at the N-terminus. Amino acid analysis showed both Lys and Met to decrease progressively, whereas furosine increased only up to day 9, indicating that the Amadori compounds were being degraded to undetermined derivatives faster than they were being formed at that stage. Lactosylation improved emulsifying activity slightly, but left the high solubility and heat stability of the macropeptide unchanged.

2.2 Reversibility of the Amadori Rearrangement

Although, generally speaking, the Amadori rearrangement is irreversible, recent evidence shows that this is not entirely true.

When 15 mM N^α-formyl-N^ε-fructosyl-lysine (fFL) in 0.2 M phosphate buffer, pH 7.4, was incubated in air for 15 d at 37 °C, not only was CML formed, but a substantial amount of lysine was recovered, which N. Ahmed et al.[50] and M.U. Ahmed et al.[51] interpreted as due to reversal of the Amadori rearrangement. The reduction of the reaction mixture with NaBH$_4$ and subsequent identification of both glucitol and mannitol supports the interpretation.

Davidek et al.[52] have studied the degradation of N-(1-deoxyfructos-1-yl)glycine (DFG) at 90 °C with pH remaining constant (pH stat) at 5, 6, 7, or 8 for up to 7 h, in water or 0.1 M phosphate buffer. The rate of release of hexose (glucose and mannose in a constant ratio of 7 : 3) increased pH 5 to 7 in the absence of phosphate. Phosphate accelerated the reaction, leading to an 18% yield at most, so the Amadori rearrangement can still be regarded as essentially irreversible. The rate of degradation of DFG increased with pH, the process being accelerated by the presence of phosphate, particularly at pH 5–7. Degradation of 25% was reached in 360, 135, 40, and 20 min in water at pH 5, 6, 7, and 8, respectively. The rate of liberation of glycine also increased with pH, most notably in presence of phosphate at pH 6–8, but the yield decreased as the reaction reached an advanced stage at all pH values, both in water and in buffer. The concentration of acetic and formic acids was also followed, their formation increasing with pH. Except at pH 8, the total amount of the two acids remained generally below the amount of sodium hydroxide added, but at pH 8, the amount exceeded it. The highest amounts of the acids formed were 0.64 and 0.14 mol mol^{-1} DFG, respectively.

3 Reaction C: Sugar Dehydration

Sugar dehydration occurs in two ways: under acid conditions, furfurals are produced, whereas in neutral or alkaline conditions and/or in the presence of amines in nearly anhydrous systems, 6-carbon and other reductones are favoured. The two pathways are illustrated in Scheme 2.3, where it can be seen that the first proceeds via 1,2-enolisation and the 3-deoxyosone, whereas the second does so via 2,3-enolisation and the 1-deoxy-2,3-dicarbonyl.

1. *Furfural formation*: Various compounds can accelerate furfural formation; for example, glycine accelerates both the conversion of xylose to furfural and

glucose to HMF. The reason seems to be that the Amadori product dehydrates more readily than the original aldose or *N*-substituted glycosylamine, giving the Schiff base of furfural, which is then hydrolysed, re-liberating part of the amine, but also condensing to melanoidins. It is generally thought that HMF is of low browning potential and does not lie on the main pathway to melanoidins.

Furfural formation is often used as a relatively simple means of following the deterioration of food during storage. Thus, Tosun and Ustun[53] showed that the HMF level in Turkish white hard grape pekmez increased progressively from 9 to 13 mg kg^{-1} on storage for 8 weeks.

2. *Reductone formation*: Reductones can be thought of as products formed from sugars by the loss of only two molecules of water, as compared with the loss of three that leads to furfurals. Reductones are compounds which contain the group $-C(OH):C(OH)-$, as in ascorbic acid, and a hexose can readily be converted on paper into the vinylogue of a reductone. Compounds such as reductones explain the reducing power that develops during browning, but they take part in browning in the dehydro form and, therefore, need oxygen to be converted into it. Similar to furfurals, they brown more readily in the presence of amines.

Feather,[54] in particular, has carried out some of these reactions in D_2O or tritiated (radioactive) water, to study the extent of H exchange:

D-glucose $\xrightarrow{\text{strong acid}\ (3\ M\ HCl)}$ HMF (no incorporation of D into C–H bonds

D-xylose $\xrightarrow{\text{strong acid}\ (3\ M\ HCl)}$ furfural no incorporation of T into C–H bonds

Amadori compounds from glucose $\xrightarrow{1\ M\ HCl}$ no incorporation at C-3, but some at times at C-1 (aldehydic C–H)

$\xrightarrow{2\ M\ HOAc}$ strong incorporation at both C-3 and C-1

It therefore seems that Amadori compounds can undergo 1,2-enolisation much more easily and under milder acid conditions than the original sugar.

Use of 1-^{14}C has shown that the methyl group is derived from C-1 of hexuronic acid or pentose.[54] If prepared in D_2O, NMR shows hydrogen exchange from both C-2 and CH_3. This accords with $CH_2:C(OH)$... intermediates (on the 2,3-enolisation route).

Maltol (see Scheme 2.5) can be formed from hexoses by the loss of three molecules of water, but is more readily formed from disaccharides, such as maltose. The precursor of maltol, the 5,6-dihydro-5-hydroxy compound, has lost one molecule of water less. Its concentration increases linearly over the range of medium roasting conditions (5–15 min at 130 °C) used commercially for cocoa beans,[55] but then tends to level off. Dimethylpyrazines, on the other hand, show a sharp increase as the beans become overroasted.

Disaccharides give quite a range of specific products. Pischetsrieder et al.[56] have examined the reaction of lactose with N^α-acetyllysine in phosphate buffer (1.28 M, pH 7.0). At 37 °C, the aminoreductone, $CH(NHR){:}C(OH).CO.CH_2.HCOH.CH_2OH$, was present after 5 d, other compounds being present in traces only. At 70 °C, it reached a maximum at 6 h, but was present still after 9 h. At 100 °C, it reached a maximum at 1.5 h, but could no longer be detected after 4 h, other compounds dominating. The Amadori compound was thought to lose galactose, leading to the 4-deoxy compound, which isomerised to the aminoreductone.

4 Reaction D: Sugar Fragmentation

The mechanism by which sugar fragmentation occurs is accepted to be principally retroaldolisation (dealdolisation), although oxidative fission is also thought to play a role.[57] Vinylogous retroaldolisation can take place in appropriate intermediates. It should be borne in mind that retroaldolisation is an important part of the EMP (Embden–Meyerhof–Parnas) glycolytic pathway, where fructose-1,6-diphosphate is split into dihydroxyacetone-phosphate and glyceraldehyde-3-phosphate.

The cleavage of hexose derivatives can be C_5/C_1, C_4/C_2, or C_3/C_3.

The sorts of reactions which occur are illustrated in Scheme 2.5. Fragments which retain the α-hydroxymethylcarbonyl grouping will undergo browning alone in aqueous solution, but this will be greatly accelerated by the presence of amines. The relative reactivities of α-hydroxymethylcarbonyl compounds and other sugar fragments are given in decreasing order in Table 2.2.

Because α-dicarbonyl compounds are particularly reactive, Weenen and Apeldoorn[57] specifically looked for these compounds by means of derivatisation with o-diaminobenzene among the butanol-soluble fragmentation products formed in 15 systems (glucose, fructose, xylose, 3-deoxyglucosone, or fructosylalanine without amine or with alanine or cyclohexylamine; 1 h, 100 °C, phosphate buffer, pH 8). Four α-dicarbonyls were obtained: glyoxal, 2-oxopropanal, butanedione, and 2,3-pentanedione.

In the absence of added alanine, the Amadori compound gave the first three in about equal amounts (ca 100 µg of each), but only ca 20 µg of 2,3-pentanedione. The four carbohydrates gave glyoxal and 2-oxopropanal only. This includes 3-deoxyglucosone, which also gave only a relatively poor yield of glyoxal, but somewhat more 2-oxopropanal, overall indicating rather modest reactivity.

In the presence of alanine, all four α-dicarbonyls are formed in each system, glyoxal being produced in largest amounts by glucose and fructose and 2-oxopropanal by xylose and 3-deoxyglucosone. Fructosylalanine gave lower amounts of α-dicarbonyls (except 2,3-pentanedione) in the presence of alanine than in its absence.

In the presence of cyclohexylamine, the yields were much higher in all cases, up to times 40, for 2-oxopropanal from fructose (1.10 mg), except for 2,3-pentanedione from glucose and 3-deoxyglucosone, where the increase was less than one-fifth. The largest yields were from xylose, fructose, and glucose, 2.23, 2.15, and 1.75 mg, respectively, each representing less than 0.1% of the carbohydrate used.

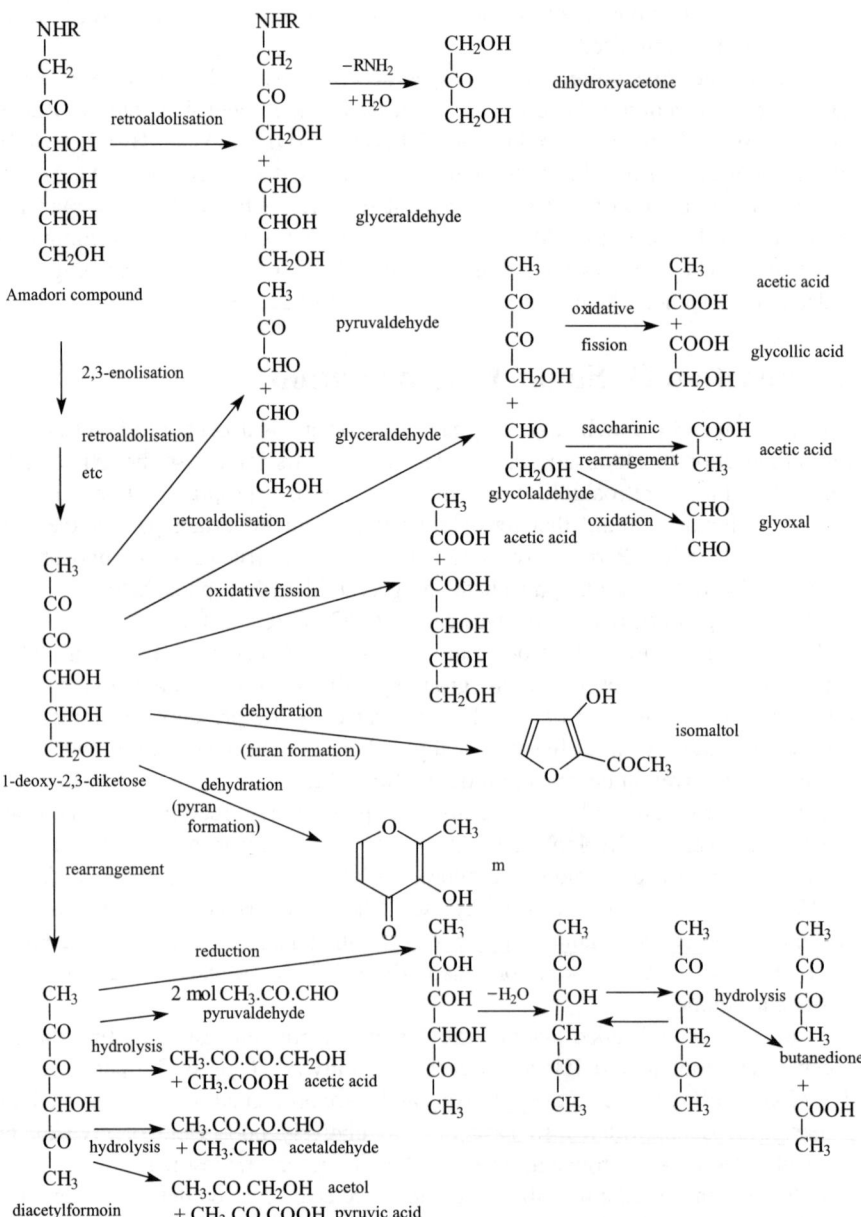

Scheme 2.5 *Examples of sugar fragmentation*

Because of the importance of dicarbonyl compounds in the Maillard reaction, Meade and Gerrard[58] are trying to elucidate their structure–activity relationships, using linear and cyclic dicarbonyls and RNase A, crosslinking being assessed by SDS PAGE.

The Chemistry of Nonenzymic Browning

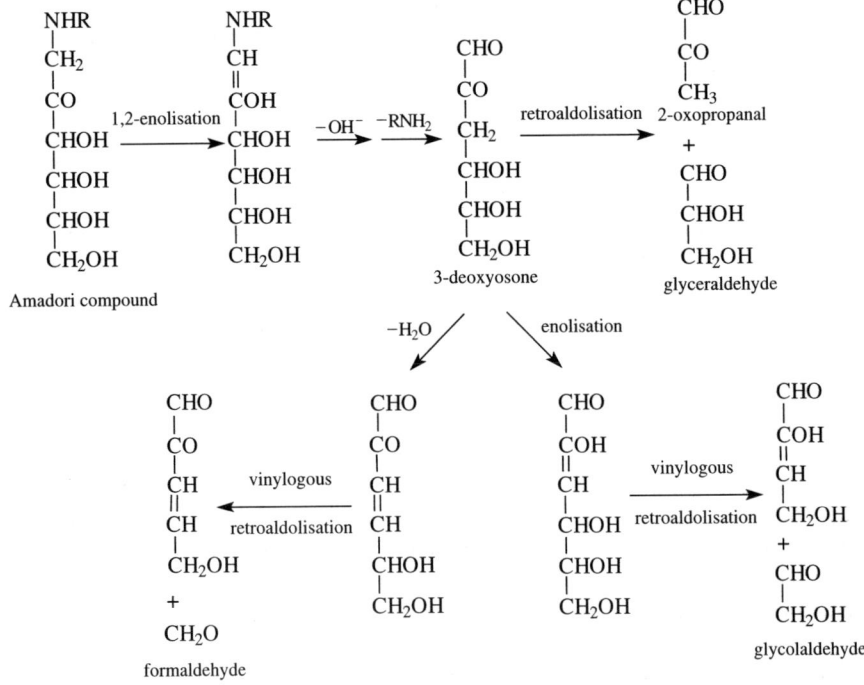

Scheme 2.5 (*Cont.*)

Table 2.2 Sugar fragmentation products in decreasing order of reactivity

Sugar fragmentation product	Formula	Reactivity
Glycolaldehyde	$CH_2OH.CHO$	Highest
Glyceraldehyde	$CH_2OH.CHOH.CHO$	
2-Oxopropanal	$CH_3.CO.CHO$	
Acetol	$CH_3.CO.CH_2OH$	
Dihydroxyacetone	$CH_2OH.CO.CH_2OH$	
Acetoin	$CH_3.CHOH.CO.CH_3$	
Butanedione	$CH_3.CO.CO.CH_3$	
Ethanal	$CH_3.CHO$	Slightly lower
Aldol	$CH_3.CHOH.CH_2.CHO$	Still lower
Propanal	$CH_3.CH_2.CHO$	Very low
Pyruvic acid	$CH_3.CO.COOH$	Even lower
Levulinic acid	$CH_3.CO.CH_2.CH_2COOH$	
Saccharinic acid	$CH_2OH.(CHOH)_2.CH_2.CHOH.COOH$ (for example)	Not at all
Lactic acid	$CH_3.CHOH.COOH$	
Acetic acid	$CH_3.COOH$	
Formic acid	$H.COOH$	
Formaldehyde	$H.CHO$	Inhibits

β-Hydroxyamino acids under pyrolysis conditions (250 °C) can give rise to α-hydroxyaldehydes and α-aminoalcohols, which can enrich the pool of sugar fragmentation products.[59]

Glyoxal can be formed by oxidation of glycolaldehyde (*e.g.*, in Scheme 2.5), but it can also be formed by autoxidation of unsaturated fats and by enzymic degradation of serine.[60] 2-Oxopropanal can be obtained by retroaldolisation of 1- and 3-deoxyglucosone or by hydrolysis of diacetylformoin (see Scheme 2.5). Butanedione can also be derived from diacetylformoin, but by reduction, dehydration, and hydrolysis (see Scheme 2.5). 2,3-Pentanedione can be formed from butanedione by aldol reaction with formaldehyde, dehydration, and reduction or by aldol condensation of hydroxyacetone and acetaldehyde, followed by dehydration.

The reactivity of glyceraldehyde is illustrated by capillary electropherograms of its interaction with N-α-acetyllysine at 37 °C in phosphate-buffered saline, pH 7.2,[61] which showed one main peak and three others, as well as much unresolved material.

5 Reaction E: Strecker Degradation

This is a reaction of α-amino acids, in which they are oxidised to the corresponding aldehyde, giving off carbon dioxide, and ammonia is transferred to other components of the system, very little being liberated as such. The reaction is initiated by compounds, such as α-dicarbonyl compounds and their vinylogues, or compounds which can give rise to them readily, such as reductones by dehydrogenation or imino analogues by hydrolysis. The reaction may therefore be represented as follows:

$$\text{R.CHNH}_2\text{.COOH} + \text{R}'\text{.CO.CO.R}'' \longrightarrow \text{R.CHO} + \text{CO}_2 + \text{R}'\text{.CHNH}_2\text{.CO.R}''$$

Studies with radioactive carbon have shown that well over 80% of the carbon dioxide liberated in the Maillard reaction does indeed originate from the amino acid, somewhat less than 10% coming from uniformly labelled glucose.[62]

The reaction is named after Strecker,[63] who reacted α-amino acids with alloxan, obtaining CO_2 and the purple murexide. The reaction was reviewed by Schönberg and Moubacher,[64] its contribution to food flavour being covered by Rizzi.[65]

The Strecker degradation enters browning reactions in two ways: on the one hand, the aldehyde formed can take part in aldol condensation leading to nitrogen-free polymers or it can react with amino compounds to give melanoidins via aldimines. However, this is not thought to be a major colour-producing reaction, because glycine can at times give more browning with sugars than alanine, yet glycine produces formaldehyde by the Strecker degradation, which has a negative effect on browning. Amino acids other than the α-ones can also produce melanoidins, but usually cannot undergo the Strecker degradation. On the other hand, the dehydroreductones, derived from Amadori products by dehydration and dehydrogenation, or the dicarbonyl fission products, pick up the nitrogen from the amino acid and go on to form melanoidins.

Dehydroascorbic acid (but see later) and quinones, such as vitamin K and those formed enzymically from polyphenols, can also act as dicarbonyl compound in the reaction.

It is known that in fish, two less commonly occurring amino acids are major contributors to the browning reaction and to the loss of free sugar, *i.e.*, anserine, a dipeptide, β-alanyl-L-methylhistidine, and taurine, $H_3N^+.CH_2.CH_2.SO_3^-$. Discoloration of fish at 0% r.h. in sunlight has been shown to be due to 1-methylhistidine without the involvement of sugars. Anserine is also prominent in chicken muscle, whereas carnosine, β-alanylhistidine, is prominent in beef extract and balenine, β-alanyl-3-methylhistidine, is characteristic of whale (except sperm whale) muscle.

Being volatile, the aldehydes formed in the Strecker degradation have often been thought to be important contributors to the aroma of foodstuffs and many patents have been granted which use the Strecker degradation to produce flavouring materials of various types, such as, maple, chocolate, coffee, tea, honey, mushroom, and bread.[66]

Strecker aldehydes can react with sugar-derived Maillard intermediates. Thus, Blank et al.,[67] using ^{13}C-labelling, have shown that, when xylose and glycine/alanine interact in phosphate buffer (0.2 M, pH 7, 90 °C, 1 h), norfuraneol constitutes more than 99% of the total 3(2H)-furanones formed. In the absence of an amino acid, the proportion falls a little (to 93%), but the total amount is more than 100 times lower. Furaneol (HDMF) and homofuraneol were detected in all samples. With labelled glycine in the system, only 45% of HDMF was labelled, indicating that the C_1 fragment can be derived from carbohydrate as well as by the Strecker degradation. With labelled alanine in the system, none of the HDMF was labelled, but 85% of the homofuraneol were. Use of labelled xylose demonstrated that norfuraneol was entirely carbohydrate-derived.

Lysine residues can be deaminated to the corresponding aldehyde (allysine) under oxidative conditions. The oxidant can be certain α-dicarbonyl compounds[68] and the reaction thus has some resemblance to the Strecker degradation. However, cupric ions are essential, and, although glucose, 3-DG, and 2-oxopropanal are effective, glyoxal itself is not. The reaction occurs under physiological conditions (pH 7.4, 37 °C) and, on hydrolysis, the blood plasma of STZ-induced diabetic rats exhibited levels of allysine more than three times as high as that of normal rats.

6 Reaction F: Aldol Condensation

Aldehydes can arise by Reactions C, D, and E, and they can then react with each other by the aldol condensation. Amines (and particularly their salts), including peptones and egg albumin, are effective catalysts. Additional carbonyl compounds which can participate in the condensation may be derived by the oxidation of lipids.

Browning has been demonstrated for 2-oxopropanal alone, furfurals, and 2-oxopropanoate plus furfural. Less browning occurs with sugars or with aldol itself.

Butanedione can undergo a similar reaction, which can lead to 2,5-dimethylbenzoquinone (Scheme 2.6).

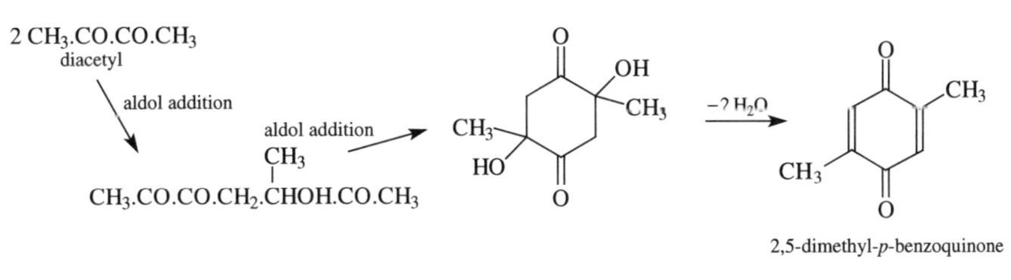

Scheme 2.6 *Formation of quinones from sugar fragmentation products*

Benzoquinones can act as dicarbonyl components of the Strecker reaction, readily form imines in the cold, and can be involved in the production of melanoidins. Benzoquinones in plant systems are usually derived from polyphenols and constitute well known intermediates in enzymic browning. They are also involved in the formation of melanin in animals, including humans.

7 Reaction G: Aldehyde–Amine Condensation

Aldehydes, particularly α,β-unsaturated ones, react readily at low temperatures with amines to give "polymeric" high molecular mass, coloured products of unknown structure, called melanoidins. Heterocyclic ring systems, such as pyridines, pyrazines, pyrroles, and imidazoles, have been shown to be present. Melanoidins usually contain 3–4% nitrogen.

The constitution of the melanoidins differs somewhat, depending on how they have been produced, for example:

furfural + glycine	high ether content
glucose + glycine	high alcoholic hydroxyl content
2-oxopropanal + glycine	high enolic hydroxyl and low ether content

As the condensations proceed, higher molecular mass products are formed, long-period products being non-dialysable.

Some work by Benzing-Purdie et al.[69] bears on this, as can be seen from Table 2.3. The following comments are worth making:

1. In relation to high-molecular-mass material, temperature seems much more important than time. The *proportion* of high-molecular-mass material increased with temperature.
2. Loss of material, presumably H_2O and CO_2, is much greater at 100 °C (54%) than at 68 °C (30%).
3. The composition of the retentate formed at 22 °C corresponds quite closely to the loss of 3 mol of H_2O: C, 49.1; H, 5.3; N, 8.2%, *i.e.*, all N appears to be retained. N is lost subsequently, some presumably as volatiles.
4. It is not clear to what extent, if any, these high-molecular-mass materials are coloured.

Table 2.3 *Composition of retentate of melanoidins from xylose and glycine (molar ratio 1 : 1)*[69]

Reaction Conditions		Yield (%)[a]		Microanalytical Data (%)		
Temperture (°C)	Time	Retentate	Diffusate	C	H	N
22	9 months	4.2	—	50.3	5.3	8.0
68	6 weeks	39.6	30.0	57.0	6.1	7.3
100	38 h	30.7	15.8	57.8	5.4	6.7

[a]Yield based on total reactants, membrane cut-off, 12 kDa.

Table 2.4 *Composition of melanoidins obtained from xylose and glycine (molar ratio 1:1) at 68 °C*[70]

Time (d)	Yield (%)[a]		Microanalytical data (%)		
	Retentate	Diffusate	C	H	N
3	4.4	58.7	51.7	5.5	7.7
7	—	—	53.4	5.2	7.3
42[b]	33.3	—	55.5	5.3	6.8
70[b]	—	—	56.2	5.5	6.9

[a]Yield based on total reactants, membrane cut-off, 12 kDa.
[b]Filtered, not dialysed.

The effect of time can be seen from Table 2.4,[70] the result for 42 d at 68 °C being close to that in Table 2.3. Here, some indication of colour was given, the 3 d melanoidin being light-brown, whereas the 42 d one was dark-brown. Whilst the infrared spectra of the melanoidins showed no significant differences, the cross-polarization-magic angle spinning (CP-MAS) ^{13}C-NMR spectra did. The main conclusions were: (i) with time the melanoidins acquired more unsaturated C, as well as more total carbonyl and carbon; (ii) the use of ^{15}N-glycine and CP-MAS ^{15}N-NMR showed that pyrrole moieties increased also and that they probably did not contain N–H bonds; and (iii) the yield of melanoidin increased asymptotically, reaching its maximum after about 45 d, when all the xylose had been consumed.

In earlier work, Olsson et al.[71] arrived at somewhat different conclusions, having found the ^{13}C-NMR spectrum of high-molecular-mass water-soluble material from glucose and glycine to be very similar to that of the corresponding Amadori compound without providing any evidence of unsaturated or aromatic carbon atoms. The material is very difficult to hydrolyse, suggesting that the glucose units are linked by C–C bonds.

Fogliano et al.[72] have prepared melanoidin [molecular mass (MM) > 12 kDa] from glucose–glycine, lactose–lysine, and lactose–N-acetyllysine; lactose–lysine proved to be the odd one out, exhibiting less than one-tenth the absorbance of the other two at 460 and 520 nm. Their CAVs (colour activity value,[73] defined as the ratio of a colorant's concentration to its visual detection threshold, both in µg kg^{-1}) were 8000, 200, and 4000, respectively. Examination by MALDI–TOF gave results only for the lactose–N-acetyllysine melanoidin, with peaks around 6, 12, and 24 kDa (the spectrum reproduced by Borrelli et al.[74] shows peaks around 6, 7.5, and 12 kDa). The antioxidant effects of the melanoidins were also examined (see p. 129).

Kato and Hayase[75] isolated a blue pigment (Blue-M1) from xylose–glycine kept in 60% ethanol (starting pH 8.1) under nitrogen for 48 h at 26.5 °C or 96 h at 2 °C. Its Structure **30** (see Chapter 4, p. 54) was determined using ^{13}C- and ^{1}H-NMR (500 MHz), as well as FAB-MS; its formula is $C_{27}H_{31}N_4O_{13}$. It was considered to be formed from 4 mol of Amadori compound by loss of 9 mol of H_2O and 1 mol each of H_2 and CO_2. It has a large peak at 625 nm and small peaks at 238, 322, and 365 nm. It fluoresces, $\lambda_{\text{ex max}} = 349$, $\lambda_{\text{em max}} = 445$ nm. Yellow and red pigments were also formed and it was proposed that the blue pigment was derived from the interaction of two yellow pigments, formed from pyrrole-2-aldehyde and the Schiff base from xylosone and 3-deoxyxylosone, respectively.

Blue-M1 showed some resemblance to a melanoidin prepared from xylose and n-butylamine neutralized with acetic acid in methanol at 50 °C for 7 d, based on sugar moiety per N, residual OH content, periodate consumption, dehydration ratio, and dehydrogenation ratio. This led Kato and Hayase to suggest Blue-M1 as a key intermediate in melanoidin formation.

Tressl et al.[76,77] have focused on 2-hydroxymethylpyrroles as melanoidin precursors, because of their remarkable reactivity. In their study of the ready browning of free or nucleic acid-bound 2-deoxy-D-ribose with compounds, such as 4-aminobutyric acid,[78] they isolated and identified 1, which they interpreted as the result of 2-hydroxymethylpyrrole being trapped by the 4-aminobutyric acid. Accordingly, they synthesised N-methyl-2-hydroxymethylpyrrole, but found it difficult to purify, because it polymerised spontaneously and exothermally to a colourless solid, which, on exposure to air, became pink and then dark red. Catalytic amounts of HCl gave a similar result. The macromolecular product had λ_{max} = 247 and 290, with a broad shoulder at 350–550 nm (chloroform). It fluoresced with λ_{ex} = 385, λ_{em} = 493 nm (chloroform). Mainly by the use of MALDI–TOF–MS, linear oligomers (or the corresponding dehydro-oligomers) of up to a dodecamer were identified. The results were supported by complementary experiments starting with N-methyl-2-hydroxy-[^{13}C]methylpyrrole (see Reaction Scheme 2.7).

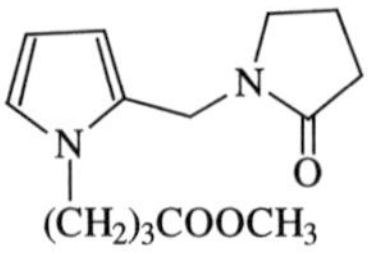

Structure 2.1

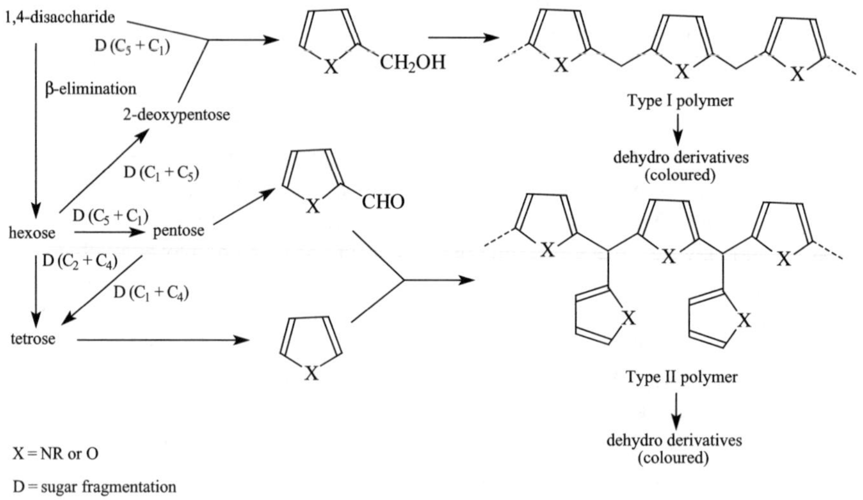

Scheme 2.7 Routes to melanoidin-type polymers in sugar–amine model systems[30]

When the starting point was a mixture of N-methylpyrrole and N-methyl-2-formylpyrrole, branched oligomers of up to 26 units were identified, the solution turning intensely red (λ_{max} = 515 nm in chloroform) and becoming brown/black. When the methylamine moiety was replaced by the methyl ester of β-alanine (non-Strecker-active), red oligomers up to a tetradecamer were identified. When N-methylpyrrole was mixed with furfural, branched oligomers of more than 30 units were identified, oligomers with the monomers combined in proportions of 2 : 1, 3 : 2, and 4 : 3, respectively, were isolated, characterised, and found to contain the expected number of centres of chirality. When N-(2-methoxycarbonylethyl)pyrrole was mixed with furfural, branched oligomers of up to 25 units were identified. When N-methyl-2-formylpyrrole was mixed with 2-methylfuran, the latter was not incorporated into the oligomers formed. These were branched, but not invariably, and could be deformylated (loss of CO).

Tressl et al.[76,77] designated the linear polymers as Type I and the branched ones as Type II. In most melanoidins, they would represent domains (or substructures), unsubstituted pyrroles and Strecker aldehydes, for example, being integrated into the melanoidin backbone, giving a complex macromolecular structure overall. Tressl et al.[77] consider the oligomerisation/polycondensation reactions described as the only experimentally established pathways by which simple Maillard products generated from hexoses and pentoses are easily and irreversibly converted into macromolecules.

Melanoidins constitute up to 25% of the dry matter of coffee beverage and they have been studied by Borrelli et al.[79] by gel filtration and MALDI–TOF. As roasting is increased, the proportion of melanoidins increased, but their MM decreases. Antioxidant activity by ABTS$^{\bullet+}$ and DMPD$^{\bullet+}$ decreased with increased roasting, yet the ability to prevent linoleic acid peroxidation was higher in dark roast coffee. On the contrary, Fraction I from the gel filtration (Sephadex G-25, > 100 kDa) of green coffee brew was most effective in scavenging DMPD$^{\bullet+}$, but was completely ineffective in inhibiting linoleic acid peroxidation. Protein (Kjeldahl) constituted > 30% of the green coffee brew, but decreased by two-thirds or more when the coffee had been roasted. All roasted coffees gave a broad peak, 1–4 kDa, on MALDI–TOF, the mean decreasing from 2.703 to 1.930 kDa, as the degree of roast increased. In contrast, the green coffee brew gave several sharp peaks, ranging up to 5752.8 a.m.u., that at 3288.2 a.m.u. being the most prominent.

There are large differences between melanoidins produced under aqueous and under dry-heating conditions, as exemplified by glucose–glycine mixtures.[80] Whereas the former produced water-soluble melanoidins with MM > 3 kDa only in trace amounts, the latter gave 13.6% (almost all with MM > 30 kDa). On the other hand, both types of melanoidin contained free radicals; in each case, the dialysed portion contained more than that in the undialysed portion, but those in the dialysed portion increased by a factor of 1.4, whereas those in the undialysed portion increased by a factor of 4 in changing from the aqueous to the dry-heated melanoidin.[81] The high-MM fraction is thus mainly responsible for the amount of free radicals detected, whereas, surprisingly, its ability to scavenge free oxygen is lower. The radicals appear to be very stable, surviving over more than 10 d dialysis.

Cämmerer et al.[82] have prepared melanoidins from glycine and a range of sugars. When prepared under water-free conditions (170 °C, 20 min), melanoidin from glucose, on hydrolysis, gave back only 3 mg glucose per 100 mg melanoidin, whereas that from maltose gave 24, from maltotriose 31.5, and from dextrose 95 mg per 100 mg melanoidin, the last having been heated for 25 h. Lactose gave only 0.7 mg glucose per 100 mg, but 16 mg galactose per 100 mg melanoidin. When melanoidin was prepared under aqueous conditions (100 °C, 10 h), glucose gave 0.3 and maltose 18.5 mg glucose per 100 mg melanoidin. Overall, these results suggest that the reaction takes place at the reducing group, leaving much of the glycosidically attached glucose free to be recovered by hydrolysis. In consequence, a structure for melanoidin was proposed, based on a backbone of 3-deoxyhexosones (still carrying glycosidically linked sugar residues), 1→3 linked by aldol reactions and with Amadori compounds attached as sidechains through their amino group.

Gel-permeation chromatography on Sephadex G25 is able to separate water-soluble melanoidins into several fractions, estimated to have MM of >12, 2.5–10, and <2.5 kDa, respectively.[80] Chromatographing the permeate from dialysis with a membrane with a 10 kDa cut-off showed that this relatively lengthy procedure had allowed a considerable proportion of material with MM >12 kDa to be formed, demonstrating the lability of melanoidins.

Examining a defatted extract of roasted coffee showed that it could similarly be separated into three fractions on Sephadex G25-fine. When the middle fraction was collected, freeze-dried, roasted at 200 °C for 3 min, and re-analysed, most of it had been converted into the fast-running high-MM peak, thus demonstrating that low-molecular-mass (LMM) melanoidins act as precursors to high-molecular-mass (HMM) ones.[83]

8 Standard Melanoidin

Since melanoidins are so complex and since their nature changes with the conditions under which they have been produced (see, for example, ref. 84), it is very difficult to compare the results of different researchers in a meaningful manner. The question of producing a standard melanoidin, or several standard melanoidins, has been discussed at some length within the framework of COST Action 919 of the Directorate-General for Research of the European Union.[85] As a consequence, the following recipe was proposed for standard melanodins from glucose and glycine:

8.1 Materials

- Glucose (Sigma; min 99.5%, Cat. No. G8270)
- Glycine (Sigma; min 99%, Cat. No. G7126)
- Glass beakers (250 ml, Pyrex, about 7 cm i.d., about 9 cm high) for heating the glucose–glycine mixture
- Freeze-drier
- Oven, heated to and stabilised at 125 °C
- Dialysis tubing (from Sigma; Cat. No. D9652), flat width 33 mm, diameter 21 mm, cellulose (retains $>90\%$ cytochrome *c*, MM 12.4 kDa, in solution over a 10 h period; supplied in rolls, dry; may contain glycerol and sulfur compounds in more than trace amounts; washing instructions supplied).

The Chemistry of Nonenzymic Browning

8.2 Method

1. Dissolve 0.05 mol of glucose and 0.05 mol of glycine in 100 ml of distilled water.
2. Place the solution in a container that is compatible with the freeze-drier and place in a freezer until completely frozen (*e.g.*, overnight).
3. Place the frozen mixture in a freeze-drier using a plate temperature of 30 °C or below. Freeze-dry until all the water has been removed (*i.e.*, to constant weight).
4. Place the glucose–glycine mixture in a 250 ml beaker and place it in an oven that has been preheated to 125 °C. Heat for exactly 2 h. Do not cover the beaker. The oven should not be opened during the heating.
5. Remove the beaker from the oven and allow it to cool to room temperature in a desiccator.
6. Take 5 g of the sample of the heated mixture and add 20 ml of distilled water. Stir to dissolve as much material as possible.
7. Filter the mixture through Whatman No 4 filter paper and collect the filtrate which contains the soluble melanoidins. Wash the residue on the filter paper with 2 × 10 ml of distilled water and mix together these washings and the original filtrate. This mixture is called *Solution A*.
8. Obtain a UV-visible absorbance spectrum of Solution A (200–600 nm) and measure the absorbance at 280, 360, 420, 460, and 520 nm.
9. Prepare a 30 cm length of dialysis tubing according to the manufacturer's instructions.
10. Close one end of the tubing with either a clip designed for the purpose or a thread. Place Solution A in the tubing and add distilled water so that the height of the solution in the tubing is about 15 cm. Close the other end of the tubing and submerge the tubing in 1 L of distilled water.
11. Leave the vessel containing the dialysis tubing for 24 h at 4 °C.
12. Replace the water surrounding the dialysis tubing with 1 L of fresh distilled water and allow to dialyse for a further 24 h at 4 °C.
13. Repeat Step 12 two times. This means dialysis in a total of 4 × 1 L of water.
14. At the end of dialysis, place the contents of the dialysis tubing in a container compatible with the freeze-drier and place in a freezer until completely frozen (*e.g.*, overnight).
15. Place the frozen mixture in a freeze-drier using a plate temperature of 30 °C or below. Freeze-dry until all the water has been removed (*i.e.*, to constant weight).
16. Store the Standard Melanoidin in a desiccator.

9 Melanoidins: Properties

9.1 Glass Transition Temperature

The glass transition temperature T_g of melanoidins has been considered by Anese et al.[86] The freeze-dried Solution A and the Standard Melanoidin showed T_g of 30 and

56 °C, respectively, compared with those of glucose, glycine, sucrose, maltose, and starch, 31, −71, 62, 87, and > 200 °C, respectively. Mixing the Standard Melanoidin with water rapidly reduced the T_g: for 4.4% (w/w, total) water $T_g = 19$ °C and for 18.3% water $T_g = -70$ °C. The DSC thermogram also showed sharp endotherm peaks for Solution A and the Standard Melanoidin at 160 and 140 °C, respectively, attributed to the degradation of carbonyl compounds. The temperature at which this peak occurs was also lowered on addition of water: 4.4% lowered it to 100 and 24% to 80 °C.

9.2 Thermal Degradation

The thermal degradation of standard glucose–glycine melanoidins has been extensively studied by Tehrani et al.[87] After heating for 2 h at 125 °C, there was a weight loss of 32%. The water-insolubles (5 g from the equimolar mixture 9 : 3.75 g originally) and the non-dialysables (119 mg) were heat-treated, either (A) successively at 100, 150, 200, and 220 °C or (B) as a fresh sample at each temperature and at 250 and 300 °C. Heat-treatment A produced about 10 times the amount of volatiles from the non-dialysables than from the insolubles. This was seen for 5-methylfurfural, furfural, 2-acetylfuran, 4,5-dimethyl-2-formylpyrrole, 2-acetyl-1-methylpyrrole, and 2-acetylpyrrole. 2-Pyrrolidone was the most abundant nitrogen compound for the non-dialysables (not reported for the insolubles), 2-formylpyrrole being the next most abundant (100 times lower for the insolubles). 2-Cyclopentene-1,4-dione was present about equally in the non-dialysables and insolubles. Heat-treatment B led to the identification of 35 compounds from the non-dialysables and 52 from the insolubles. 5-Methylfurfural from the non-dialysables exhibited a maximum at 200 °C, whereas for the insolubles its level was maintained till 250 °C. Oxazole derivatives, such as the 4,5-dimethyl, benzo, and 2,5-dimethylbenzo, were mainly formed at higher temperatures. Pyrroles only started to be formed at 150 °C and tended to peak at 200–220 °C. 2-Acetyl-1-methylpyrrole increased remarkably between 250 and 300 °C. N-Methylsuccinimide was detected in the non-dialysables at 200 and 220 °C and at a much lower level in the insolubles at 250 °C. It has been found in pyrolysates of polyglycine at 500 °C, thereby hinting at the presence of chains of glycine in the melanoidins. 2,6-Dimethylpyridine, which had not been previously recognized as a Maillard product, was detected quite irregularly in the non-dialysables. It was formed from the insolubles in lower amount, but increased with temperature. Up to 250 °C, the amount of pyridine formed from the insolubles was greater than that from the non-dialysables, but the latter increased sharply, being three times that of the former at 300 °C. The insolubles gave 12 pyridines and 12 pyrazines, whereas the non-dialysables gave 5 and 10, respectively. The pyrazines from the non-dialysables tended to peak at 200 °C, whereas those from the insolubles continued to increase with temperature.

10 Reaction H: Free-Radical Reactions

The presence of stable free radicals in melanoidins has been known for some time: Mitsuda et al.[88] have detected a relatively stable free radical as a broad singlet by ESR in the solid melanoidin from glucose–glycine (1 h, 100 °C). Namiki and his colleagues[89,90] went on to show that free radicals developed at an early stage of the

The Chemistry of Nonenzymic Browning

Maillard reaction. Of the range of sugars and sugar degradation products used, glycolaldehyde gave the most intense ESR signal, both with α- and β-alanine, these combinations also producing the highest level of colour. Furfural browned as much, but gave no ESR signal. Namiki and Hayashi[91] summarized their results, reporting that the products of the interaction of α- or β-alanine with arabinose gave rise to striking ESR spectra with 17 and 23 lines, respectively, and that similar spectra had been obtained with a range of sugars and sugar degradation products. Most other amino acids and primary amines examined also gave ESR spectra with hyperfine structure, but secondary and tertiary amines did not; neither did amides. These signals were attributed to the presence of N,N'-dialkylpyrazine cation radicals. The radicals are detected before the Amadori compound and so led to the conclusion that a new pathway (H, see Scheme 1.1, p. 3) for sugar fragmentation and thus for browning had been discovered. The hyperfine structure of the ESR spectra degraded with heating time to a broad singlet, similar to that observed by Mitsuda *et al.*[88]

Hayashi and Namiki[92] allowed glucose (or another hexose or pentose) to react with an alkylamine in 95% ethanol under reflux. They examined the product by TLC on silica gel after spraying with 2,4-dinitrophenylhydrazine, detecting glyoxal. They then showed that glyoxal was an artefact, produced by the action of silica gel on its precursor, its bisalkylimine. The bisalkylimine reached a maximum in about 20 min, just about when 3-deoxyglucosone, the free radical, and absorbance at 420 nm began to be detectable. The free radical concentration reached its maximum at about 90 min, but 3-deoxyglucosone and browning were still increasing at 120 min. Hayashi and Namiki concluded that the sugar or the glycosylamine was the most likely starting point, with formation of the Schiff base and oxidation subsequent to retroaldolisation, as shown in Scheme 2.8.

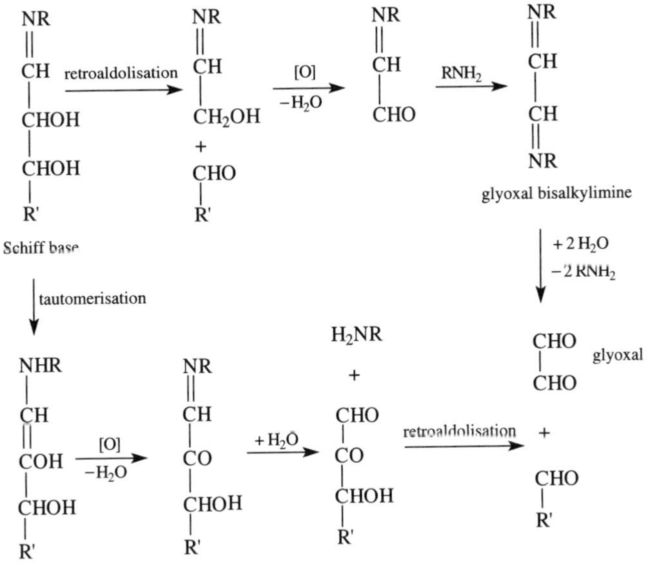

Scheme 2.8 *Formation of glyoxal according to Hayashi and Namiki[92] and Hofmann et al.[94]*

In similar work, Hayashi et al.[93] followed up the additional spots seen on the sprayed TLC plates and traced these to 2-oxopropanal. Glucose without amine (pH 9.3, 95 °C, 10 min) gave no 2-oxopropanal. Glucose and n-butylamine (1 M each, 80 °C) had formed C_2 and C_3 imines already when the products were first analysed (at 5 min) and reached maxima at about 15 and 18 min, respectively, the latter being at about twice the concentration of the former. Their formation correlated well with the decrease in glycosylamine and the increase in Amadori compound. Browning increased progressively. The Amadori compound was shown to be able to degrade to the C_3 imine and was therefore thought to lie on the path of formation.

Using their colour dilution factor, Hofmann et al.[94] obtained results similar to those of Namiki and Hayashi[95] (see Table 2.5). Whereas both glycolaldehyde and furfural led to the highest dilution factors, only the former gave free radicals at the highest intensity observed. The conclusion from this is that two distinct mechanisms of browning must be in operation, a radical and an ionic one. It should be noted that glyoxal gave only about one-tenth of the colour and 4% of the radical formation of glycolaldehyde. Table 2.5 also shows that the Amadori compound gives rise to only half the colour derived from the corresponding glucose/α-alanine mixture, from which it also follows that colour cannot be formed exclusively via the Amadori compound.

Although the pyrazinium doubly charged cations, produced on oxidation of the radical cations, are very unstable, Namiki and Hayashi[95] considered them very likely to be active intermediates in the browning reaction. This point was re-examined, using the diethylpyrazinium dication.[94] Its extraordinary lability in water was confirmed, intense colour developing instantly on dissolution and being accompanied by free-radical formation. The ESR spectrum was that of the radical cation and decreased rapidly to half its intensity in 10 min. During this period, colour (at 420 nm) increased quickly, but continued to increase more slowly thereafter.

Clearly, there is a redox equilibrium among the 1,4-dialkyl-1,4-dihydropyrazine, the radical cation, and the dication, as shown in Scheme 2.9. LC-MS of an aqueous

Table 2.5 *Colour development and radical formation in binary mixtures of alanine and sugars or sugar degradation products (from Hofmann et al.[94])*

Carbonyl compound	CD factor[a]	Radical formation (rel. %)
Glucose	16	4
Xylose	64	8
N-(1-Deoxy-D-fructos-1-yl)-L-alanine	8	1
Glycolaldehyde	1024	100
Glyoxal	128	4
Furfural	1024	0
Pyrrole-2-carboxaldehyde	256	0
Pyruvaldehyde	256	0
Butanedione	128	n.d.
HMF	2	n.d.
Glyceraldehyde	2	n.d.
2-Hydroxy-3-butanone	2	n.d.

n.d., not determined.
[a] The colour dilution (CD) factor was used to compare the colour intensities of reaction mixtures heated for 15 min at 95 °C.

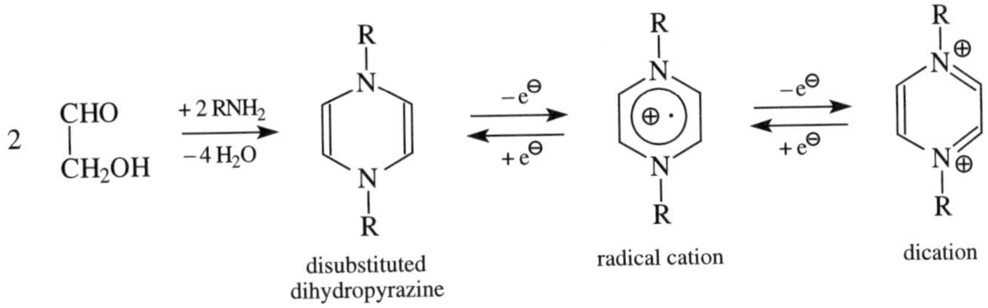

The Chemistry of Nonenzymic Browning

Scheme 2.9 *Redox equilibria between 1,4-dialkyl-1,4-dihydropyrazines and the corresponding radical cations and dications.*[95]

solution of freshly prepared dication surprisingly gave no evidence of the doubly charged ion at *m/z* 69.[94] The base peak was at *m/z* 155 (100%) and there was a prominent peak at *m/z* 138 (35%), corresponding to the radical cation as well as a peak at *m/z* 171 (8%). When a methanolic solution was submitted to LC-MS, the base peak was at *m/z* 169 (100%) and there still was a prominent peak at *m/z* 138 (32%), but also a new peak at *m/z* 199 (17%). This was interpreted as the dication being converted rapidly into its dihydrohydroxy and dihydrodihydroxy derivatives in water and into the corresponding methoxy and dimethoxy derivatives in methanol. These dihydro derivatives are potential reducing agents, capable, for example, of reducing dications to radical cations.

Namiki and Hayashi[95] proposed that glycolaldehyde imine or the isomeric alkylaminoethanal is formed prior to the Amadori rearrangement by a retroaldolisation of the Schiff base, and is then readily oxidised by atmospheric oxygen to glyoxal monoimine. When an aqueous mixture of glucose and alanine was heated[94] the reverse was found to be the case, namely, glyoxal was formed rapidly, reaching a maximum at 15 min (100 μg mmol^{-1} glucose) and decreasing to 20 and 10 μg mmol^{-1} glucose after 60 and 240 min, respectively, whereas glycolaldehyde increased only slowly, approaching a maximum value of about 20 μg mmol^{-1} glucose by 240 min. 3-Deoxyhexosone and 1-deoxyhexo-2,3-diulose reached maxima of about 1000 and 300 μg mmol^{-1} glucose, respectively, at about 40 min, *i.e.*, subsequent to the formation of glyoxal.

Where does the glyoxal originate? Refluxing aqueous pH 7.0 solutions of glucose, of glucose and alanine, and of the corresponding Amadori compound for 10 min gave 18, 96, and 253 μg glyoxal mmol^{-1}.[94] The last two experiments were repeated, but oxygen was removed by passing nitrogen through the solutions prior to heating. The amount of glyoxal formed decreased to 8 and 14 μg mmol^{-1}, respectively. Oxygen is clearly involved.

The Amadori compound was the most effective glyoxal precursor, even though the yield was only 0.4%. Under the same conditions, glucose and alanine give the Amadori compound in only 1.1% yield and so most of the glyoxal from glucose/alanine cannot originate from the Amadori compound. Since the presence of the amino acid seems to be a prerequisite for effective glyoxal formation, Hofmann *et al.*[94] proposed a mechanism in which oxidation of the Schiff base is the key step (see Scheme 2.8).

From the data presented above, it was concluded that glyoxal needs to be reduced to glycolaldehyde as a prerequisite for cation radical formation. This led Hofmann et al.[94] to follow the time course of the formation of reducing substances by titration with 2,6-dichloroindophenol. On heating an aqueous solution of glucose and alanine, both free radicals and reducing substances started to be formed after 10 min, the former peaking at 30 min and the latter at 45 min.

Free radicals were believed to be involved when green coffee beans were roasted.[96] Radical formation increases strongly with time and temperature of roasting, running in parallel with the colour intensity of the roasted coffee. Recent work has traced the nature of the radical to peptide-bound 1,4-bis-(5-amino-5-carboxy-1-pentyl)pyrazinium radical cations.[94] Pascual et al.[97] found that the ESR spectra of soluble coffee display single-line free-radical signals in both the solid state and in aqueous solution, along with signals from paramagnetic Fe^{3+} and Mn^{2+}. The intensity of the signal corresponded to about 7.5×10^{16} unpaired electrons g^{-1}. In aqueous solution, the signal declined rapidly and was attributed to pyrazinium radical cations; however, in coffee solutions, radicals are being generated as well as destroyed. Although the decrease was not caused by oxygen, oxygen is involved in the radical generation, which continued over extended periods under aerobic conditions. Although the free radicals responsible for the single-peak ESR signal did not react with any of the spin traps tested, adducts of C-centred radicals were detected in coffee solutions in the presence of PBN (phenyl-*N*-t-butylnitrone) and 4-POBN (α-(4-pyridyl-1-oxide)-*N*-t-butylnitrone). No adducts were detected when DEPMPO (5-(diethoxyphosphoryl)-5-methyl-1-pyrroline *N*-oxide) was used as a spin trap, but its hydroxyl-radical adduct was unstable in the presence of coffee, illustrating the strong free-radical scavenging power of coffee solutions.

Free radicals in the Maillard reaction are the subject of a recent in-depth review by Rizzi.[640]

CHAPTER 3

Recent Advances

1 Introduction

Knowledge of the Maillard reaction is being extended very actively in many different ways. The participation of free radicals has already been dealt with in Chapter 2 and work on colour and flavour aspects is being deferred to Chapters 4 and 5, respectively. This chapter deals with a number of relatively disparate topics, namely, the effects of pH, high pressure, T_g, and the use as reactants of amines other than amino acids, of lipids, and of oligo- and polysaccharides, as well as the determination of α-dicarbonyl intermediates, control of aldol/retroaldol reactions, fluorescence, kinetic aspects, and sites of protein glycation.

2 Effect of pH

pH exerts a crucial effect on the Maillard reaction when the Amadori compound has been formed. At that point it determines the extent to which reactions proceed by its enolisation, either 1,2 at low pH or 2,3 at higher pH.

A good example of the effect of pH[98] is that observed on a xylose–lysine model system (1 M each, refluxed 1 h with diethyl ether in a Likens and Nickerson apparatus, initial pH 4.9, either kept at pH 5 with NaOH additions or left, when the final pH is 2.6); 54 and 28 volatiles were identified, respectively, 2-furaldehyde dominating with 52.2 and 99.9% (w/w). Total yield and number of nitrogen-containing compounds were greater at higher pH values of the former system, and monocyclic pyrroles, pyridines, and 2,3-dihydro-1H-pyrrolizines were identified only in that system.

4-Hydroxy-5-methyl-3(2H)-furanone (norfuraneol) is an important contributor to cooked beef flavour, although it has a caramel-type odour. It can be obtained by heating xylose, ribose, or ribose-phosphate with amine salts. It is thought to be formed by 2,3-enolisation, probably as in Scheme 3.1.

Therefore, furanone can be taken as symptomatic of 2,3-enolisation similar to furfural and HMF being symptomatic of 1,2-enolisation (see Table 3.1).[54]

Table 3.1 shows clearly that low pH favours formation of 2-furaldehyde (1,2-enolisation), whereas higher pH favours formation of the furanone (2,3-enolisation). It also illustrates how the basicity of the amine can affect the route; the least basic (dibenzylamine) is less likely to be protonated and therefore more liable to undergo 2,3-enolisation.

The work reported by Arnoldi and Boschin[99] with equimolar aqueous solutions of xylose–glycine, heated at 100 °C for 2 h at different pH values, maintained by additions

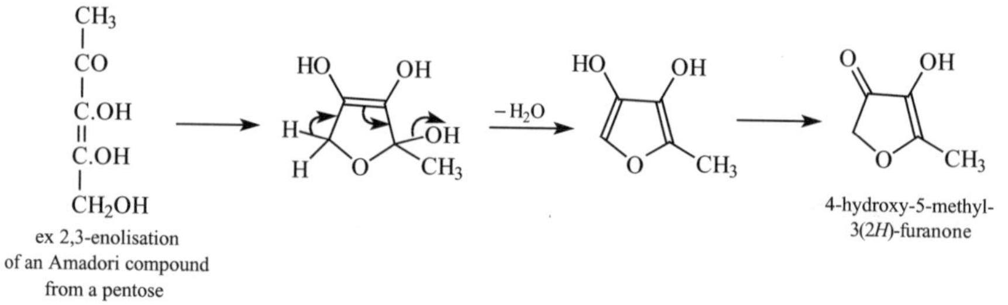

Scheme 3.1 *Formation of 4-hydroxy-5-methyl-3(2H)-furanone from a pentose via an Amadori compound and 2,3-enolisation*

Table 3.1 *Degradation of Amadori compounds under acidic and neutral conditions*

Amadori from	M H_2SO_4/100 °C		pH 7/100 °C	
	2-Furaldehyde	Furanone	2-Furaldehyde	Furanone
Glucuronic acid+1 mol of dibenzylamine (derivative of fructuronic acid)	16%	Traces	0	Present
Glucuronic acid+1 mol of benzylamine (derivative of fructuronic acid)	23%	0	0	Present
Xylose+1 mol of benzylamine (derivative of xylulose)	41%	0	0	Present

of alkali without buffers, supports this work up to a point. They showed that free furfural was obtained only at the lowest pH (3) and that furfural combined in the furfurylidenefuranone decreased progressively as the pH was increased to 7. However, the free furanone increased only up to pH 5 and then decreased, suggesting that it must be combining to form compounds other than the furfurylidene derivative.

In studying the degradation of fructosylglycine in unbuffered and buffered aqueous solutions at 90 °C over 7 h, the pH being kept constant at 5, 6, 7, or 8, HMF was detected by Davidek et al.[100] only at pH 5, whereas 2,3-dihydro-3,5-dihydroxy-6-methyl-4H-pyran-4-one was formed over the whole pH range studied. The phosphate buffer accelerated the formation of the latter compound at pH 5 and 6, but the formation of HMF was practically unaffected, suggesting that phosphate catalyses 2,3-enolisation at lower pH values.

As the final pH of meat increases above the normal range of 5.6–5.8, there is a decrease in meat flavour.[101,102] The reaction between norfuraneol and cysteine, which plays a crucial role in meat flavour, is very sensitive to pH.[103] At pH 4.5, the major volatile products are mercaptoketones, furan- and thiophenethiols, 2-methyltetrahydrothiophenone, and 3,5-dimethyl-1,2-dithiolan-4-one, whereas, at pH 6.5, these compounds were detected, if at all, only in traces, except for the thiophenone; however, pyrroles, pyrazines, thiazoles, and oxazoles had achieved prominence. The sensory observations are a clear reflection of such differences. Similar effects of pH had been found for the ribose–cysteine and other systems.[104]

Recent Advances

3 Effect of High Pressure

The use of high pressure is one of the more recently developed methods of food preservation. The primary interest has been the inactivation of microorganisms; however, the effect of high pressure on the functional properties of food ingredients is clearly also of relevance, as is its effect on the chemical reactions occurring during processing and storage.

Physical principles imply that high pressure will increase the rates of reaction where the volumes of reaction are negative, *i.e.*, where the number of molecules is reduced, *e.g.*, through addition. High pressure thus encourages addition, but discourages the formation of small molecules, such as flavour compounds (see Table 3.2).[105]

Imine formation, the Amadori rearrangement, and aminoketose decomposition, with glucose or xylose and tryptophan, led to activation volumes of -14, $+8$, and $+17$ mL mol^{-1}, pressure thus accelerating the first, but retarding the other two reactions.[106] As a consequence, the formation of certain volatiles, such as norfuraneol at pH 7 and 2-methylpyrazine at pH 10 in lysine–xylose systems, is greatly decreased.[107]

The condensation of glyceraldehyde with di- and tripeptides was little affected with increase in pressure (up to 500 MPa), and the reaction of glyceraldehyde, glycolaldehyde, or xylose with amino acids was suppressed very weakly. Browning in the latter systems was suppressed significantly, with activation volumes of $+13$ to 28 mL mol^{-1}.[108]

The yields of the coloured 2-furfurylidene-4-hydroxy-5-methyl-3(2*H*)-furanone and 2-furfurylidene-5-methyl-3(2*H*)-pyrrolinone from xylose and alanine at pH 5 were strongly affected, depending on whether the reaction was carried out at atmospheric pressure or at 400 MPa.[109] Formation of both colorants increased progressively

Table 3.2 *Some of the volatiles isolated from a glucose–lysine system, pH 10.1, heated at 60 °C at atmospheric pressure and 600 MPa*[105]

Volatile	Amount (μg) formed at AP^a	600 MPa	RPY^b
2-Methyl-3-(2*H*)-furanone	14	4	29
Furaneol	108	17	16
Methylpyrazine	257	3	1
2,5- and/or 2,6-Dimethylpyrazine	3951	110	3
2,3-Dimethylpyrazine	127	6	5
Trimethylpyrazine	758	20	3
3-Ethyl-2,5-dimethylpyrazine	28	–	0
3-Methyl-1,2-cyclopentanedione	25	4	16
2-Acetyl-1,4,5,6-tetrahydropyridine	30	9	30
2-Acetylpyrrole	38	–	0
2,5-Dimethyl-2,5-cyclohexadiene-1,4-dione	7	–	0
2,3-Dihydro-5-hydroxy-6-methyl-(4*H*)-pyran-4-one	51	4	8
2,3-Dihydro-3,5-dihydroxy-(4*H*)-pyran-4-one	442	5	1
5-Formyl-6-methyl-2,3-dihydro-(1*H*)-pyrrolizine	17	5	29
7-Acetyl-5,6-dimethyl-2,3-dihydro-(1*H*)-pyrrolizine	78	12	15

^{a}AP, atmospheric pressure.
^{b}RPY, relative percentage yield at 600 MPa to AP.

with time at atmospheric pressure; however, under high pressure, much lower amounts were formed at longer reaction times (>2 h), with maximum formation observed at 2 h. In the case of the pyrrolinone, almost twice as much was formed at 2 h under high pressure than at atmospheric pressure.

For a glucose–lysine system, the different stages of the Maillard reaction were affected differently. Thus, Moreno et al.[110] found that, unbuffered at an initial pH ≤ 8, the formation of the Amadori compound was not greatly affected by pressure (400 MPa), whereas the intermediate and advanced stages were suppressed, suggesting that the degradation of the Amadori compound was being retarded. In buffered solutions, pH ≤ 8, pressure slowed down the Maillard reaction from the start. These effects were attributed to the pH drop caused by the pressure-induced dissociation of acidic groups. In solutions at initial pH 10.2, pressure accelerated the formation and the subsequent degradation of the Amadori compound, leading to increased levels of intermediate and advanced reaction products.

Moreno et al.[111] have also studied the effect of high pressure (400 MPa, 60 °C, 3 h) on the degradation of lactose (10 g L^{-1}) in alkaline media. As can be seen from Table 3.3, formation of isomers (lactulose, epilactose) and galactose is decreased in 4 and 8 mM NaOH and in sodium carbonate buffer (pH 10.0). Colour development (420 nm) is almost negligible in the NaOH systems, but high pressure markedly reduces the appreciable colour formed in the carbonate buffer system.

When β-casein was heated with glucose or ribose, the fluorescence emission spectra did not differ significantly, but, with increased pressure, the spectra from the former system became less pronounced, whereas those from the latter increased in intensity and developed a shoulder at 385 nm, corresponding to that of pentosidine (see Chapter 8). An equimolar mixture of N^{α}-acetyllysine, N^{α}-acetylarginine, and ribose was therefore subjected to increasing pressure.[112] The pentosidine content increased about sixfold at 600 MPa when compared with that at atmospheric pressure. The pentosidine content of β-casein similarly treated with ribose at atmospheric pressure and at 600 MPa increased from not detectable (<32 μg per 100 g protein) to 4.8 mg per 100 g protein, respectively.

Table 3.3 *Effect of high pressure on isomerisation and degradation of lactose in alkaline media for 3 h at 60 °C[111]*

Medium	pH^a	Lactulose[b]	Epilactose[b]	Galactose[b]
4 mM NaOH				
AP^c	8.2	823	18.6	30.3
HP^d	9.6	633	10.9	26.6
8 mM NaOH				
AP	7.9	1329	45.7	104.4
HP	9.1	1046	21.4	86.7
Carbonate buffer				
AP	9.7	1876	80.2	633.5
HP	10.0	780	21.9	90.0

[a]Final pH.
[b]Expressed as mg per 100 mL.
[c]AP = atmospheric pressure.
[d]HP = high pressure (400 MPa).

Recent Advances

Pyrraline (see also Chapter 8) can be assessed fluorometrically ($\lambda_{ex}/\lambda_{em}$ = 350/430 nm), allowing it to be determined in mixtures undergoing the Maillard reaction. When N^α-acetyllysine was treated with 3-DG, an appreciable amount of pyrraline was formed at atmospheric pressure, but none could be determined at 600 MPa.

The contrasting results above serve to emphasise that individual steps in protein–carbohydrate reactions differ significantly in their pressure dependence and, therefore, the behaviour at high pressure in complex reactions, such as the Maillard reaction, cannot be deduced easily from that at atmospheric pressure.[112]

4 Fluorescence

Fluorescence characteristically develops during the intermediate stage of the Maillard reaction. For analytical purposes, colour has always been the method of choice, even though it is far from ideal. Fluorescence, partly because of its greater sensitivity, particularly in the presence of colour due to sources other than the Maillard reaction, has frequently been found useful, especially for biological samples.

The fluorescence of some molecules, derived from crosslinks and related structures formed in Maillard reactions, are given in Table 3.4.

Table 3.4 *The fluorescence of some molecules derived from crosslinks and related structures*

	$\lambda_{ex\,max}$ (nm)	$\lambda_{em\,max}$ (nm)	Reference
Argpyrimidine	320	382/384	Al-Abed et al.;[415] Shipanova et al.[626]
Imidazolonylornithine	320	398	Lo et al.[627]
Pentodilysine, penK$_2$	320/366	440	Graham et al.[369]
Pentosidine	325/335	385	Westwood and Thornalley;[370] Hayase et al.[359]
Arg-hydroxy-triosidine	331	380	Tessier et al.[357]
C-Pentosidine	335	385	Miyazaki et al.[376]
MRX^a	340	402	Osawa et al.[383]
Blue-M1	349	445	Kato and Hayase[628]
Other AGE	350/370	430/440	Westwood and Thornalley;[370] Dawney et al.;[629] Ruggiero-Lopez et al.[630]
Lys-hydroxy-triosidine	354	440	Tessier et al.[357]
Coffee-foam fraction	361	440	Petracco[505]
Fluorophore LM-1	366	440	Nagaraj and Monnier;[631] Tessier et al.[368]
Lysylpyrropyridine	370	448	Watanabe et al.[390]
Acetyl-lysylpyrropyridine	376	450	Hayase et al.[359]
Crosslines etc.	379	463	Nakamura et al.[632]
N-Methyl-2-hydroxymethyl-pyrrole polymer	385	493	Tressl et al.[76]

^{a}The conjugate of the oxidation product of glucose with a cysteine residue of BSA, formed in the presence of arginine (or a guanidino-type compound).

When a xylose–glycine system is heated under reflux in 0.07 M phosphate buffer, pH 8.2, cooled and extracted first with light petroleum (b.p. 60–80 °C), and then with ether, the latter extract contained many fluorescent compounds amongst the large number of components separated by HPLC.[113] Two types of fluorescence spectra stood out because of their repeated occurrence:

	$\lambda_{ex\,max}$ (nm)	$\lambda_{em\,max}$ (nm)
A	270/285	415/425
B	270/310	390/425+480/540

Unfortunately, interpretation of fluorescence spectra is rudimentary. Comparing the above data with those in Table 3.4, it is noteworthy that none of the latter exhibits the two-peaked emission of Type B. However, fluorescence is such a spectacular phenomenon, intimately involved with the Maillard reaction, that more attention devoted to it should rapidly improve its diagnostic value.

5 Determination of α-Dicarbonyl Intermediates

Clearly, α-dicarbonyl intermediates are very important and, as a consequence, their detection and quantification is of great interest. Glomb and Tschirnich[114] list the following seven methods:

- as such
- as the corresponding alcohol after reduction
- as O-aryl or O-alkyl hydroxylamines, such as O-methylhydroxylamines
- as hydrazines, e.g., with Girard Reagent T
- as cysteamines
- as azines by reaction with o-phenylenediamine
- as triazines by reaction with aminoguanidine

The stability of 3-DG is much greater than that of 1-DG. Thus, 3-DG can be recovered quantitatively, using aminoguanidine or o-phenylenediamine, after incubation in phosphate buffer for 24 h, whereas 1-DG is completely degraded within 5 h to compounds no longer accessible to the trapping reagents.[114] Glucosone is only 50% recovered through reaction with aminoguanidine for 24 h. Glyoxal and 2-oxopropanal react much more rapidly, their reaction even with aminoguanidine being almost complete within 5 h.

Recently, Biemel et al.[115] have demonstrated, by means of aminoguanidine and o-phenylenediamine derivatives, the lability of carbonyl groups along the carbohydrate backbone, showing that compounds such as 1,4-dideoxy-5,6-hexosones are very important Maillard intermediates.

6 Control of Aldol/Retroaldol Reactions

Although aldolisation and retroaldolisation play a role in many aspects, particularly in the intermediate stage, of the Maillard reaction, virtually no attention has been paid to manipulating even the direction of the reaction. The problem was raised some years ago,[116] but little effort has gone towards solving it.

7 Kinetics of the Maillard Reaction

The kinetics of the early-stage Maillard reaction has been considered by Ge and Lee.[117] Their basic equation was as follows:

$$A + S \underset{k_{-1}}{\overset{k_1}{\rightleftarrows}} [AS] \overset{k_2}{\rightarrow} AP \quad (1)$$

where A represents an amino acid, S a sugar, AS the Schiff base complex, and AP the Amadori compound. For the early-stage reaction, it was assumed that (i) A is not subject to the Strecker degradation, (ii) caramelisation of S is negligible, and (iii) conversion of AP to AMRP as well as the reverse reaction is also negligible. Using phenylalanine, because of its ready UV determination, it was shown that k_{-1} and k_2 are 103 times greater than k_1, indicating that Schiff base formation, but not AP formation, is rate-limiting. E_{a2} (33.5, 8.01) was found to be slightly greater than E_{a1} (27.3, 6.52) and E_{a-1} (31.3, 7.49 × 10^4; units in kJ mol^{-1}, kcal mol^{-1}, respectively), showing AP formation to be more sensitive to temperature and favoured by high temperature. k_{-1} and k_2 increase from about 70 °C, and very sharply from 90 °C. The rate of formation of AP roughly levelled between pH 2 and 8, but increased at pH 10, and very significantly at pH 12.

Colour formation, which lies well beyond the early-stage Maillard reaction, has been described by Wedzicha and Leong[118] by the following scheme (DH = deoxyhexosone):

$$\text{Aldose} \overset{k_1}{\longrightarrow} \text{DH} \overset{k_2}{\longrightarrow} \text{Int} \overset{k_3}{\longrightarrow} \text{melanoidins}$$

Here the first, slow step (rate constant k_1) covers Reactions A, B, and part of C of Hodge's scheme (see Chapter 1). The second, slow step lies also within Reaction C, whereas the fast, third step takes the system through Reactions F and G. Wedzicha and Leong[118] have shown clearly that the rate constants for Step 3 are much greater for lysine, glycine, and serine than those for the other steps. Their data further demonstrate clear differences among amino acids, glycine being relatively the slowest in Step 2, but almost the fastest in Step 3, whereas arginine is relatively the fastest in Step 2, but seven times slower than glycine in Step 3.

Mundt et al.[119] have adapted the above scheme for the maltose–glycine system (pH 5.5, 70 °C). Multiresponse modelling gave excellent fits for the time course of the concentrations of glucose, melanoidins (A_{470}, number of maltose molecules incorporated from U-^{14}C-labelled maltose), 3-DH (as quinoxaline derivative), maltose, and S^{IV} (in maltose–glycine–S^{IV} systems).

Martins and van Boekel[120] have considered the key intermediates in the early stages of the above scheme, which they modified to

$$\begin{array}{c} \text{AS} \xleftarrow{k_1} \text{ARP} \xrightarrow{k_2} \text{1-DG} + \text{A} \\ k_4 \downarrow \qquad\qquad\qquad k_3 \downarrow \\ \text{3-DG} + \text{A} \qquad\qquad \text{S*} \\ k_5 \downarrow \\ \text{M} \end{array}$$

AS is a reactive intermediate, formed from the amino acid (A) and sugar (S) prior to the Amadori product (ARP). Its nature, say, Schiff base or glucosylamine/enaminol, was not studied. However, its inclusion was of major importance in obtaining fits by multiresponse modelling of the decomposition of fructosylglycine on its own (pH 5.5, 100 °C), when glycine and 1- and 3-deoxyosones (DG) were the only products detected. Because of the reactivity of DGs, further products were expected, hence the inclusion of k_3.

The derived rate constants were applied to the glucose–glycine system, where a good fit was obtained. In accord with the above scheme, ARP and 3-DG were formed simultaneously, while there was a lag phase for colour (melanoidins, M) and 1-DG. 3-DG proved to be the main pathway for colour formation.

For the model system glucose–glycine (0.2 M each in 0.1 M phosphate buffer, pH 6.8), van Boekel and Martins[121] explain that the glucose reacts to a greater degree than glycine, because (1) glucose isomerises to fructose as well as undergoing the Maillard reaction, and (2) part of the glycine is recycled, being released from intermediate-stage Maillard products. From 80 to 100 °C, the Amadori product and melanoidins completely account for the loss of glycine, but, at higher temperatures, 10–20% of the glycine input (up to 66% of the reacted glycine) becomes unaccounted for. Some of the explanation will lie in those parts of the Strecker degradation products that are not taken up in the melanoidins, but which go on to form low-molecular-mass nitrogen heterocycles. It is clear from this that the behaviour of glycine is such that the kinetics of its decrease should not be taken as a measure of the Maillard reaction.

Using a laboratory reaction cell to model Maillard colour development, Bates et al.[122] examined a starch–glucose–lysine system at 100–155 °C, 13–18% H_2O, and pH 2.9–10.7. The tristimulus Z values allowed a quadratic response surface regression equation to be developed, which fitted well except at the higher pH values. Colour development followed Arrhenius kinetics over the range 120–155 °C, with activation energies lying in the range 41.1–110.6 kJ (9.8–26.4 kcal) mol^{-1}. The activation energy generally decreased as the moisture content increased and as the pH became more alkaline, 8.5 to 10.7.

Brands and van Boekel[123] have been concerned with the kinetics of the interaction of monosaccharides and casein for some time. It is not surprising that the model they have developed is complex. The current form is depicted in Scheme 3.2. It incorporates unidentified intermediates, C_n and C_5, which react with lysine residues and form advanced Maillard products (AMPs). This model fitted lysine

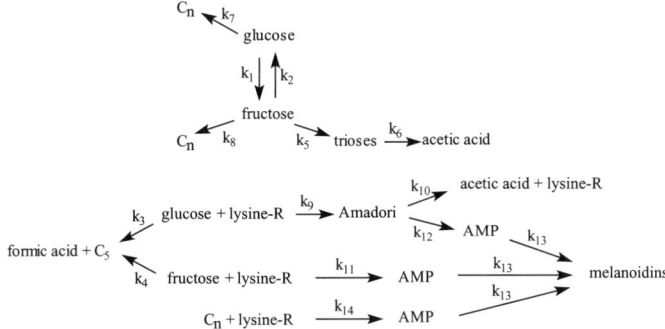

Scheme 3.2 *Model of sugar–casein reactions*[123]

loss in fructose–casein systems heated at 100 °C much better than previous ones, without affecting the goodness of fit at higher temperatures. The activation energies, calculated for every step, were around 120 kJ (28.7 kcal) mol^{-1}, as expected for chemical reactions. However, the reaction between fructose and casein proved to be much more temperature-dependent [E_a = 175 kJ (41.8 kcal) mol^{-1}] and, therefore, less important at lower temperatures than the reaction between the intermediates and the protein.

Brands and van Boekel[124] noticed that there were significant differences between aldoses and ketoses, the latter seeming more reactive in sugar degradation reactions. Whereas Amadori compounds were detected in aldose–casein systems, no such intermediates were found for ketose–casein systems.

The considerable differences in the slopes of the Arrhenius plots for Maillard volatiles have led Reineccius[125] to point out that high temperatures of processing favour, *e.g.*, 2,3-dimethylpyrazine (nutty), whereas lower processing temperatures allow, *e.g.*, methylcyclopentenolone (caramel), to dominate. However, the rates of formation of both of these types of components become negligible at storage temperatures (say 30 °C). The low temperature favours off-flavours, *e.g.*, hydroxymethylfurfural (stale), and extending the time compounds the problem, rather than compensating for an increase in temperature. Reineccius[125] concludes that, at some particular temperature–time combination, a flavour is perceivable, but it is not at all probable for the same flavour to be perceived under other temperature–time combinations.

Stahl and Parliment[126], too, examined higher temperature reactions. They studied the glucose–proline system (0.2 M in each) in a continuous-flow reactor either at different times (0.25–5.0 min) at 200 °C or for 1.0 min at different temperatures (160–220 °C). The main volatile components identified at 200 °C after 0.25 min were only three: 2-acetyltetrahydropyridine, 5-acetyl-2,3-dihydro-1*H*-pyrrolizine, and maltoxazine (see Scheme 5.3, p. 66). After 2 min the pyrrolizine dominated. The kinetics for each of the three compounds was pseudo-zero order. The activation energy of the pyrrolizine was 188 kJ (45 kcal) mol^{-1}, much higher, as expected, than those of the other two compounds [25–63 kJ (6–15 kcal) mol^{-1}]. The low value for maltoxazine fits with its presence in malted barley and its formation in aqueous glucose–Pro and maltose–Pro systems at boiling point.[127]

Stahl and Parliment[126] also determined the loss of glucose and proline with time and found that the rate constant at 200 °C for the former was about three times that of the latter (0.633 and 0.227 min^{-1}, respectively). This difference may be partly due to regeneration of the amine and partly due to caramelisation independent of the amine. The activation energies for glucose and proline were 61.5 (14.7) and 86.7 (20.7) kJ (kcal) mol^{-1}, respectively, and that for browning products (A at 300–304 nm) was 112.6 kJ (26.9 kcal) mol^{-1}.

The kinetics of flavour formation during the Maillard reaction has been reviewed recently by Reineccius.[128] So far, only little work has been carried out and he warns that the formation of flavour may be very system-dependent. The initial period of flavour formation tends to be linear and can be modelled using pseudo-zero-order kinetics. For longer heating times, flavour concentration in the system will generally tend to form a plateau, and first-order kinetics will become a better basis for modelling. Activation energies for flavour formation range from 50 to 190 kJ (12 to 45 kcal) mol^{-1}, the higher values being for the linear phase of pyrazine formation. Values for many other volatiles are close to 84 kJ (20 kcal) mol^{-1}, more in line with the consumption of Maillard reactants or melanoidin formation. 2-Acetyl-1-pyrroline has an activation energy of about 59 kJ (14 kcal) mol^{-1}, a very low value, which, together with its low threshold (0.1 µg L^{-1}), may explain its frequent detection in foods by sniffing.

Recently, Peleg et al.[129] have made a case for applying non-Arrhenius and non-Williams-Landel-Ferry kinetics to the Maillard reaction, since its reactivity is only noticeable above a certain temperature. They show that a relatively simple, empirical log-logistic relationship,

$$Y = \ln\{1 + \exp[c\,(T-T_c)]\}^m$$

where Y is the rate parameter, T_c is the marker of the temperature range, and c and m are constants, gives a highly satisfactory fit to a range of published data sets.

Jousse et al.[130] have surveyed the data available on the kinetics of aroma formation by the Maillard reaction and have assembled a simplified, but broadly based, kinetic scheme, as outlined in Scheme 3.3.

The scheme is summarised by the following equations, based on the abbreviations given in Scheme 3.3:

$$S' = -R_1 S - R_2 S.AA$$
$$AA' = -R_2 S.AA + R_4 ARP - R_8 C.AA$$
$$ARP' = R_2 S.AA - R_3 ARP - R_4 ARP$$
$$PY' = R_3 ARP - R_{11} PY$$
$$RS' = R_4 ARP - R_5 RS - R_6 RS$$
$$FU' = R_5 RS + R_7 C.C - R_{11} F$$
$$C' = 2R_6 RS - R_7 C.C - R_8 C.AA - R_{11} C$$
$$I' = R_8 C.AA - R_9 I - R_{10} I.I$$
$$SA' = R_9 I - R_{11} SA$$
$$PZ' = R_{10} I.I - R_{11} PZ$$

The rates are expressed as s^{-1} for unimolecular reaction steps R_1, R_3, R_4, R_5, R_6, R_9, and R_{11}, and M^{-1} s^{-1} for bimolecular reactions R_2, R_7, R_8, and R_{10}. The abbreviations, such as S for sugar, represent its molarity, whereas S' is the time derivative. The

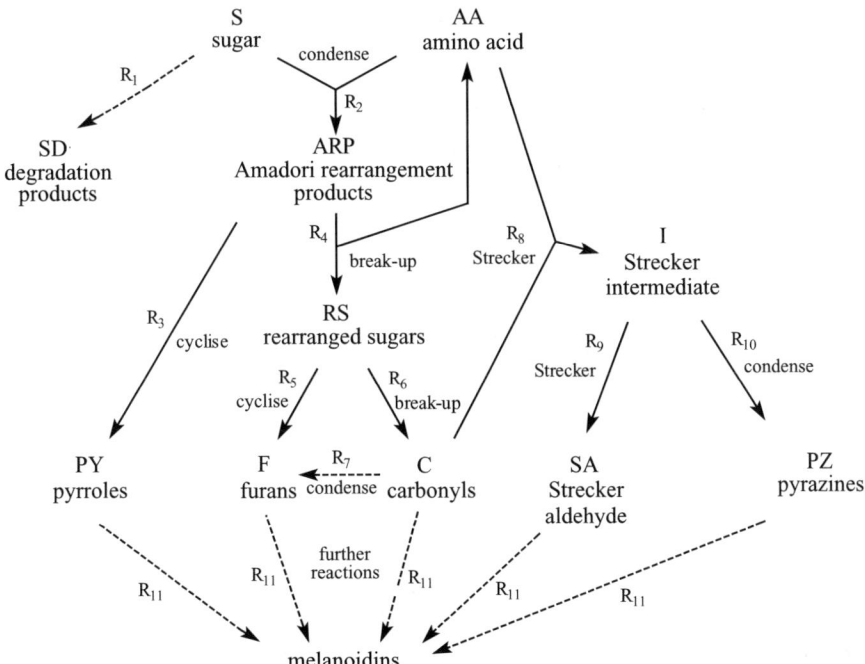

Scheme 3.3 *Simplified kinetic scheme for flavour generation by the Maillard reaction*[130]

Table 3.5 *Arrhenius rates used to fit the experimental concentrations of pyrroles, furans, Strecker aldehydes, and pyrazines from a glucose–alanine system as a function of time and temperature*[130]

Process	R_0	E (kJ mol^{-1})
R_1 (s^{-1})	0	0
R_2 (M^{-1}s^{-1})	5.0×10^{12}	120.5
R_3 (s^{-1})	6.0×10	35.6
R_4 (s^{-1})	1.5×10^5	52.9
R_5 (s^{-1})	2.0×10^{11}	109.3
R_6 (s^{-1})	5.0×10^5	66.5
R_8 (M^{-1}s^{-1})	5.0×10^{11}	83.1
R_9 (s^{-1})	1.0×10^{15}	116.3
R_{10} (s^{-1})	Fast	–
R_{11} (s^{-1})	1.0×10^{10}	99.7

changes of the system with time are completely determined by the values of the rates and the initial molarities of sugar and amino acids.

The rates used to fit the experimental data for PY, FU, SA, and PZ from a glucose–alanine system are given in Table 3.5. These values can be used to give a first approximation of the profile of volatiles to be expected to result from a Maillard reaction between specific components under particular conditions.

8 Effect of Glass Transition Temperature T_g

Bell et al.[131] were the first to try to correlate the effect of T_g on browning with that on reactant consumption. They studied three poly(vinylpyrrolidone) systems (PVP, MM ~40, ~30, and <3.5 kDa), containing glycine and glucose (0.1 molal each) at $a_w = 0.11$–0.76. At $a_w = 0.54$, the plot for loss of Gly was second-order and that for browning had pseudo-zero order. At $a_w = 0.11$, where all systems were almost 30 °C or more below the T_g, the rate constants both for Gly loss and for browning were very low. At $a_w = 0.33$, both had increased ($P < 0.05$), even though all the systems were still glassy. At $a_w > 0.54$, the rate for Gly loss decreased, which was attributed to matrix collapse, whereas the rate of browning remained steady. The low-MM PVP system changed from glassy to rubbery, as a_w was increased from 0.11 to 0.44, and the rate constant for browning increased by more than 30 times. For the other PVP systems, the rate constants hardly changed as they altered from glassy to rubbery. Other factors must be playing a role. Also, there was no correlation between the rate constant for Gly loss and for browning. In some cases, the rate of browning was high, but the loss of Gly was low; conversely, the highest rate of Gly loss did not correspond with the highest rate of browning.

Craig et al.[132] determined the rates of reactant consumption for the Maillard reaction between glucose and lysine in the T_g region of a non-crystallising matrix of trehalose–sucrose–water. Above the T_g, the rates showed Arrhenius temperature dependence with E_a of 135 and 140 kJ (32.3 and 33.5 kcal) mol^{-1} for lysine and glucose, respectively. For the glassy samples, reaction rates were finite and faster when compared with the non-glassy ones. Consideration of the results indicated that the reaction controlled the rates above the T_g, but below the T_g the systems approached diffusion control.

The effect of T_g on the rate of browning of mixtures of amorphous lysine and xylose, containing maltodextrin (MD) or PVP and equilibrated to $a_w = 0.23, 0.33$, or 0.44, has been studied.[133] Rates of browning increased with temperature, were greater for PVP than MD systems, and increased with a_w, but did not correlate with T_g. Following this work, Lievonen et al.[134] compared xylose with fructose and glucose in aqueous, MD, and PVP systems (the last two at $a_w = 0.33$). In water, NEB exhibited a lag phase followed by a linear relationship, whereas, in MD and PVP systems, there was no lag phase and the relationship was linear at first, but then levelled off. Xylose browned fastest in all systems and fructose generally browned faster than glucose. Comparing the matrices, browning decreased for fructose in the order water > MD > PVP and for glucose water > PVP > MD, but for xylose PVP > water > MD. Phase separation in the PVP system was considered to provide a possible explanation. The ratios of browning in solid/aqueous systems were generally low below T_g and increased significantly only about 10–20 °C above T_g. Again, xylose proved an exception, the ratios for the PVP system beginning to increase immediately, well below T_g. This clearly links to the fact that the rate constant for the PVP system is greater than that for the aqueous system, even at temperatures where the PVP system is glassy.

Lievonen et al.[134] analysed the temperature dependence of NEB, using the Arrhenius equation. All systems could be analysed, five of the six giving a visually observable break, usually 2–12 °C above T_g. For glucose/MD, the break was observed below T_g and, for glucose/PVP, no break was apparent. E_a calculated for below and above the break did not differ much, so changes in physical state had only a moderate effect on NEB kinetics. Application of the Williams–Landel–Ferry equation did not prove superior.

9 Amines Other than Amino Acids as Reactants

9.1 Peptides

It has been appreciated for a long time that, in model systems, the reactivity of amino acids in Maillard reactions does not necessarily remain the same when present as a component of a peptide. Thus, van Chuyen et al.[135,136] showed that glyoxal reacted with amino compounds at 80 °C in the order tetraglycine > triglycine > diglycine > dialanine > glycine > alanine, the reactivity of the peptides being much greater than that of the amino acids. From the reaction mixtures, they were able to isolate and characterise a series of pyrazinones (see Structure **2**).

The reactivity of peptides has also been studied by de Kok and Rosing,[137] who have reviewed earlier work. In addition, they undertook several series of experiments to determine the kinetics of glucose degradation in mixtures of glucose with peptides (ca 0.4 M in each reactant, 100 °C, pH 5.6, up to 6 h). Sugar degradation in the absence of peptide was negligible, as was the hydrolysis of peptide in the absence of glucose. The reactivity in terms of glucose degradation was found to decrease in the order diglycine > triglycine > glycine, even though the pK_2 values were 8.25, 7.91, and 9.77, respectively, showing that the terminal amino group is not the only factor operating. Intramolecular catalysis by the carboxy group (see Scheme 3.4) seems to be an acceptable explanation.

In the case of dipeptides GlyX (X = Gly, Val, Thr, Pro, Phe, His, Lys, Asp, or Glu), the reactivities were similar, except for X = Glu, where there was a strong relative

	R = e.g.,	–CH(COOH).CH$_2$.CH(CH3)$_2$	from glycylleucine
		–CH$_2$.CO.NH.CH$_2$.COOH	from triglycine
2		–CH$_2$.CO.NH.CH$_2$.CO.NH.CH$_2$.COOH	from tetraglycine

Structures 3.1   1-Substituted pyrazin-2-ones from the reaction of glyoxal with peptides at 100 °C and pH 5.0[135,136]

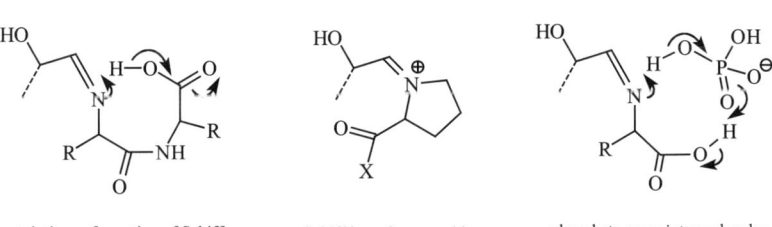

catalytic conformation of Schiff base from an aldose and a dipeptide

Schiff base from an aldose and a proline dipeptide, Pro-X

phosphate as an *intra*molecular protontransfer agent

Scheme 3.4   *Mechanisms to explain the reactivity of dipeptides and phosphate catalysis*[137]

enhancement. This was true both in the absence and in the presence of phosphate, which overall gave a reactivity about three times as high.

In the case of dipeptides MetX (X = Met, Gly, Glu, Lys, Pro), the reactivities were again similar, except for X = Glu, where again there was strong relative enhancement, both with and without phosphate, which overall gave a reactivity about twice as high.

For the dipeptides ProX (X = Gly, Glu, Lys, Pro), surprisingly, the 'glutamic acid effect' was not evident, but X = Lys gave a strong relative enhancement, both with and without phosphate, which overall gave a reactivity about three times as high. Intramolecular catalysis by the carboxy group in the sidechain of glutamic acid is not possible here, as the nitrogen of the Schiff base is quaternary and so has no proton to transfer to the carboxy group. Instead, the ε-amino group of the lysine is able to pick up the proton from C-2 of the carbohydrate moiety (see Scheme 3.4).

The degree of phosphate catalysis (DPC = ratio of reaction rate in the presence and absence of phosphate) varies with the nature of the dipeptide. The plots suggest that, within each series, DPC is negatively correlated with the relative rate in the absence of phosphate. Phosphate is postulated to act as a proton transfer mediator between the dipeptide carboxy group and the reactive imine centre (see Scheme 3.4). If direct transfer from the dipeptide carboxy group is already sufficiently efficient, the enhancement is decreased.

The dipeptide carnosine, β-alanyl-L-histidine, is one of the most abundant N compounds present in the non-protein fraction of vertebrate skeletal muscles. It constitutes, for example, 50, 150, and 276 mg per 100 g of muscle tissue from chicken leg, bovine leg, and porcine shoulder, respectively. Chen and Ho[138] examined its effects on volatile generation in a model system of ribose and cysteine (180 °C, 2 h, pH 5 and 8.5). These were complex; the levels of thiophenes and some meaty compounds, such as 2-methyl-3-furanthiol, 2-furfurylthiol, and their associated dimers, were generally lowered, but those of important N compounds, such as pyrazines and thiazoles, which are known to elicit roasty and nutty flavours, were enhanced.

9.2 Phospholipids

Phospholipids, bearing an amino group, are potentially able to act as amino components in Maillard reactions, the involvement of phospholipids being first suggested by Bucala et al.[139] In vitro, dioleylphosphatidylethanolamine is 40% glycated by an excess of glucose in 2 weeks at 37 °C, mostly as the Schiff base, only 18% of the product being stable at pH 5 for 1 h at 37 °C, i.e., was in the form of the Amadori compound, not the Schiff base,[140] as determined by GC-MS. The phospholipids isolated from red blood cell membranes were glycated at a level of 0.48 mmol mol^{-1} ethanolamine, almost all as the Amadori compound. In samples from diabetic subjects, this value was three times as high. In earlier work,[141] using LC-MS, 1.2 and 2.3% of phosphatidylethanolamine from red blood cell membranes and from plasma, respectively, were found to be glycated, with values increasing to 10 and 16%, respectively, for samples from diabetic subjects. It should be noted that the ratio of diabetic/nondiabetic is here about 10.

Carboxymethylethanolamine (CME) has been shown to be present in red blood cell membrane hydrolysates at ca 0.14 mmol mol^{-1} ethanolamine. Carboxymethyllysine was present in the membrane proteins at ca 0.2 mmol mol^{-1} lysine, in both cases there being no difference between samples from diabetic (free of complications) and nondiabetic subjects.[142] CME was also detected in fasting urine at 2–3 nmol mg^{-1} of creatinine from samples of both diabetic and nondiabetic subjects. Requena et al.[142] calculate a value of the 2.8 μmol CME excreted daily, $>$99% arising from the normal turnover of membrane lipids of cells other than RBC.

Miyazawa et al.[143] determined the Amadori compounds from dioleylphosphatidylethanolamine by normal-phase HPLC as the 3-methyl-2-benzothiazolinone hydrazones (MBTH) through their UV absorption (318 nm) with detection limits of 4.5 and 5.3 ng for the fructosyl (F) and the lactulosyl (L) compound, respectively, and linear ranges of up to about 2 μg. Using their method, they obtained the following results: infant formula, F 32–112, L 49–88 μg g^{-1}; mayonnaise, F 12.2 μg g^{-1}; chocolate, F 3.9, L 1.5 μg g^{-1}; cow's milk, L 0.079 μg mL^{-1}; soybean milk, F 0.24, L 0.13 μg mL^{-1}; and rat plasma, F 0.23 μg mL^{-1}. Significant amounts were not detected in human milk.

Above, N-(carboxymethyl)phosphatidylethanolamine was reported to be present in red blood cell membranes to the extent of 0.14 mmol mol^{-1} ethanolamine,[142] yet N-(carboxymethyl)phosphatidylserine has been obtained in 0.44 mmol mol^{-1} serine,[140] even though the ratio of phosphatidylethanolamine to phosphatidylserine is about 2:1. Such preferential modification of phosphatidylserine was unexpected.

Much attention has been focused on low-density lipoprotein (LDL) owing to its importance to health. Glycation of both its protein and its lipid component is believed to be involved when LDL is oxidised. Using glucosylated phosphoethanolamine (PE), Ravandi et al.[144] have identified glycated ethanolamine phospholipids in LDL. LDL specifically enriched with glucosylated PE showed increased susceptibility to lipid oxidation, the presence of glucosylated lipid resulting in a 5- and 4-fold increase in the production of phospholipid-bound hydroperoxides and aldehydes, respectively. Inclusion of glucosylated PE in the surface lipid monolayer of the LDL resulted in a rapid loss of polyunsaturated cholesteryl esters from the interior of the particles during oxidation. Glycated ethanolamine phospholipids were also isolated and identified from atherosclerotic plaques collected from both diabetic and nondiabetic subjects.

Atherosclerosis develops rapidly in patients with diabetes or renal insufficiency. Plasma lipoprotein profiles are often abnormal in these cases, with higher levels of apoprotein B-containing components VLDL and LDL. High levels of circulating AGEs also occur in diabetes and end-stage renal disease. Bucala et al.[145] therefore investigated the possibility of circulating AGEs' reacting directly with plasma lipoproteins to prevent their recognition by tissue LDL receptors, thus hindering clearance. AGE-specific ELISA showed increased AGE-modified LDL in the plasma of such patients, and AGE-LDL formed readily in vitro when native LDL was incubated with either synthetic AGE peptides or AGE peptides isolated from the plasma of patients. LDL, which had been modified to the same level as that in the plasma of diabetic patients with renal insufficiency, exhibited markedly impaired clearance kinetics when injected into transgenic mice expressing human LDL receptors. Administration of the advanced glycation inhibitor aminoguanidine to diabetic patients lowered circulating LDL by 28% over a 4-week period, supporting the above reasoning.

9.3 Fumonisin

These toxins were first identified as produced by *Fusarium verticillioides* Sacc. (Nirenberg) growing on maize. They are commonly found in fodder, but also occur in maize products consumed by humans. A range of toxic responses has been reported for animals, and fumonisin has been associated with a high incidence of human oesophageal cancers in South Africa and China.[146] Fumonisin B_1 (**3**) has been shown to possess a primary amino group and so it has been suggested that nonenzymic browning could explain losses in samples exposed to increased temperatures and durations.

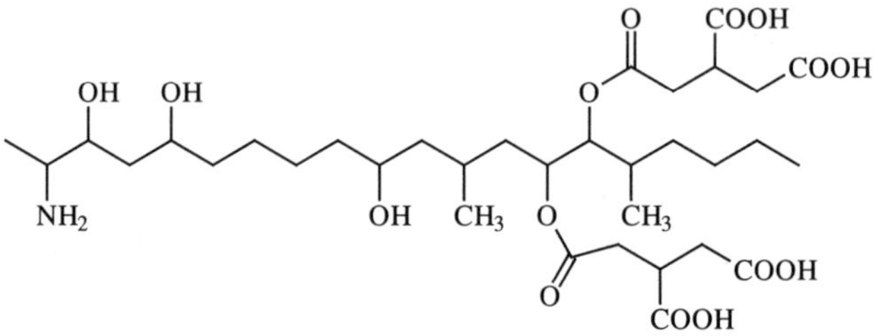

3

fumonisin B_1

Accordingly, Lu *et al.*[147] have shown that **3** heated with glucose at 65 °C for 48 h leads to four primary products: the *N*-methyl, *N*-carboxymethyl, *N*-(3-hydroxyacetonyl), and *N*-(2-hydroxy-2-carboxyethyl) derivatives. The *N*-(1-deoxyfructos-1-yl) compound was detected when heating was limited to 60 °C. Heating at 80 °C for 48 h led to more than 10 products. The reaction kinetics were apparently first-order and led to an activation energy of 105.7 kJ (25.3 kcal) mol^{-1}, implying that heat is necessary to achieve a useful degree of reaction with feasible processing times. Heating maize with glucose at 80 °C caused gelatinisation, interference from which could be avoided by addition of α-amylase; heating at 60 °C did not cause gelatinisation and heating at 50 °C or below gave no reaction. At 60 °C, 8 d were needed to reduce **3** by 50%, whereas at 80 °C only 2 days were required, but the extra cost of α-amylase needs to be borne in mind.

Costelo *et al.*[148] baked maize muffins with glucose at 200 °C, obtaining a reduction of 45–70% in **3**, whereas extrusion at 160 °C gave 90% reduction.

9.4 Chitosan

When chitosan is heated for 1 h at 25 °C, followed by 15 min at 175–180 °C in the presence of water, yeast, and reducing sugar, it becomes largely nonrecoverable because of the Maillard reaction.[636]

10 Effect of Lipids

The effect of lipids on the volatiles formed in foods has been reviewed by Whitfield.[149]

Lipid oxidation products react with proteins and other amino compounds to form brown substances, similar to melanoidins. The formation of such brown substances was reviewed already at the first Maillard Symposium.[150] The pigments formed are partly soluble in chloroform–methanol and partly insoluble, whereas true melanoidins are largely water-soluble. As most brown pigments of fish muscle are soluble in benzene–methanol and only to a lesser extent in water, the implication is that here oxidised lipid–protein interactions are more important than Maillard browning due to ribose–amino acid interactions.

Of the different types of lipids in foods, the phospholipids, being more unsaturated, are particularly important in relation to aroma formation in meat.[151] The aroma of cooked meat was not affected by the prior extraction of triglycerides with hexane, but the use of a more polar solvent (chloroform–methanol), which extracts all lipids, including phospholipids, resulted, after cooking, in the replacement of the meaty aroma by a roast or biscuit-like one. This was reflected in the volatiles, the dominant aliphatic aldehydes and alcohols being replaced by alkylpyrazines. This implies that the participation of the lipids in the Maillard reactions inhibited the formation of heterocyclic compounds.

Thiazoles with long alkyl chains in the 2-position provide evidence of the interaction of lipid oxidation products with Maillard intermediates. The presence of such thiazoles has been reported for fried chicken, roast beef, and fried potatoes (see Mottram[152]). Thiophens with long alkyl chains in the 2-position have also been encountered. They were obtained when phospholipid was added to a cysteine–ribose system.[153] Lipid degradation to the 2,4-dienal, followed by reaction with H_2S, was thought to be responsible.[154]

Browning due to interaction of oxidised lipids and proteins occurs *in vivo*, the brown pigment deposited in tissues being termed lipofuscin or ceroid. The appearance of brown discoloration in the adipose tissues of, *inter alia*, pig, mink, and chicken, has been called the 'yellow fat disease'. Ceroid accumulates slowly and has therefore been described as the 'age pigment'.

When polyunsaturated fish oil esters were mixed with an excess of egg albumin, heated to 140 °C for 1 h, and extracted with chloroform–methanol, the liposolubles could be chromatographed on silica gel. Fractions containing only traces of nitrogen were light in colour, but the deep-brown fraction contained 0.9%.[150] The effect of water depends on the system; the rate of browning of methyl linoleate-casein mixtures[155] at r.h. of 0, 33, and 75% was highest under the last conditions.

The brown products are much less intensely coloured when formed from less unsaturated lipids.[156] The introduction of novel cultivars, such as SunOleic peanuts with only 3–6% linoleic acid, compared with the 30% of a traditional cultivar, such as Virginia, is therefore advantageous as regards the manufacture of, say, snacks, where a light colour is desirable.

11 Sites of Protein Glycation

Starting with a synthetic heptapeptide, N^α-Lys-Lys-β-Ala-Lys-β-Ala-Lys-Gly, and equimolecular amounts of glucose or lactose (2 h, 110 °C, pH 6.7), the sites and nature of glycation were determined using MALDI–TOF–MS, collision-induced decomposition ESI-MS/MS, and [^{1}H]- and [^{13}C]-NMR on fractions obtained on RP-HPLC.[157] Glucose gave the Amadori compound at Lys1 or Lys2 or both. Other fractions gave Lys1 Amadori and Lys2 furosine (-2H) or formylpyrrole, and Lys1 furosine (-2H) or CML (-2H). Similarly lactose gave the Amadori compound at Lys1 or Lys2 or both, as well as Lys1 Amadori and Lys2 formyl or CML and Lys1 formyl or a C_3 compound. There was no glycation at Lys3 or Lys4.

Hasenkopf et al.[158] studied the glycation of various proteins, using glucose, ascorbic acid, or related compounds, followed by enzymic hydrolysis, silylation, and GC-MS. Polylysine and dehydroascorbic acid led to carboxymethyl-, carboxyethyl-, oxalyl-, and formyl-lysine (CML, CEL, OL, and FoL, respectively). CML was formed in the highest amount, with only minor amounts of CEL and OL obtained. Only OL was formed during ascorbylation, but it can also be obtained from sugars, provided oxidative conditions are used. Ascorbylated polylysine gave an additional compound, identified as 1-carboxy-3-hydroxypropyllysine (CHPL). β-Lactoglobulin and sugars gave CML, CEL, OL, and CHPL. Note that synthetic OL is converted into FoL on GC-MS, but is stable to enzymic hydrolysis, followed by derivatisation with phenyl isothiocyanate, and can be determined by HPLC with UV detection. Polylysine and sugars gave OL and FoL in different ratios: $\geq 7:1$ for ascorbic acid, dehydroascorbic acid, and maltose, 5:2 for glucose, 3:2 for ribose, 2:3 for fructose, and 1:3 for lactose. For all the proteins investigated, CML was detected in the highest amount, ranging from 3.0 with lactoglobulin and lactose to 167.2 mmol mol^{-1} Lys for AGE-lactoglobulin (glucose for 60 d at 37 °C), indicating 0.05–2.7 CML mol^{-1} protein (16 Lys residues). OL was obtained in 9.0–35.8, FoL in 1.2–24.2, and CEL in 1.1–3.5, but CHPL in only <0.25 mmol mol^{-1} Lys. Ribose gave the highest sum of products of glycation, namely, 209.1, lactose the lowest, 3.0 mmol mol^{-1} Lys. Very different results have been reported when other markers of glycation have been used, and so, at this stage, generalisations are to be avoided.

Kislinger et al.[159] have used MALDI–TOF–MS for the relative quantification of CML, imidazolone A, and Amadori product (AP) in glycated chicken-egg lysozyme. The enzyme (0.7 mM) was treated with glucose (500, 250, 100 mM) for 1, 4, 8, and 16 weeks in phosphate-buffered saline (pH 7.8) at 50 °C. The product was digested with endoproteinase Glu-C, and the three Maillard products were assayed in the resultant N- and C-terminal peptide fragments. They constituted the main glycation products under the conditions used. Their formation was dependent on glucose concentration and on contact time. The kinetics were similar to that obtained by competitive ELISA, an established method for quantifying CML and imidazolone A. Inhibition experiments showed that incubation with N^α-acetylarginine suppressed the formation of the imidazolone, but not of the AP or CML. In the presence of N^α-acetyllysine, the modification of Lys was inhibited, but the concentration of imidazolone A increased. Treatment with o-phenylenediamine decreased the yield of AP and completely inhibited the formation of CML and imidazolone. The endoproteinase Glu-C specifically cleaves peptide bonds C-terminal to glutamic or aspartic acid, thus giving amino acid fragments 1–7 (KVFGRCE) and 120–129 (VQAWIRGCRL) as the N- and C-terminal peptides from lysozyme. Samples (500 pmol μL^{-1}) were diluted 1:50 in saturated α-cyano-4-hydroxycinnamic acid in 0.1% TFA with 33% acetonitrile and 1 μL of the solution was spotted onto the target of the MS. Under the conditions used, only singly charged ions were detected. The monoisotopic peak integrals of modified peptides were measured in comparison with the monoisotopic peak integrals of the unmodified peptides, defined as 1.0, thus producing relative integrals. The formation of AP was dependent both on glucose concentration and on time. The higher glucose concentration seemed to lead to excessive glycation and thus to incomplete digestion. AP was detected after 1 week with 100 mM glucose, but CML was not. AP plateaued after 16 weeks. CML was detected after 4 weeks. Being an advanced glycation endproduct (AGE), it was expected to continue to increase with time, but, in fact, it decreased, probably again due to incomplete digestion. The imidazolone, formed on Arg-125, increased with glucose concentration and with time. In contrast with CML, its curve for relative % was almost linear both for 100 and 250 mM glucose, as expected for an AGE, not flattening until 24 weeks were reached. With 500 mM, the curve flattened after 8 weeks.

Brock et al.[160] have used ESI-LC-MS to compare the sites of glycation and carboxymethylation of ribonuclease A (RNase), a 13.7 kDa enzyme with a known sequence of 124 amino acid residues, including 9 Lys residues. The enzyme (1 mM) was incubated with glucose (400 mM) at 37 °C for 14 d in phosphate buffer (0.2 M, pH 7.4) under air, followed by tryptic digestion. The major sites of glycation were Lys41, Lys7, Lys1, and Lys37, the degree of glycation decreasing in that order. Three of these, in decreasing order, Lys41, Lys7, and Lys37, were also the major sites of CML formation. When RNase was incubated under anaerobic conditions (1 mM diethylenetriaminepentaacetic acid, N_2 purged) to form glycated protein, which was then incubated aerobically to allow CML formation, the major sites of glycation and CML formation remained the same. When RNase was incubated with up to 5 mM glyoxal, substantially more than is formed by autooxidation of glucose under the conditions used, lysine residues were modified in traces only, primarily at Lys41. It follows that the primary route to CML formation is by autooxidation of glycated protein, rather than by glyoxal derived by glucose autooxidation, and carboxymethylation, like glycation, is site-specific.

A method of assaying a wide range of early and AGEs was developed by Ahmed et al.[50] It involves hydrolysis, derivatisation with 6-aminoquinolyl-N-hydroxysuccinimidylcarbamate (AQC), and HPLC with fluorometric detection (structural isomers of hydroimidazolones from glyoxal, 2-oxopropanal, and 3-deoxyglucosone and THP, N^δ-(4-carboxy-4,6-dimethyl-5,6-dihydroxy-1,4,5,6-tetrahydropyrimid-2-yl)ornithine). AGEs with intrinsic fluorescence (argpyrimidine and pentosidine) were assayed without derivatisation. The limits of detection are 2–17 pmol and the recoveries 50–99%. Except for pentosidine, the compounds detected had short half-lives of only 0.7–12 d at pH 7.4. Contrary to in in vitro[161], in in vivo samples, GOLD, MOLD, DOLD, CML, CEL, and pyrraline could not be determined because of lack of sensitivity and/or resolution.

The assay was applied to HSA modified with 0.5 or 100 mM MGO in 100 mM phosphate buffer (pH 7.4) for 24 h, when 1.45 and 1.92 MG-H1, 0.27 and 1.72 THP, and 0.23 and 3.19 mol mol^{-1} protein ARGPy were found, respectively.[161] Similar experiments with 40 and 1670 mM glucose led mainly to fructosyllysine and CML, with minor amounts of THP, MG-H1, G-H1, 3-DG-H1, ARGPy, and DOLD. With the higher concentration of glucose, pyrraline was found in addition and the fructosyllysine reached about 8 mol mol^{-1} protein (36 lysine residues). The arginine-derived AGEs accounted for 80% of the arginine residues modified, but the corresponding value for lysine residues was only about 25%. It is worth noting that commercial batches of HSA contained high and variable amounts of G-H1 (up to 3.15 mol mol^{-1} protein) and small amounts of fructosyllysine, 3-DG-H1, MG-H1, THP, and ARGPy (all formed 0.28 mol mol^{-1} protein or less).

Capillary electrophoresis is a powerful separation technique. Hinton and Ames[162] subjected BSA (1 mM) to capillary electrophoresis, incubating it alone or with glyoxal (25 mM) in phosphate buffer (pH 7.5) at 37 °C for 14 d. Electropherograms of the < 5 kDa fractions from tryptic digests exhibited > 70 peaks with some unique to each sample.

Chevalier et al.[163] glycated β-lactoglobulin (0.217 mM) with equimolar amounts of ribose, arabinose, galactose, glucose, rhamnose, or lactose at 60 °C in 0.1 M phosphate buffer (pH 6.5) for 72 h, strictly anaerobically, the average number of amino groups modified being 11.0, 8.8, 6.7, 6.6, 6.5, and 5.5, respectively.

Lactosylation of caseins was studied by Scaloni et al.[164] Combined MALDI–MS/Edman degradation on enzymic digests showed that moderate heating involved the sites Lys34 (α_{S1}-casein) and Lys107 (β-casein), more severe heating leading to additional attack on lysines at positions 7, 83, 103, 105, 132, and 193 and 32, 48, 113, and 176, respectively.

12 Effect of Oligo- and Polysaccharides

In Maillard reactions with lactose, the 4-hydroxy group is substituted and hence β-dicarbonyl formation is blocked, thus preventing the production of 2-acetylpyrroles.[30]

Hollnagel and Kroh[165] compared glucose, maltose, and maltotriose (MT) under 'caramelisation' and 'Maillard' conditions in aqueous solution (0.5 mL, 0.25 M, up to 240 min at 100 °C in sealed tubes), i.e., without and with glycine (0.5 mL, 0.25 M), respectively. Under both conditions, 3-deoxypentosone (3-DP) was the predominant α-dicarbonyl formed from maltose and MT, but it was not formed from glucose. In consequence, a special 'peeling-off' pathway was proposed for the formation of 3-DP (see Scheme 3.5), 'peeling-off' because the (n − 1) oligosaccharide eliminated can undergo a further cycle of reactions. The presence of the amino acid favours such a pathway, which is significant neither in the Maillard reaction of monosaccharides nor in the caramelisation of mono- and oligosaccharides.

In model systems of lysine with maltose, maltotriose, or maltotetrose, heated for 15 min at 110 °C, glucosyl- (GP), diglucosyl-, or triglucosyl-β-pyranone were identified by HPLC–MS–MS, respectively.[166] When the sugar component was amylose and β-amylase was added after the heat treatment, only GP and diglucosyl-β-pyrone were detected. Formation of GP, as well as of the derived glucosylisomaltol (GIM) and glucosyl-lysylpyrrole (GLP), was studied in heated systems containing maltose and increasing amounts of lysine. GIM was detectable when the ratio of Lys/maltose exceeded 1:100 and increased quickly as GP degradation began. GLP behaved similarly (see also p. 11, ref. 45).

Recent Advances

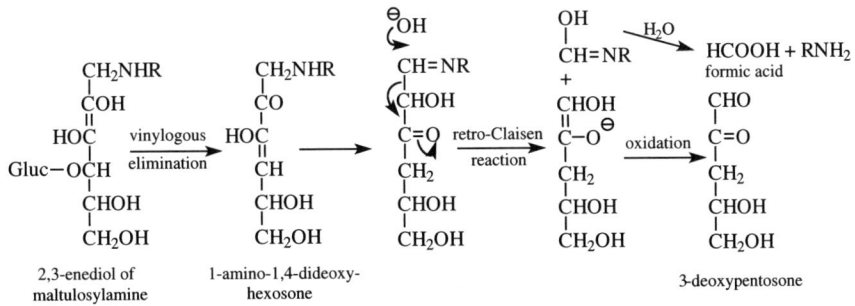

Scheme 3.5 *'Peeling-off' mechanism proposed for the formation of 1-amino-1,4-dideoxy-hexosone from an oligosaccharide, such as maltose*[165]

Dried spaghetti contains 3–600 GP and 0.02–6 μmol kg^{-1} GIM, depending on the heat involved in drying.[166] Due to the exposure of crackers and bread crust to higher temperatures, lower amounts of GP (50–100) and higher amounts of GIM (10–30 μmol kg^{-1}) are formed.

CHAPTER 4

Colour Formation in Nonenzymic Browning

1 General

Colour formation is the primary characteristic of the Maillard reaction, yet even now our knowledge of the coloured moieties responsible for the coloration is only rudimentary.

The colour produced can be readily measured by reading the absorbance in the visible region of the spectrum, typical wavelengths used being 360 and 420 nm. Since the spectra of crude Maillard reaction products have few characteristic features, absorption being strong below 400 nm but gradually falling away at higher wavelengths, it is not too important, as a first approximation, which wavelength is chosen for instrumental quantification, provided measurement remains consistent.

In many ways, colour, which is a sensation, should be assessed visually. On analogy with the treatment of odorants, Hofmann[73] has taken a crucial step in this area by defining a colour dilution factor (CD) and a colour activity value (CAV). CD is the factor required for any solution of a colorant x to be diluted to its colour threshold. CAV_x is the ratio of the concentration of x ($\mu g\ kg^{-1}$) to its threshold concentration ($\mu g\ kg^{-1}$). The colour contribution of a component colorant to the colour of a mixture can then be defined as

$$\text{colour contribution of colorant } x = \frac{CAV_x}{CD_{total}} \times 100\%$$

where CD_{total} is the colour dilution factor for the mixture as a whole.

Structures 4.1 show some low-molecular-mass coloured compounds formed in model Maillard systems. The organic chemistry underlying colour in melanoidins was not clear until 1972, when a specific compound, **4a**, was isolated by Severin and Krönig[167] in 0.07% yield from a heated aqueous mixture of xylose and isopropylamine acetate. It is yellow with $\lambda_{max} = 365$ nm. When the amine was replaced by glycine or lysine, the yield dropped further. Similar results were obtained when arabinose was substituted for xylose. Subsequent work led to the identification of related compounds (see **4b** and **4c**), in some of which the ring oxygen of the furfural is replaced by NR.

Colour Formation in Nonenzymic Browning

Even though **4a** and its analogues are relatively simple compounds, they come to assume an importance which is quite disproportionate, when their origin is considered. Their two parts are derived differently: the furfural resulting from 1,2-enolisation of the Amadori compound and the furanone from 2,3-enolisation (see Chapter 2), the former being favoured by low pH and the latter by moderate pH. This means that for both components to be available for interaction, the reaction conditions, including pH, need to be carefully balanced. pH optimisation of the reaction will also be affected by the acid/base properties of the amino acid moiety of the Amadori compound.

Structures 4.1 *The structures of some low-molecular-mass coloured compounds formed in model Maillard systems*

Structures 4.1 (*cont.*)

The methylene group of the furanone is clearly sufficiently reactive to undergo aldol condensation with aldehydes, such as furfural. Thus, other aldehydes will also condense, giving products coloured yellow to red. 3-Hydroxy-2-butanone will also condense.[168] Accordingly, dicarbonyl compounds should lead to crosslinking. The use of 2-oxopropanal, a well-known cleavage product of 3-deoxyglucosone, gave intractable mixtures, but its dimethylacetal did give rise to a yellow derivative of **5**.

The methyl group of **4a** is vinylogously activated by its carbonyl group and is therefore capable of undergoing aldol condensation too, leading to the trinuclear compound, **6a**, which is deep orange. Analogues **6b** and **6c** have also been isolated.

Hexoses can undergo similar reactions.[169] Thus, the Amadori compound from glucose and piperidine gives the yellow **7** (λ_{max} = 365) on heating with furfural. Glucose and butylamine acetate gave the yellow-orange **8** (λ_{max} = 453 nm), in some ways, a hexose equivalent of **1c**,[170] without the involvement of furfural.

Additional study uncovered further elaboration in the form of compounds, such as **9** to **11,** from the reaction of glucose with methylammonium acetate.[171] Compound **9** is yellow-orange, **11** (the structure of which is not certain) is yellow (**11a,** λ_{max} = 352 nm; **11b,** λ_{max} = 354 nm), and **10** has λ_{max} = 406 nm. It should be noted that **9** has a C_8 chain, which must be the result of the cleavage of the C_6 chain of glucose, followed by a recombination of fragments. Compound **10** may be formed similarly, but the pathway to **11** has not been clarified.

Exchanging the aqueous medium for an organic one makes reaction conditions more remote from those used in practical food technology, but nevertheless offers more insights into the Maillard reaction. Thus, methanolic xylose and diethylammonium acetate give the orange **12,** demonstrating the ability of pentoses to cyclise to a cyclopentane. Similarly, in ethanol, glucose and piperidine acetate yield **13,** which is also orange. Here, not only is a hexose converted into a methylcyclopentane, but it also forms a methylfuranone with the loss of a carbon atom, probably as formaldehyde by a retroaldolisation.[172] Methanolic xylose and glycine allowed both the yellow **14** and **15** (λ_{max} = 386 nm) to be isolated, and glucose and glycine similarly gave **16,** isolated as its yellow acetate with λ_{max} = 384 nm.

Painstaking separations of the product from xylose and glycine in buffer at pH 8.2 or in water at an initial pH of 6.0 led to 17 coloured compounds, ranging in molecular mass from 149 to 448 Da,[113] **4a** being predominant. A yellow compound with λ_{max} = 292 nm was assigned the tricyclic structure **17** on the basis of high-resolution MS, PMR, and UV data. It clearly exhibits its origin from three xylose units. The compounds at the top of the molecular range isolated are likely to be tetracyclic.

Similarly, a mixture of xylose and lysine monohydrochloride in water at an initial pH of 4.6 led to the yellow bicyclic compound **18,**[173] in which a pyrrole ring is linked to a cyclopentenone, the latter being produced here in an aqueous medium (*cf.* above).

Spray drying is extensively used in industry, and therefore spray drying a glucose-glycine model system has specific technological relevance. Compounds **19** and **20** were tentatively identified in the product.[174]

Sulfite can be used as an inhibitor to trap intermediates in model systems.[175] Its use (0.2 M) has been explored with mixtures of 6-aminohexanoic acid (0.38 M) and xylose or glucose (2.8 M) in phosphate buffer, (pH 7.35) at 37 °C. Ion-exchange chromatography and HPLC allowed yellow compounds **21** and **22** to be isolated, respectively. They exist in equilibrium with their hemiacetals.

Other inhibitors and blocking agents have also been explored. Furfural is an aldehyde which lacks α-hydrogen atoms and so cannot act as the nucleophile in aldol condensations (see ref. 176). It thus behaves as a chain terminator in polymerisation, yet it is an important Maillard product. Hofmann[177] has exploited its blocking properties to discover several novel types of coloured Maillard products. Aqueous furfural and proline heated for 15 min at 50 °C led to the intensely yellow, but unstable **22.**[177] Similarly, aqueous furfural and alanine heated for 1 h at 70 °C led to the red isomers **24.**[171] In these, carbon atom 4, as well as those designated as 1, 5, and 16 on structure **24,** originate from the carbonyl group of the furfural, as demonstrated using a ^{13}C label. The mechanism of formation proposed derives the aldehyde group of **24** from C-4 of a furfural molecule, which remains intact and stretches to form carbon atom 4 with its original carbonyl group. In a further experiment, aqueous xylose

(66 mmol) and alanine (16 mmol) were heated for 15 min under reflux in phosphate buffer (pH 7.0). Then furfural (100 mmol) was added and heating continued one more hour. Elaborate separations led to orange **25a** and **25b** as well as to **4a, 6a, 24a,** and **24b**. When the preparation was repeated in a medium of buffer/methanol 2:1 (v/v), **26a** and **26b** were obtained in addition to the six compounds mentioned already and the methyl ethers of **25a** and **25b**.[73]

Using the blocking agent in an alcoholic medium led to the isolation of the red **27**,[169] which was thought to have been formed via diacetylformoin, a hexose Maillard intermediate with progressively increasing significance. Compound **27** has been found to be a potent inhibitor of human tumour cell growth *in vitro*[178] (see Chapter 8). Hofmann[179] has shown that **27** results from diacetylformoin when heated in the presence of secondary amines (proline), whereas primary amines (glycine methyl ester) lead to yellow compounds of the type **28**, a very important switch. Compound **28** is hexose-derived, yet it bears considerable resemblance to the pentose-derived **4c**.

Much attention has recently been paid to arginine as a melanoidin precursor. In a model system, N^α-acetylarginine was first allowed to react with glyoxal in aqueous solution at pH 7, followed by furfural. Repeated chromatography of the aqueous residue after ethyl acetate extraction led to a 1% yield of the deep red **29**.[180] Such a compound provides an example of a potentially coloured crosslink in proteins derived by the Maillard reaction (see Chapter 8).

The effect of the medium has been explored by Mustapha *et al.*[181] Xylose and lysine (1 g of each) were added to water/secondary liquid mixtures (10 g; the amount of secondary liquid being varied from 0 to 10 g) and the samples were stored for 1 week at 20 °C; A_{480} was then determined. Results when the medium contained about 16.5% water are given in Table 4.1.

Colour formation in the Maillard reaction normally reaches its maximum at intermediate a_w, but here the data for water/glycerol are out of step. At higher levels of water, each of the systems exhibited the highest colour intensity (in the range A_{480} = 96–122) at 25–33% water. The stored water/glycerol and water/propyleneglycol systems were analysed for residual lysine, which was found to have a good inverse relation with A_{480}. With the water/polypropyleneglycol mixtures, complications arose due to phase separation.

So far, a wide range of coloured, relatively small molecules, has been isolated, mainly from model systems. Because of the complexity of the mixtures of substances produced, their separation and isolation have at times needed to be facilitated

Table 4.1 *Absorbance (at 480 nm) and a_w for samples containing different secondary liquids after storage at 20 °C for 1 week*[181]

Secondary liquid	a_w	A_{480}
Propyleneglycol	0.868 ± 0.009	59.9 ± 2.3
Glycerol	0.012 ± 0.006	23.0 ± 2.8
Water	0.983 ± 0.010	15.4 ± 1.5
Polypropyleneglycol 425	0.135 ± 0.011	0
Polypropyleneglycol 1200	0.175 ± 0.014	0

by the use of trapping agents, such as sulfite or furfural. Overall, there still seems to be quite a large gap in terms of intensity of colour between the compounds identified and melanoidins.

2 Coloured Polymeric Maillard Products

According to Rizzi,[182] the types of systems studied are conveniently divided into simple model systems of reducing sugars and amines or amino acids, sugars and proteins, and sugars by themselves (*i.e.*, caramelisation). Each of these will now be considered in turn, bearing in mind, in particular, the key question of whether the backbone of the polymeric material itself is coloured or whether the colour resides in moieties attached to an essentially colourless backbone.

2.1 Model Systems of Reducing Sugar and Amines or Amino Acids

Chemical degradation of polymers formed in xylose/glucose-ammonia/butylamine systems indicated that the repeating units were unsaturated $C_{5/6}$ compounds derived via Amadori compounds; however, since conjugation was not extensive, the implication is that the backbone is not coloured.[183] This was supported by ^{13}C-NMR spectra of water-soluble glucose–glycine melanoidin,[71] which provided very limited evidence of olefinic and aromatic unsaturation. A similar polymer from glucose/fructose-glycine (>16 kDa) at pH 3.5 gave ^{13}C-NMR spectra, which suggested some aromatic unsaturation.[184] Use of [1-] and [2-^{13}C]glycine, as well as [1-^{13}C]glucose, in conjunction with ^{13}C-NMR, showed that both atoms of glycine are incorporated into the melanoidin and that the C-1 of glucose is present as a substituted methyl group. Melanoidin prepared with HMF differed from those obtained with glucose or fructose:

Glycine +	%C	%H	%N
glucose	51.2	5.3	6.2
fructose	50.5	4.9	6.9
HMF	47.6	3.7	3.5

Melanoidins (>12 kDa) from xylose-glycine have been examined by ^{13}C and ^{15}N cross polarisation-magic angle spinning NMR.[70] At 68 °C, there was a definite increase in unsaturation with time, which levelled off at about 45 d, when the xylose had been used up. This increase was accompanied by a decrease in total carbonyl and carbohydrate carbon signal. Pyrrole and/or indole nitrogen increased, with the nitrogen not bearing hydrogen. As expected, xylose-alanine gave rise to extra aliphatic-C resonances, whereas xylose-urea gave few. A comparison was made of the melanoidins produced from xylose-glycine at 68 °C in 3 d with those produced at 22 °C in 150 d, starting material not being limiting in either case. For ^{13}C-NMR, the main difference lay in the aliphatic-C region, the resonance at 35 ppm being displaced towards 48 ppm. ^{15}N-NMR showed that, whereas the higher temperature product gave only a single amide resonance, the product obtained at the lower temperature gave two. The low-molecular-mass fractions gave ^{15}N-NMR evidence of the presence of the C_{12} enaminol.

Electron-spin resonance (ESR) indicated free radicals in a brown polymer (mean ca 1 kDa by field-desorption MS) from a glucose-4-chloroaniline model system.[185] Such free radicals (see Chapter 2) could be a source of visible colour. Exposure of the melanoidin to nitric oxide, a radical-trapping agent, diminished the ESR signal by 48% and changed the colour to red-brown.

Blue-M1 is the name given to a pigment isolated from D-xylose-glycine.[75] It was identified as the pyrrolopyrrolium compound **30** and assumed to be a dimer of the yellow pyrrolopyrrole-2-carboxaldehyde.[186] It is capable of polymerising further to melanoidins and, similar to melanoidins, it has strong antioxidant activity. There is evidence of another pigment, Blue-M2, with molecular mass (MM) = 0.942 kDa.

Melanoidins based on polymeric pyrroles were dealt in Chapter 2.

2.2 Model Systems of Reducing Sugar and Proteins

Early work by Hannan and Lea[187] on pigment formation was concerned with glucose in relation to casein, poly-L-lysine, and N^α-acetyl-L-lysine. Similarity in behaviour fitted the expectation that the group primarily involved is the ε-amino group of protein-bound lysine. Chemical tests suggest that colourless N-glycosides are formed first, followed by Amadori compounds.

Melanoidin, prepared from glucose-casein [1:2 (w/w), dissolved at pH 7.0, and freeze-dried] by keeping at 36 or 55 °C, 75% r.h., was washed at the isoelectric point of the protein, filtered off, and hydrolysed with proteolytic enzymes.[188] The so-called 'limit-peptide pigment' (LPP) was purified by gel and ion-exchange chromatography. It was water-soluble (10–50 g per 100 mL) and exhibited a typical featureless electronic spectrum. The purest fraction based on absorbance at 420 nm had a mass of 1.5 kDa, but still contained some protein. There were no free radicals present by ESR. Using glucose-6-t, the loss of tritium correlated directly with the amount of brown colour produced, strongly suggesting dehydration at C-6 to be a key reaction leading to crosslinking and colour generation.[189] The LLP obtained at 55 °C after 37 d had lost 26% of its tritium. LPPs derived from glucose-U-^{14}C and insulin (ca 1.5 and 7.8 kDa) contained a high ratio of sugar residues to available amino groups, about 1.5 and 8, respectively.[190]

Hofmann[73] also allowed an aqueous solution of glucose and β-casein to react (4 h at 95 °C) and fractionated the resulting intensely brown product by ultracentrifugation (Table 4.2). High-molecular-mass compounds (>50 kDa) constituted 61.5% by weight and accounted for most of the colour, whereas the lowest molecular mass fraction (<1 kDa), although constituting 23.7% by weight, exhibited hardly any colour. Bearing in mind that β-casein has a molecular mass of 24 kDa, the results imply that much of the β-casein has been crosslinked to tetramers or larger oligomers by the action of the glucose.

Casein heated in aqueous solution with furfural rapidly turns orange-brown. Melanoidins (>10 kDa), isolated by ultracentrifugation, were enzymically hydrolysed and the product was separated by HPLC. Two peaks led to red compounds in the ratio 1:7, which were unequivocally identified by Hofmann[191] as the lysine analogues of **24a** and **24b** (MM = 0.476 kDa), respectively. This is the first demonstration of the attachment of a specific type of coloured Maillard product to a protein

Table 4.2 *Yield and browning intensity of fractions obtained from a heated glucose/casein solution by sequential ultracentrifugation with different cut-off filters*[191]

Fraction	Molecular mass kDa	Amount mg	%	CD factor
1	>100	895.3	43.4	2048
2	100–50	373.8	18.1	512
3	50–30	93.1	4.5	8
4	30–10	76.5	3.7	<1
5	10–3	9.1	0.4	<1
6	3–1	36.0	1.7	<1
7	<1	488.5	23.7	2
	Fractions 1–7	1972.3	95.5	n.d.
	Complete mixture	2061.0	100.0	4096

n.d. = not determined.

backbone, and supports the hypothesis that the colour of proteins modified by the Maillard reaction resides, not in the backbone, but in sidechain(s). 'Melanoprotein' is a term that has been applied to such entities.[72]

With a glucose–casein system, dialysis with Visking tubing with a 12 kDa cut-off allows about 70% retention of the colour (420 nm), whereas glucose–amino acid systems retain only about 10% of the colour, even with a 3.5 kDa cut-off. Using uniformly labelled [^{14}C]glucose and plotting the absorbance of the melanoidin fraction against ^{14}C incorporation, Brands *et al.*[192] obtained ε values of 477 ± 50 and 527 ± 35 cm^2 mol^{-1} for glucose-casein and fructose-casein melanoidins, respectively. The difference is not statistically significant. From microanalytical data, the number of glucose or fructose molecules attached per casein molecule is 2, 5, and 10 or 2, 7, and 13, respectively, after 10, 30, and 60 min of heating at 120 °C. Casein has 13 lysine sidechains.

In a gluten–glucose system, colour is mainly due to low-molecular-mass compounds formed by interaction of glucose with ammonia derived by deamidation of glutamine residues.[193]

Glycolaldehyde has been identified as a key precursor in the formation of pyrazine radical cations (see Chapter 2). Heating glycolaldehyde with BSA led to rapid browning.[179] Formation of pyrazine radical cations should result in crosslinking and so the fraction with molecular mass > 100 kDa (BSA has molecular mass of 67 kDa), produced by ultracentrifugation, was examined. It was orange-brown and gave an ESR spectrum without fine structure, but with a single broad band ($g = 2.0038$; free electron spin value = 2.0023), in accordance with the presence of immobilised free radicals. No ESR signal was observed when the reaction was carried out in the presence of 20% sulfite. When the glycolaldehyde was replaced by glyoxal in the absence of sulfite, the ESR signal was absent, but reappeared on addition of ascorbic acid (see Chapter 2), with the addition of a doublet ($g = 1.8$), attributed to the ascorbate radical.

When light-brown wheat bread crumb and dark-brown crust of toasted wheat bread were ground and submitted to ESR, no signal was obtained with the former, but the latter gave a broad singlet ($g = 2.0038$), identical with that observed for glycolaldehyde/BSA. When wheat bread crumb was sprayed with 20% sulfite, dried at 35 °C, and then toasted, browning was strongly suppressed and no ESR signal was obtained. The ESR signal ($g = 2.0038$) was given by dark-brown roasted cocoa and coffee beans, but not by unbrowned, heated milk.[179] It seems that the radical mechanism of browning (see Chapter 2) is operative in foods equally well as in model systems with amino acids or with proteins.

The contrary argument that the polymer backbone itself contributes substantially to the colour has been put forward most strongly by Tressl et al.,[76] namely, that high-MM coloured molecules are produced from low-MM Maillard intermediates, such as pyrroles.

The red pigment, obtained from the interaction of amino acids and dehydroascorbic acid and shown to have structure **31** (p. 54) by Kurata et al.,[194] provides another model chromophore for Maillard reaction products.

2.3 Caramelisation

Theoretically, caramelisation is the browning reaction undergone by sugars on heating by themselves or, at least, in the absence of amino compounds. Cooks consider the product to be caramel.

Caramel is also the designation attached to certain food colour additives, which are coded E150(a-d) by the European Parliament and Council Directive, Colours for Use in Foodstuffs,[195] corresponding to the four caramel classes I to IV of the International Technical Caramel Association. Caramels are produced by heating edible sugars, Class I requiring sodium hydroxide, Class II sodium hydroxide and sulfur dioxide, as such or in the form of sodium sulfite or metabisulfite, Class III ammonia, and Class IV ammonia and sulfur dioxide, as such or in the form of ammonium sulfite, ammonium hydrogen sulfite, sodium sulfite, sodium hydrogen sulfite, or sodium metabisulfite.

According to the Directive, the term caramel relates to products of a more or less intense brown colour, which are intended for colouring. It does not correspond to the sugary aromatic product obtained from heating sugars and which is used for flavouring food (e.g., confectionery, pastry, and alcoholic drinks). Caramel is the only colour permitted in malt bread, vinegar, and alcoholic drinks, such as beer, whisky, and liqueurs.

Clearly, Class III and IV caramels are close to melanoidins produced by the Maillard reaction, but detailed structures for the coloured components cannot be given for any of these commercial caramels.

Caramel colours are the most widely used food colouring agents, contributing about 90% by weight of the total colouring agents supplied to the UK food industry. World-wide, 80% has been quoted.[196] In some cases, caramel colours also contribute to the flavour and/or stability of the food. Total production/usage in the UK is equivalent to 550 mg person^{-1} d^{-1}, which is well below the daily intake acceptable to the Joint Expert Committee on Food Additives (200 mg caramel per kg body weight for Classes III and IV; no limit was set for Classes I and II). Annual global consumption exceeds 200 kt.[196]

The colour intensity of caramel colours has been defined as the absorbance of a 0.1% (w/v) solution of caramel colour solids in water in a 1 cm cell at 610 nm, the range of values for the four classes of colour being 0.01–0.12, 0.06–0.10, 0.08–0.36, and 0.10–0.60, respectively. Linner has been reported[196] to have developed a Hue index or measure of redness for caramel colours:

$$\text{Hue index} = 10\log(A_{510}/A_{610})$$

The Hue index for caramel colours is usually in the range 3.5–7.5.

Primary usage of the four classes is in spirits and desserts, ice cream and liqueurs, beers and baked goods, and soft drinks, respectively. In the UK, Class III accounts for almost 70% of total caramel consumed, Classes IV and I contributing 25 and 5%, respectively. World-wide usage is very different, Class IV constituting 70%, Class III 28%, Class I 2%, and Class II 1%.

Pyrolysis–GC–MS has given some interesting results.[197] More than 100 volatile pyrolysis products were obtained. Class III caramels gave relatively large amounts of nitrogenous heterocycles, including isomeric methylpyrazine-2-carboxaldehydes. Separating the commercial colours by membrane ultrafiltration into five fractions ranging from <0.5 to >10 kDa showed the specific colour intensity (610 nm) to increase rapidly with molecular mass.[198] Class III caramel gave the most intense colour according to the absorption at 610 nm and it is noteworthy that a significant proportion of the colour resided in lower mass compounds (1–5 kDa).

Capillary electrophoresis at pH 2.5 and 9.5, combined with ultrafiltration through a regenerated cellulose membrane with a nominal cut-off of 5 kDa, showed clear differences among the caramel classes.[199] The high-MM Class I and IV caramel peaks migrate with a negative charge at both pH values, but the Class IV caramel also has several sharp peaks from low-MM components, whilst Class I caramel gives only one broad neutral peak. Class III caramels have one broad high-MM peak which is positively charged at pH 2.5. Its migration time is inversely related to the caramel's nitrogen content, which determines the charge. For high-MM Class IV caramel, it is the coloured broad peak at pH 9.5, the migration time of which is related to the caramel's sulfur content.[200] Capillary electrophoresis at pH 9.5 can thus be used to identify and quantify Class IV caramel in soft drinks.

Determining Class III caramel in food products is more difficult. Semiquantitative results have been achieved with some food products by ion-pair reversed-phase HPLC.[201]

Aspartame (α-L-aspartyl-L-phenylalanine methyl ester) is widely used as an intense sweetener, particularly in diet soft drinks. In colas, Class IV caramel is the predominant ingredient, a typical concentration being 1400 ppm. Such a concentration has been shown to affect the stability of aspartame at the typical pH of 3.0–3.2.[202] Thus, at 55 °C, about 90% of the aspartame in a simulated beverage (4 mM phosphate, pH 3.1) has been lost in 27 d by peptide hydrolysis, rearrangement, ester hydrolysis, and cyclisation to the diketopiperazine. The degradation of aspartame was not affected by 250 ppm caramel, but started at 700 ppm.

Heating sucrose (e.g., at 125 °C for 46 h or at 170 °C for 100 min) gives a fructoglucan with a DP of about 25 by gel permeation chromatography in almost 40% yield.[203] The ratio of fructose to glucose units in it is about 1:2 and it has appreciable absorbance at 450 nm.

CHAPTER 5

Flavour and Off-Flavour Formation in Nonenzymic Browning

1 Flavour

Flavour is a complex sensation, made up principally of smell and taste, but touch and hearing contribute as well. The human senses of smell and taste differ in sensitivity, between each other and depending on the nature of the component eliciting the sensation. Substances may have no impact at all (such as oxygen or carbon monoxide) or exhibit very low thresholds (such as 2×10^{-14} g mL^{-1} water for the odour of bis-2-methyl-3-furyl disulfide).[204] In general, odour thresholds are much lower than taste thresholds and so flavour tends to be dominated by odorous components, *i.e.*, by substances able to reach the olfactory epithelium high up in the nose, that is, substances with at least some volatility. Hence, the emphasis on the volatile compounds derived from the Maillard reaction.

2 Volatile Compounds

The formation of volatile compounds by the Maillard reaction has been most recently reviewed by Mottram[152] and Tressl and Rewicki.[30]

The volatile products of the Maillard reaction can be classified into the following three groups:[205]

1. 'Simple' sugar dehydration/fragmentation products:
    - Furans   (*e.g.*, HMF)
    - Pyrones   (*e.g.*, maltol)
    - Cyclopentenes (*e.g.*, methylcyclopentenolone)
    - Carbonyls   (*e.g.*, CH$_3$COCOCH$_3$)
    - Acids   (*e.g.*, CH$_3$COOH)
2. 'Simple' amino acid degradation products:
    - Aldehydes, *cf.* Strecker degradation
        Gly   CH$_2$O
        Ala   CH$_3$CHO

Flavour and Off-Flavour Formation in Nonenzymic Browning 63

    Val $(CH_3)_2.CH.CHO$
    Leu $(CH_3)_2.CH.CH_2.CHO$
    Ile $CH_3.CH_2.CH(CH_3).CHO$
    Phe $C_6H_5.CH_2CHO$
    Tyr   insufficiently volatile
    Asp   insufficiently volatile
    Glu   insufficiently volatile
    Lys   insufficiently volatile
    Arg   insufficiently volatile
    His   insufficiently volatile
    Try   insufficiently volatile
    Ser   [$CH_2OH.CHO$]
    Thr   [$CH_3CHOH.CHO$]
    Cys   [$CH_2SH.CHO$]
    Met $CH_3S.CH_2CH_2CHO$, methional
    Pro   unable
    Hypro unable
- Sulfur compounds
    Cys H_2S
    Methional $CH_3SH + CH_2:CH.CHO$

3. Volatiles produced by further interactions:
- Pyrroles
- Pyridines
- Imidazoles
- Pyrazines
- Oxazoles
- Thiazoles
- Compounds ex aldol condensations

Much has been written on the isolation, separation, and identification of aroma volatiles, and the techniques involved have been reviewed recently by Teranishi *et al.*[206] and Reineccius.[207] Solid-phase micro-extraction (SPME) has received much attention because of its convenience and because it samples the important headspace directly; however, it is not without problems. Thus, Coleman III[208] has shown that CW/DVB fibres tend to give higher recoveries than PDMS fibres, but there are exceptions, such as 4-ethyl- and 5-ethyl-2-methylpyridine. There are also matrix effects, recovery being affected by the presence of other components. For pyrazines, carboxen/PDMS fibres tended to give the highest recoveries. They were the only fibres to allow 2-methoxy-3 (or 6)-methylpyrazine to be detected. The Mixxor™ device for solvent extraction (New Biology Systems Ltd, Israel) is worth considering.[209]

4-Hydroxy-2,5-dimethyl-2-furanone (HDMF; Furaneol™) is an important odorant, which can be difficult to purify and quantify because it is reluctant to partition into organic phases. Recently, it has been quantified along with its ethylmethyl homologue by GC-orthogonal acceleration time-of-flight MS;[210] xylose-Gly and xylose-Ala mixtures were heated in phosphate buffer (pH 6.0) at 90 °C for 1 h, giving HDMF and its homologue in 16.6, 0.3 and 4.7, 8.5 μg mmol^{-1} sugar ($n = 6$), respectively.

It has been shown that the flavour activity of diacetylformoin lies in the open-chain form, the cyclic isomer being completely odourless.[211]

2.1 Mechanisms of Formation

Sugar dehydration and fragmentation were dealt with in Chapter 2, as was the Strecker degradation. Here, the focus will be on the formation of the different heterocyclic volatiles.

2.1.1 Furans

Wnorowski and Yaylayan[212] have compared pyrolysis (250 °C, 20 s) with reaction in aqueous solution (120 °C, 3 h, sealed tube) for glucose–glycine mixtures, the glucose being singly labelled at C-1 to C-6 in a series of experiments. Although more products are formed than in aqueous media, most of the products identified in aqueous systems are present in pyrolysates with identical label distribution, even though the proportions may differ. Thus, the labelling of the 5-methylfurfural formed corresponded 100% to the incorporation of glucose intact with C-1 at the aldehyde group for both sets of reaction conditions. For the 2-acetylfuran formed, incorporation for pyrolysis was 100% intact with C-1 in the methyl group, but for the aqueous system it was only 70%, 30% of the methyl group of the 1-acetylfuran having come from C-2 of the glycine. For the 2-hydroxymethylfuran formed, incorporation of the glucose skeleton, C-2 to C-6, was 100% for the aqueous system, but only 90% for pyrolysis, 10% of the 2-hydroxymethylfuran being derived from C-1 to C-5. For the 2-methyltetrahydro-3-furanone [2-methyldihydro-3(2H)-furanone] formed, incorporation of the glucose skeleton for the aqueous system was 100% intact, with C-3 missing (as shown in Scheme 5.1), the methyl group being derived from C-1, but this did not occur for pyrolysis, which was not explained.

Diacetylformoin is an important intermediate in the formation of HDMF (mixed with silica gel, 180 °C, 10 min), the yield being increased substantially by the presence

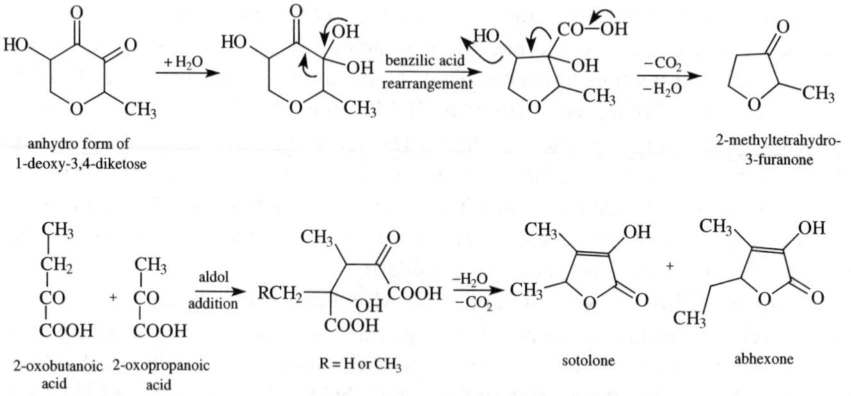

Scheme 5.1 *Proposed formation of 2-methyltetrahydro-3-furanone from 1-deoxy-3,4-diketose[212] and of sotolone and abhexone from 2-oxoacids[214]*

Flavour and Off-Flavour Formation in Nonenzymic Browning 65

of reducing agents, such as ascorbic and methylenereductinic acid (from 748 to 1959 and 1485 μg mmol^{-1}, respectively).^{213} Proline is also effective in this reaction (yield, 1765 μg mmol^{-1}), presumably by a Strecker-type process. In all these reactions, labelling shows that the diacetylformoin skeleton is incorporated intact into HDMF. However, when proline is used in aqueous solution, labelling indicates that only about one-third of the HDMF comes from intact diacetylformoin, the remainder being formed via C$_3$ fragments. Of such compounds, a mixture of 2-oxopropanal and dihydroxyacetone proved to be the most effective. These experiments demonstrate clearly how sensitive the Maillard reaction is to conditions under which it is carried out.

Sotolone and abhexone to a lesser extent make important contributions to roast coffee aroma. They have been shown to be formed in model systems of 2-oxobutanoic acid and 2-oxopropanoic acid by Kobayashi^{214} (see Scheme 5.1).

2.1.2 Pyrroles

Pyrroles and related compounds were reviewed by Maga;^{215} 37 pyrroles from about 20 model systems have been tabulated by Vernin and Párkányi.^{216}

Just as furfural can be derived as an offshoot of the 1,2-enolisation pathway to melanoidins from the Amadori compound (see Scheme 5.2), so can pyrroles, as first shown by Kato and Fujimaki.^{217}

In comparing pyrolysis (250 °C, 20 s) with reaction in aqueous solution (120 °C, 3 h, sealed tube) for glucose–glycine mixtures, the glucose being singly labelled at C-1 to C-6 in a series of experiments, Wnorowski and Yaylayan^{212} found that the glucose skeleton was incorporated intact 100% into the 2-acetylpyrrole and the

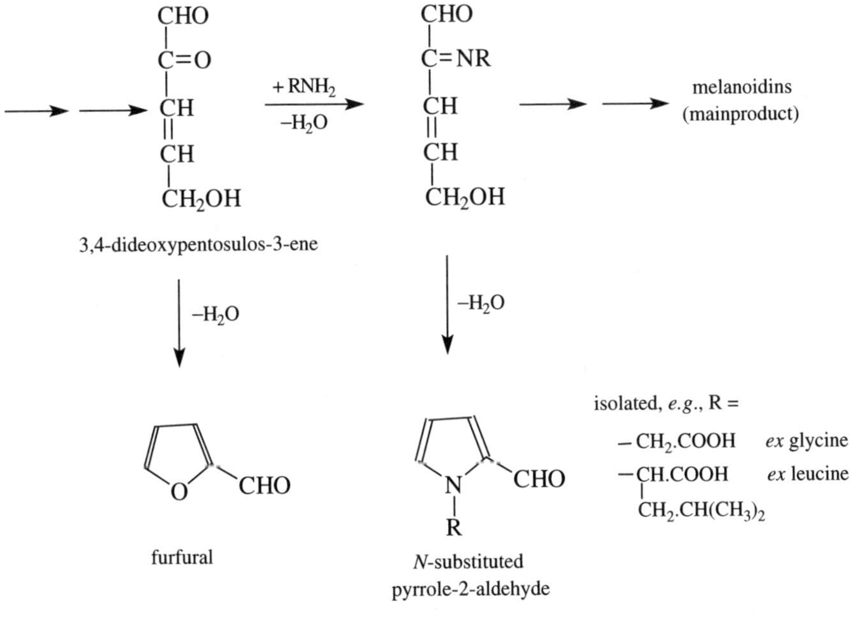

Scheme 5.2 *Pyrrole formation*^{217}

5-methylpyrrole-2-aldehyde formed, with C-6 in the methyl group of each, under both sets of conditions. It is worth noting that the glucose is incorporated into 2-acetylfuran with C-1 in the methyl group (see above).

The pyrrolidine system is already present in proline. With proline and hydroxyproline, transamination and thus the Strecker degradation is not possible, yet a whole range of characteristic Maillard products results. Shigematsu et al.[218] had already identified 5-acetyl-2,3-dihydro-1H-pyrrolizine, its 6-methyl homologue, and the corresponding 5-formyl compound, as well as 5,6,7,8-tetrahydroindolizin-8-one and its 2-methyl homologue, from glucose–proline mixtures (0.1 mol each), heated at 200 °C for 6 min. By 1985, Tressl et al.[219,220] had identified more than 120 proline-specific compounds. Proline gives rise to maltoxazine and N-5-hydroxy-methylfur-furylpyrrolidines, and hydroxyproline to N-alkylpyrroles and indolizinones (see Scheme 5.3).[30]

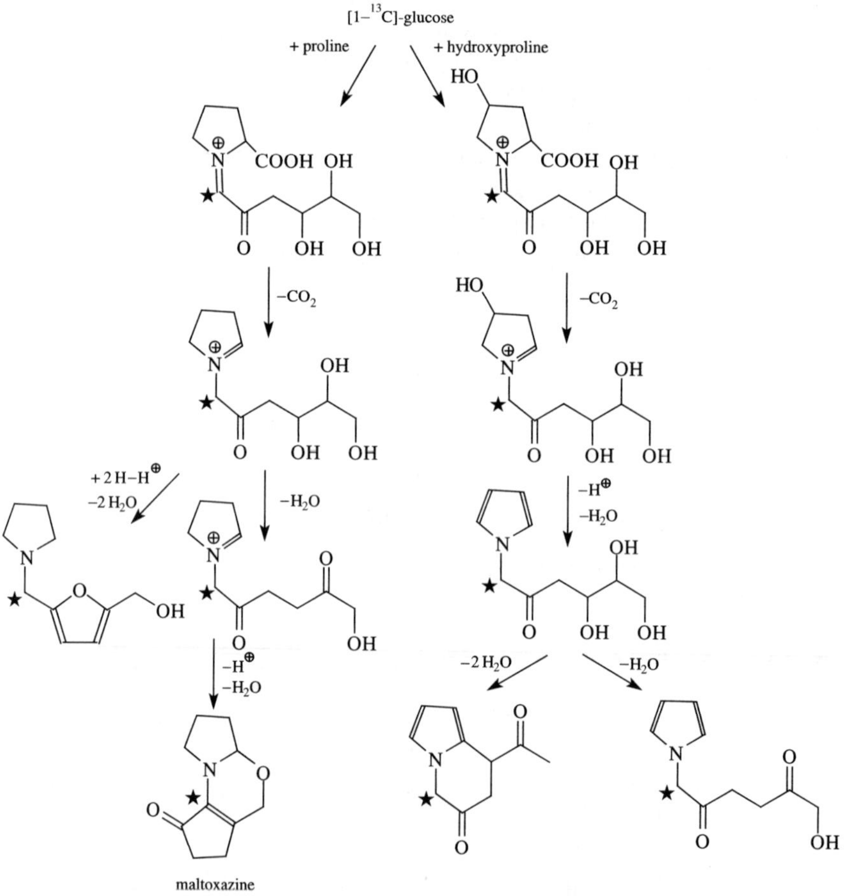

Scheme 5.3 *Formation of proline- and hydroxyproline-specific heterocyclic volatiles from [1-^{13}C]glucose via 3-deoxyglucosone*[223]

Dark malt contains 10 mg kg^{-1} maltoxazine, as well as 15, 60, 3, and 3 mg kg^{-1} of maltol, its dihydrohydroxy derivative, cyclotene, and HDMF, respectively.[127] A model system of maltose–proline leads to 40 mg kg^{-1} sugar maltoxazine, whereas the model system of glucose–proline gives only slightly less.

2-Acetyl-1-pyrroline (ACPY) is a very important odorant with a roast and sweet aroma and a very low threshold ($T = 0.1$ µg L^{-1}). Its origin has therefore been of great interest. Often, the situation is complicated by the concomitant formation of 2-acetyltetrahydropyridine (ACTPY; see the next section), which also has a roast aroma, but with a somewhat higher threshold ($T = 1.6$ µg L^{-1}). Schieberle[221] showed, by stable isotope dilution analysis, that in reaction with 2-oxopropanal, both proline and ornithine form ACPY, but only proline forms ACTPY. Ornithine gives slightly higher yields of ACPY than proline. 4-Aminobutyraldehyde, the Strecker aldehyde from ornithine, not unexpectedly, also reacts with 2-oxopropanal to give ACPY. 1-Pyrroline in place of 4-aminobutyraldehyde quadruples the yield (it is the internal Schiff base of the latter). Table 5.1 illustrates some of the above. It is noteworthy that fructose can replace 2-oxopropanal for ACPY formation from ornithine, but not from proline, as long as the buffer is phosphate. With fructose in place of 2-oxopropanal, the yield of ACTPY from proline is three times as high.

The free amino acids in bakers' yeast were determined, ornithine (318) ranking third and proline (89 mg per 100 g) 11th. None of the amino acids ranked above proline, other than ornithine, was an effective precursor for ACPY or ACTPY.

Schieberle, according to Kerler et al.,[222] had proposed that 1-pyrroline is converted into ACPY as shown in Scheme 5.4. Kerler et al. raise two questions in relation to this scheme, neither of which has yet been answered:

1. Would substitution at the vinylogous Position 4 not be expected?
2. What are the precedents for obtaining formaldehyde by α-cleavage?

Experiments using [1-^{13}C]D-glucose led Tressl et al.[223] to propose the mechanism presented in Scheme 5.5, which explains the 50% labelling of ACPY observed. Degradation of proline with diacetylformoin results in 1-pyrroline and a 1,6-dideoxydiketose, which

Table 5.1 *Formation of 2-acetyl-2-pyrroline (ACPY) and 2-acetyltetrahydropyridine (ACTPY) in model systems*[221]

4 mmol	Reactants 0.1 mmol	ACPY (µg)	ACTPY (µg)
Ornithine	2-Oxopropanal^a	43	< 0.3
Proline	2-Oxopropanal^a	41	160
Ornithine	Fructose$^{a',b}$	53	< 0.3
Proline	Fructose$^{a',b}$	< 0.3	478
Ornithine	Fructose^{b,c}	0.5	< 0.3
Lysine	2-Oxopropanal^a	< 0.3	< 0.3

^{a} 100 mL of 0.1 M phosphate, pH 7.0.
$^{a'}$ Presumably the same as in footnote a.
^{b} 2 mmol.
^{c} Phosphate replaced by malonate.

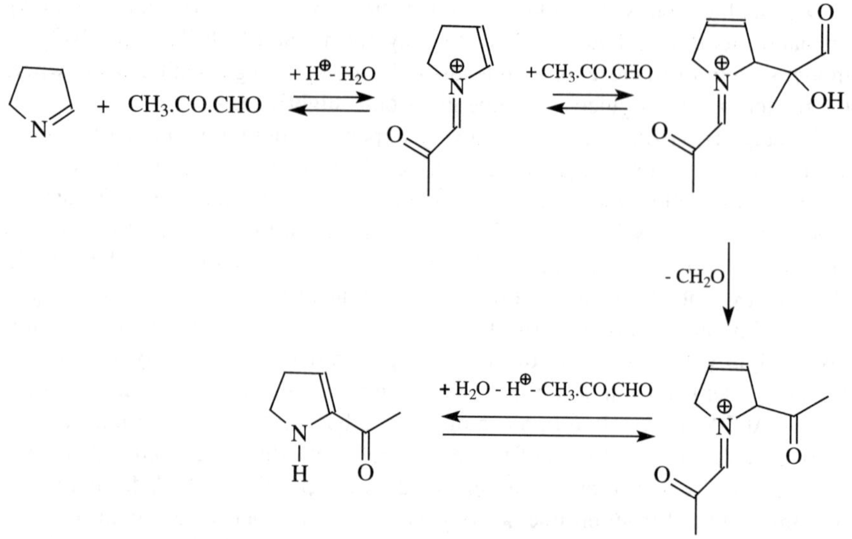

Scheme 5.4 *Formation of 2-acetyl-2-pyrroline from 1-pyrroline and 2-oxopropanal (after Schieberle[227])*

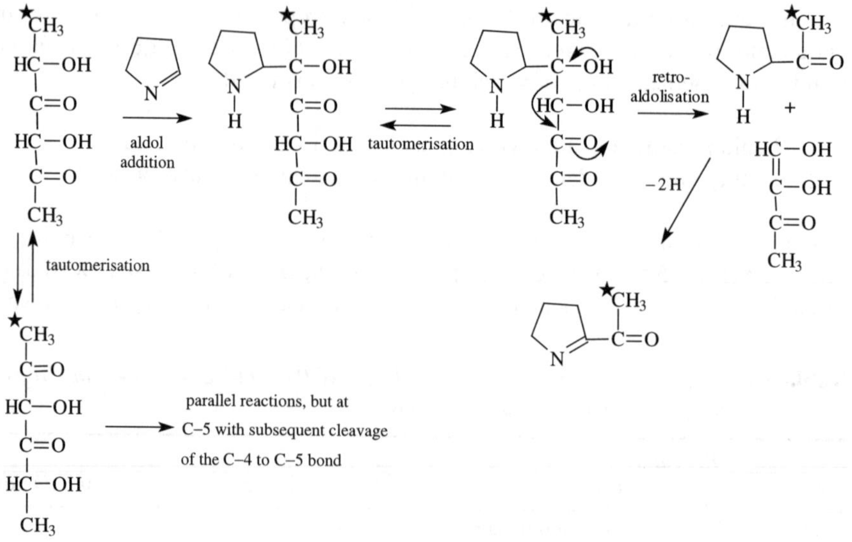

Scheme 5.5 *Formation of 50% labelled 2-acetyl-1-pyrroline from [1-^{13}C]glucose and proline via 1-deoxy-2,3-diketose and dihydrodiacetylformoin[223]*

may proceed to an aldol addition, followed by a C_2/C_4 retro-aldol cleavage and dehydrogenation, to form 2-acetyl-1-pyrroline and a C_4 reductone (*cf*. ref. 30)

It is worth noting that ACPY was found to be the character-impact compound of cooked rice[224] and was subsequently shown to play an important role in the more aromatic varieties of rice.[225]

2.1.3 Pyridines

More than 20 pyridines from 13 systems have been tabulated by Vernin and Párkányi.[216] They arise mainly during the thermal degradation of sulfur-containing amino acids alone or in the presence of glucose. They are also formed in glucose–proline systems, on degradation of Amadori intermediates of the glucose–glycine system, and on pyrolysis of α- and β-alanine.

In comparing pyrolysis (250 °C, 20 s) with reaction in aqueous solution (120 °C, 3 h, sealed tube) for glucose–glycine mixtures, the glucose being singly labelled at C-1 to C-6 in a series of experiments, Wnorowski and Yaylayan[212] found that the glucose skeleton was incorporated intact 100% into the 3-hydroxy-2-methylpyridine formed, with C-6 in the methyl group, under both sets of conditions.

Recently, 2-acetylpyridine has been found in Xiangjing scented rice and in the spice, yahonkaoluo.[226]

The important odorant, 2-acetyltetrahydropyridine (ACTPY), has already been mentioned in the previous section. ACTPY and ACPY play key roles in the aroma of popcorn.[227] Freeze-dried maize contains relatively high amounts of proline (155), whereas ornithine is not detectable (< 5 mg kg^{-1}). Schieberle treated a low-molecular-mass fraction of an aqueous extract of maize in different ways and determined ACTPY and ACPY by isotope dilution assay (Table 5.2). Steam-distillation extraction gave 130 times as much ACTPY than ACPY; however, in the presence of added 2-oxopropanal, the amount of the former was multiplied by 4, but that of the latter by 29. Dry-heating, as in popping, increased the latter further, but the former became undetectable.

ACTPY was first identified in a proline–dihydroxyacetone model system by Hunter et al.[228] The mechanism for its formation, shown in Scheme 5.6, is due to Hodge et al.[229] De Kimpe et al.[230] tried to validate this mechanism by synthesising doubly protected N-acetonyl-4-aminobutanal, but, on hydrolysis, no ACTPY was detected.

The mechanism,[30] shown in Scheme 5.7, has diacetylformoin, formed from glucose via the 1-deoxy-2,3-diketose, as its starting point. In diacetylformoin, tautomerism can move the central keto group from the 3- to the 4-position, leading to a 50% dilution of the ^{13}C-label in the C-1 methyl group.

Related results were obtained by Hofmann and Schieberle,[213] who heated precursors (1 mmol of each) in a phosphate buffer (10 ml, 0.5 mM, pH 5.0) for 20 min at

Table 5.2 *Determination of 2-acetyltetrahydropyridine (ACTPY) and 2-acetyl-2-pyrroline (ACPY) in low-molecular-mass, water-soluble fractions of maize treated in different ways*[227]

	A	B	C
ACTPY(μg kg^{-1} maize)	40	166	n.d.
ACPY(μg kg^{-1} maize)	0.3	8.7	12.7

Method A: Steam-distillation extraction with diethyl ether, 2 h.
Method B: As in A, but prior addition of 4 mg of 2-oxopropanal.
Method C: Freeze-dried extract mixed with silica gel and heated, 10 min at 150 °C.
n.d., not detected.

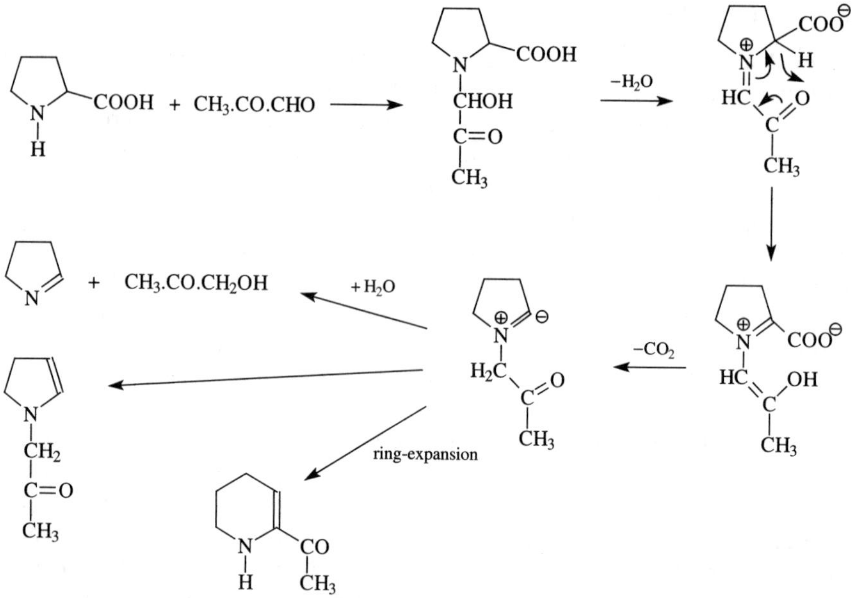

Scheme 5.6 *Conversion of proline into 1-pyrroline, N-acetonyl-2-pyrroline, and 2-acetyl-1,4,5,6-tetrahydropyridine (based on Hodge et al.[229])*

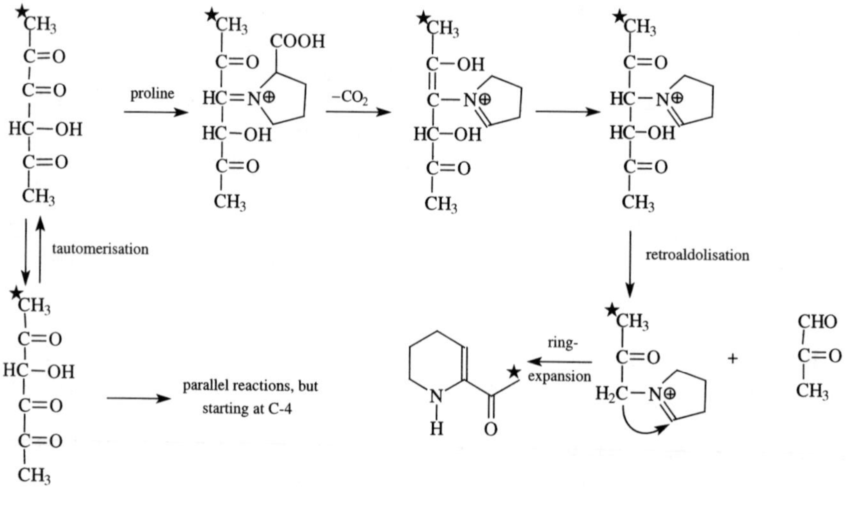

Scheme 5.7 *Formation of 50% labelled 2-acetyl-1,4,5,6-tetrahydropyridine from [1-^{13}C]glucose and proline via 1-deoxy-2,3-diketose and diacetylformoin (based on Tressl et al.[223])*

145 °C, proline with glucose, as the Amadori compound, and with diacetylformoin giving 3.2, 2.0, and 62.5 µg of ACTPY, respectively. The 20-fold yield of ACTPY from diacetylformoin clearly supports its key role. Under these aqueous conditions, ACTPY is largely formed via C_3 fragments, rather than by intact incorporation of diacetylformoin (*cf.* HDMF above).

Hofmann and Schieberle synthesised the intermediate, 2-(1-hydroxy-2-oxopropyl)pyrrolidine, and refluxed its solution in a 0.5 M phosphate buffer for 30 min, the yield of ACTPY increasing with pH 3–9 up to 35%, underlining the pyrrolidine's key role.

2.1.4 Pyrazines

The formation of some pyrroles, pyrrolines, pyrrolidines, pyridines, and tetrahydropyridines were considered above. Next come the pyrazines, a very important group of odorants. Pyrazines have been reviewed periodically by Maga.[231–233] Vernin and Párkányi[216] have tabulated 26 pyrazines, as well as 11 6,7-dihydro-(5H)-cyclopentapyrazines and 9 5,6,7,8-tetrahydroquinoxalines from 15 systems.

The formation of pyrazines is generally linked to the Strecker degradation (see Chapter 2), in which the dicarbonyl reagent undergoes transamination, leading to an α-aminocarbonyl. Two molecules of this readily condense to a dihydropyrazine, as shown in Scheme 5.8.

It should be noted that this mechanism involves a final oxidation. The need for this is avoided when amino acids with an α-hydroxy group in the sidechain (serine, threonine) are involved, as pointed out by Shibamoto and Bernhard[234] (see Scheme 5.9). Baltes and Bochmann[235] obtained as many as 123 pyrazines, both mono- and bicyclic, by interacting sucrose with serine and threonine under coffee-roasting conditions.

Koehler and Odell[236] already studied a number of factors affecting the formation of pyrazines, based on a system of glucose and asparagine (0.1 mol of each), heated for 24 h at 120 °C in diethyleneglycol/water (10:1, v/v). The dichloromethane extract of a distillate was analysed by GC. Below 100 °C, essentially no pyrazines were formed, but the yield increased steeply as the temperature was increased. At 120 °C, the yield increased rapidly with time, up to about 24 h, when it levelled off. Methylpyrazine was the major product in the first 3 h, but the ratio of dimethyl/methyl derivatives continued to increase for about 9 h, when it remained essentially constant at about 3. Altering the proportions of the reactants to 3:1 decreased the yield of methylpyrazine 10-fold and that of the dimethyl compounds 25 times, whereas a proportion of 1:3 decreased the former only by about a one-fourth and hardly affected the latter. Adding

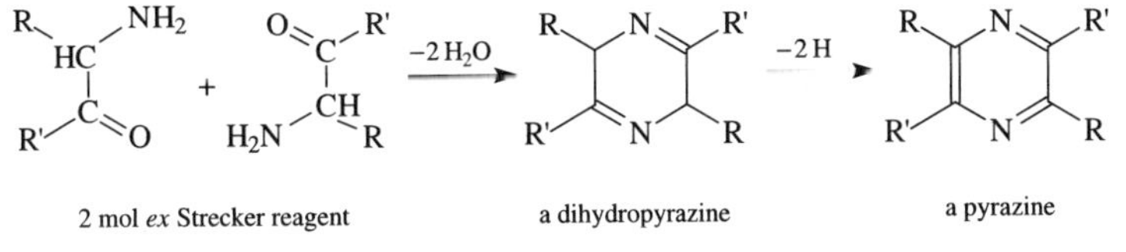

Scheme 5.8 *Pyrazine formation from 1,2-aminocarbonyl derived from the Strecker degradation*

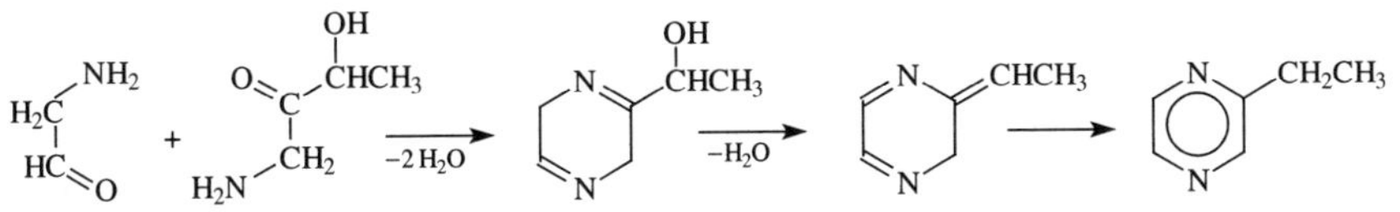

Scheme 5.9 *Pyrazine formation without the need for an oxidation step*[234]

0.1 mol of sulfuric acid decreased the yields virtually to 0, whereas adding 0.1 mol of NaOH multiplied the yield of methylpyrazine by 10 and that of the dimethyl compounds by 5. The yields with asparagine were at least five times those with glycine, alanine, lysine, or aspartate. Use of ammonia multiplied the yield of methylpyrazine by about 13, but decreased that of the dimethyl compounds by more than seven times, lowering the dimethyl/methyl ratio to 0.04. Replacing glucose with fructose almost tripled the yield of dimethylpyrazines, but increased that of the methyl homologue only by a quarter, resulting in a ratio of 8:4. On the other hand, arabinose lowered the yield of dimethylpyrazines about eight times, but that of the methyl compound by less than one-fifth, thus decreasing the ratio to 0.5. Replacing glucose by potential fragmentation products gave some interesting results. Thus, glyoxal gave mainly unsubstituted pyrazine, but also some methyl derivative. Butanedione gave relatively large amounts of tetramethylpyrazine and hydroxyacetone was outstanding at yielding dimethylpyrazines, both doing so essentially without the formation of homologues.

Rizzi[237] studied some relevant model systems. When heated in diethyleneglycol dimethyl ether at 160 °C, butanedione and alanine gave 0.5% tetramethylpyrazine in 2 h, 3,4-hexanedione and alanine gave tetraethylpyrazine, and benzil and alanine gave 14% of tetraphenylpyrazine in 7 h. However, 2-oxopropanal and glycine gave not only 2,5- and 2,6-dimethylpyrazine, but also the trimethyl homologue. To throw light on this unexpected result, aminoacetone and glycine were kept at 25 °C, when the pyrazines produced consisted of 50% 2,5-dimethyl, 33% trimethyl, and 17% 3-ethyl-2,5-dimethyl derivatives. Rizzi put up Scheme 5.10 as a plausible hypothesis to explain his results. Aldol condensations of appropriate intermediates with formaldehyde or acetaldehyde lead to the extra methyl or ethyl group.

Weenen et al.[238] have examined the processes occurring in more detail, using ^{13}C glucose and fructose with asparagine. There was little difference in incorporation of ^{13}C from [1-^{13}C]glucose and fructose, although the yield from the latter was somewhat higher (Table 5.3). The pathways differ, depending on whether the Amadori or the Heyns compound is involved, but both lead to 1- and 3-deoxyglucosone (1- and 3-DH – see Reaction Scheme 5.11). Both deoxyglucosones undergo retroaldolisation to yield 2-oxopropanal and glyceraldehyde, but, whereas the 2-oxopropanal from 1-DH carries the label in the methyl group, that from 3-DH carries it in the aldehyde group. The 1- and 3-DH pathways thus lead to the labelling of the pyrazines in the methyl and methine groups, respectively. The two pathways contribute about equally for fructose, but with glucose the former is favoured slightly. Since the formation of 2,3-dimethylpyrazine is negligible, the pathway via 4-DH seems to have little importance. Similarly, C_4 retroaldolisation products do not seem to play significant roles.

As expected, [2-^{13}C]glucose leads to the labelling of the methylpyrazines almost exclusively in the quaternary C atoms. There was evidence of a small amount (about 5%) of label in the methine groups and this was attributed to retroaldolisation of glucose or the glucosylamine (see ref. 92).

Retroaldolisation of the Heyns compound or α-cleavage of 1-DH can give acetaldehyde, which could lead to ethylmethyl or ethyldimethylpyrazines, but significant amounts of these were not observed; however, see Rizzi.[237]

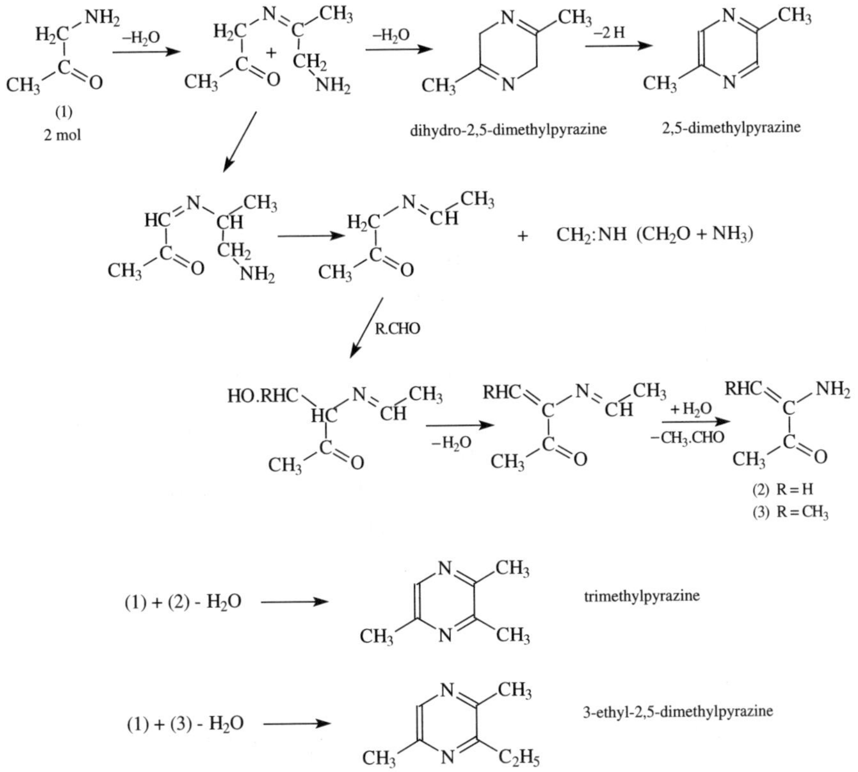

Flavour and Off-Flavour Formation in Nonenzymic Browning

[Scheme 5.10 reaction diagram showing formation pathways:]

(1) 2 mol: $H_2C(NH_2)-C(=O)-CH_3$ (aminoacetone) → $-H_2O$ → dimer with $H_2C-N=C(CH_3)$ and $C(=O)(CH_3)-CH_2-NH_2$ → $-H_2O$ → dihydro-2,5-dimethylpyrazine → $-2H$ → 2,5-dimethylpyrazine

Alternative pathway via $HC=N$ intermediate → $H_2C=N-CH(CH_3)$ + $CH_2:NH$ $(CH_2O + NH_3)$

With R.CHO: $HO\cdot RHC-HC(N=C(CH_3))-C(=O)CH_3$ → $-H_2O$ → $RHC=C(N=CH(CH_3))-C(=O)CH_3$ → $+H_2O$, $-CH_3\cdot CHO$ → $RHC=C(NH_2)-C(=O)CH_3$

(2) R = H
(3) R = CH$_3$

(1) + (2) − H$_2$O → trimethylpyrazine

(1) + (3) − H$_2$O → 3-ethyl-2,5-dimethylpyrazine

Scheme 5.10 *Hypothetical formation of trimethyl- and 3-ethyl-2,5-dimethylpyrazine from aminoacetone*[237]

Table 5.3 *Yield (%) of pyrazines from various carbon sources and asparagine*[238]

Carbon source	Pyrazine	MMP[a]	DMP	TMP
Glucose	—	1.8	3.8[b]	0.3
Fructose	—	2.0	6.2[c]	0.4
Glycolaldehyde	0.1	0.35	0.25[d]	—
2-Oxopropanal	—	—	6.2	1.5
Glyceraldehyde	—	—	6.3	0.55
Glycolaldehyde + 2-oxopropanal	—	1.05	2.35	0.45
Glyceraldehyde + 2-oxopropanal	—	—	7.5	1.15

[a] MMP, monomethylpyrazine; DMP, dimethylpyrazines; TMP, trimethylpyrazine.
[b] Ratio 2,5-DMP/2,6-DMP = 4.6.
[c] Ratio 2,5-DMP/2,6-DMP = 4.1.
[d] Ratio 2,3-DMP/2,5+2,6-DMP = 5.

The amount of 2,5-dimethylpyrazine was 4–5 times that of its 2,6-isomer. Weenen et al.[238] attribute this to the preferential pick up of ammonia or amino groups at the

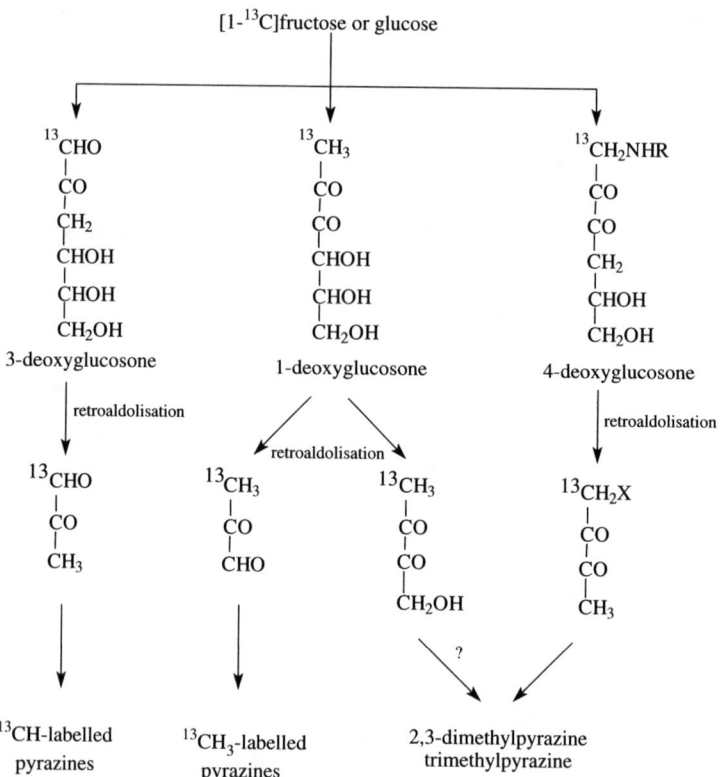

Scheme 5.11 *Formation of pyrazines from [1-^{13}C] fructose or glucose via deoxyhexosones*[238]

aldehyde group of 2-oxopropanal and glyceraldehyde, whereas one molecule of a C$_3$ compound with a 2-NH$_2$ or 2-NHR is essential for forming 2,6-dimethylpyrazine.

Data for reactions with asparagine of C$_2$ and C$_3$ precursors of pyrazines are included in Table 5.3. Glycolaldehyde should lead to a good yield of unsubstituted pyrazine, but only 0.1% was obtained. 2-Oxopropanal, as expected, gave no unsubstituted or monomethylpyrazine, yielding mainly the dimethyl compounds, but also some trimethyl derivative. This was attributed to the formation of formaldehyde, which reacted with the dihydrodimethylpyrazine intermediates (*cf.* Rizzi[237]).

Glutamine also carries an amide sidechain, which constitutes an additional source of nitrogen and so leads to a greater formation of pyrazines with glucose or fructose in aqueous solution at 160 °C, as compared with serine and threonine.[239] Glucose and fructose systems also gave greater amounts of acetylpyrazine than the corresponding ribose systems.

In comparing pyrolysis (250 °C, 20 s) with reaction in aqueous solution (120 °C, 3 h, sealed tube) for glucose–glycine mixtures, the glycine being labelled at C-2, Wnorowski and Yaylayan[212] found trimethylpyrazine to be unlabelled, labelled at the 2-methyl group, and doubly labelled at the 2- and the 5-methyl group to the extent of 20, 70, and 10 and 30, 60, and 10%, respectively.

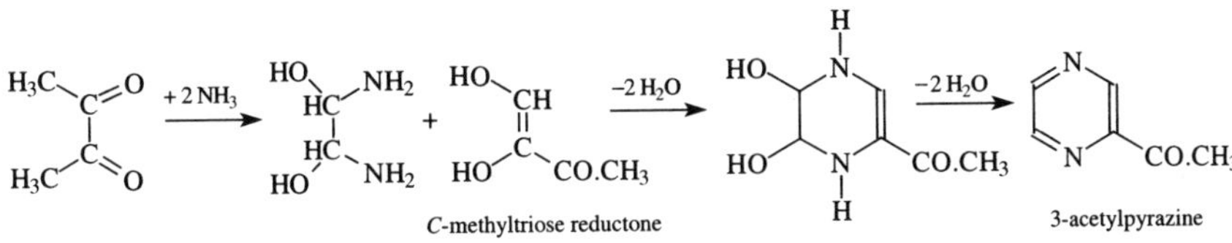

Scheme 5.12 *Formation of 2-acetylpyrazine (after Scarpellino and Soukup[240])*

2-Acetylpyrazine (roast, $T = 62$ ppb) is thought to be formed from glyoxal, ammonia, and C-methyltriose reductone (Scheme 5.12).[240]

Dipeptides and glyoxal give rise to some interesting pyrazinones (see Chapter 3).

The amount and diversity of pyrazines formed on extruding a wheat flour system can be substantially enhanced by the addition of ammonium bicarbonate and 2-oxopropanal.[241]

2.1.5 Oxazoles

The α-aminocarbonyls are not only precursors of pyrazines, but can also lead to pyrroles,[242] as well as imidazoles and oxazoles.[243] Pyrolysis–GC–MS is relatively readily available and provides a productive technique. As mentioned previously, Wnorowski and Yaylayan[212] had shown that, although more products are formed on pyrolysis than in aqueous media, most of the products identified in aqueous systems are present in pyrolysates with identical label distribution, even though the proportions may differ. Pyrolysates (250 °C, 20 s) of model systems of carbonyl and [2-^{13}C]Gly or Ala were analysed. In the butanedione–Gly system, acetaldehyde and formaldehyde formed by decomposition of the carbonyl are unlabelled, but formaldehyde by Strecker degradation is labelled. 2,4,5-Trimethyloxazole was found to be unlabelled, being formed from acetaldehyde, but 4,5-dimethyloxazole was 15% mono-labelled, i.e., 15% of the precursor formaldehyde had been derived by Strecker degradation.

2.1.6 Imidazoles

Vernin and Párkányi[216] have tabulated 14 imidazoles from four model systems.

The imidazole system is already present in histidine. Reacting histidine with glucose under the conditions of roasting (220 °C) or autoclaving (120, 150, or 180 °C) led to the identification of 2-acetyl- and 2-propionylpyrido[3,4-d]imidazole, as well as the corresponding tetrahydropyrido derivatives.[244] The first of these compounds was also obtained on heating tuna flesh with glucose, tuna flesh being known to contain relatively high amounts of histidine (up to 1%). The same compound also resulted when histamine was heated with 2-oxopropanal (see Scheme 5.13).

Similar products were obtained on autoclaving 1-, 2-, and 3-methylhistidine with glucose (cf. Chapter 2).

Imidazole formation runs closely parallel to oxazole formation.[243] That the N-alkyl groups are derived from the amino acid was shown by the use of [2-^{13}C]Gly and Ala.

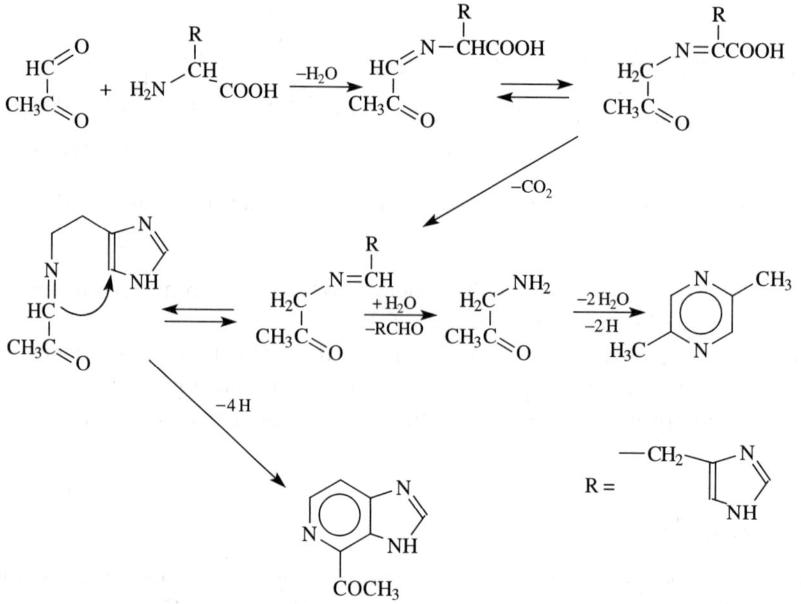

Scheme 5.13 *Mechanism proposed for the formation of 2-acetylpyrido[3,4-d]imidazole via Strecker degradation of histidine with 2-oxopropanal*[244]

Imidazole rings appear in a number of physiological modifications of arginine residues (see Chapter 8). Some imidazoles elicit undesirable physiological responses (see Chapter 6).

2.1.7 Heterocyclic Sulfur Compounds

The sulfur-containing amino acids also provide special cases. By 1986, MacLeod[245] was able to list seven aliphatic and 65 heterocyclic sulfur compounds (the sulfur is not always in the heterocyclic ring) with meat-like flavours.

H_2S is often the source of the sulfur, either directly or indirectly. Zheng and Ho[246] found the activation energy (E_a) of its formation over the pH range 3–9 to be 123.0–134.7 kJ mol^{-1} (29.4–32.2 kcal mol^{-1}) for cysteine and 78.9–128.9 kJ mol^{-1} (18.8–30.8 kcal mol^{-1}) for glutathione. At pH 5, the values were almost the same, but at other pH values, the E_a for cysteine was about 44 kJ mol^{-1} (about 10 kcal mol^{-1}) higher. The lower activation energies for glutathione imply that the molecular environment plays a significant role in the release of H_2S. The isoelectric points for cysteine and glutathione are also quite different, 5.07 and 2.83, respectively.

Cysteine can undergo the Strecker degradation, transamination, and β-elimination, as shown by Tressl *et al.*[247] using [1- or 6-^{13}C]glucose (equimolar aqueous solution, 160 °C, 1.5 h). 2-Furylmethanethiol ($T = 0.005$ ppb), very important in providing the aroma of roasted coffee and roasted meat, is formed as shown in Scheme 5.14 from [1-^{13}C]glucose via the 3-deoxy-1,2-dicarbonyl, which loses

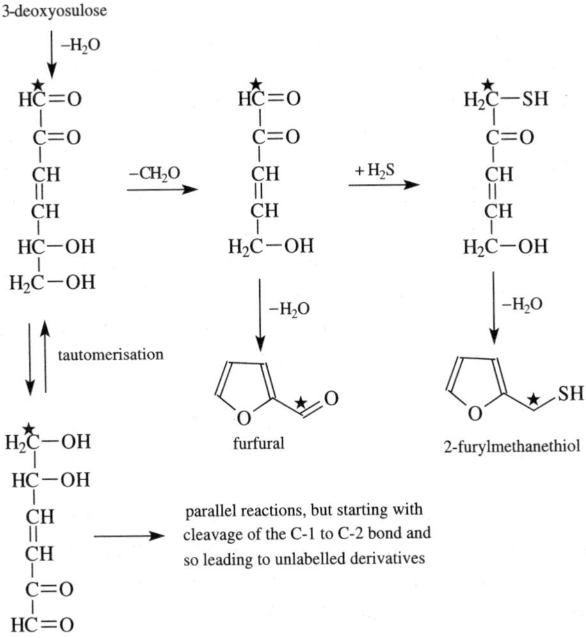

Flavour and Off-Flavour Formation in Nonenzymic Browning

Scheme 5.14 *Formation of 50% labelled furfural and 2-furylmethanethiol from [1-^{13}C]glucose and cysteine via 3-deoxyosone*^{223}

water, allowing C-1 and C-6 to equilibrate, thus resulting in the observed 50% dilution of the label in the CH$_2$ group.^{30} This experiment constitutes an excellent example of the efficacy of ^{13}C-labelling.

Kahweofuran (Structure **32**) is also a sulfur derivative of furan and occurs in coffee, together with its dimethyl and ethyl homologues. It provides a roasted, smoky aroma at high dilution. Its formation in coffee has not been elucidated.^{248} Kahweofuran and its dimethyl and ethyl homologues have also been found in bread volatiles.^{249}

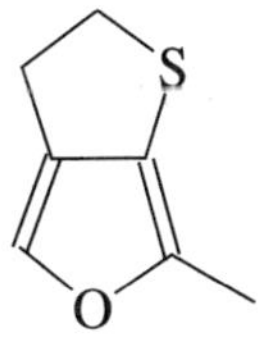

Structure 5.1 *Kahweofuran*

Cerny and Davidek^{250} used [^{13}C$_5$]ribose to show that, when heated 3:1 with cysteine at 95 °C for 4 h in a phosphate buffer (0.5 M, pH 5), the ribose is incorporated intact not only into furylmethanethiol, but also into 2-methyl-3-furanthiol (meat-like,

$T = 0.0004$ µg L^{-1}) and 3-mercapto-2-pentanone. The methylfuran moiety of 2-methyl-3-(methylthio)furan also originates in the ribose, but the methylthio carbon comes partly from ribose and partly from cysteine. All the carbons of 3-mercapto-2-butanone come from ribose, one carbon of which is split off. All the carbons of 3-thiophenthiol stem from cysteine. When norfuraneol was incorporated into the reaction mixture, virtually all the 2-methyl-3-furanthiol continued to be formed from ribose. However, 2-mercapto-3-pentanone was formed from it, whereas the 3,2-isomer was derived both from norfuraneol and from ribose.

To explain the formation of 3-mercapto-2-pentanone without its 2,3-isomer, a new pathway was proposed (Scheme 5.15), in which 1,4-dideoxyosone and 5-hydroxy-3-mercapto-3-penten-2-one are intermediates. Both 2-methyl-3-furanthiol and 2-methyl-3-hydroxyfuran are readily derived from this pathway without involving norfuraneol.

When the reaction was carried out in the presence of 2-furaldehyde (equimolar to the ribose), it turned out to be a much more efficient precursor for 2-furyl-methanethiol (92:8).

For the thiophen mentioned above, all the carbon atoms stemmed from cysteine, carbohydrate apparently not having a role. On the other hand, theoretically, thiophens can be derived from furans simply by reaction with H$_2$S. Belitz and Grosch[251] postulate derivation from 2-mercaptoethanal (Strecker aldehyde from cysteine) and acrolein or butenal (aldol condensation product from acetaldehyde). Vernin and Párkányi[216]

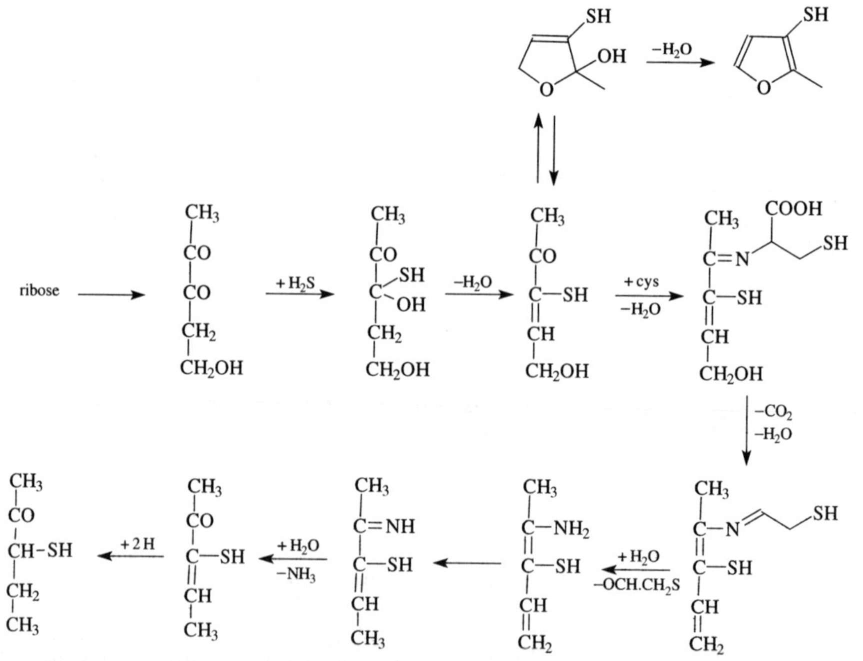

Scheme 5.15 *Proposed formation of 3-mercapto-2-pentanone and 2-methyl-3-furanthiol from ribose and cysteine via 1,4-dideoxy-2,3-diketose*[250]

have listed the following four pathways to obtain 2-formylthiophens: (a) from 3,4-dideoxypentulosene and H_2S, (b) 2-furfural and H_2S, (c) attack on C-3 of 3-substituted acroleins by mercaptoacetaldehyde (giving 5-substituted 2-formylthiophens), and (d) as (c), but attack on C-1 (giving 3-substituted 2-formylthiophens). Thiophens have been extensively reviewed by Maga[252] and Vernin and Vernin (1982).[253] 2-Acetylthiophen ($T = 0.1$ ppb) has an onion- or mustard-like aroma, but can develop a malty, roast aroma note in coffee.[152] 2-Formylthiophen and its 5-methyl homologue have $T = 2$ ppm and 1 ppb, respectively, with benzaldehyde- and cherry-like odour, respectively.

When ribose and cysteine are heated at 140–145 °C for 20–30 min, both 3,2- and 2,3-mercaptopentanone are formed, with the former predominating.[254,255] The higher temperature/shorter time favours a mechanism different from that in Scheme 5.15.

As far as 2-methyl-3-furanthiol is concerned, it can be formed from cysteine, but thiamine constitutes its more important precursor by far.[256] When the reaction is carried out in the presence of thiamine in an aqueous medium at 120 °C for 1 h, only about 8% of the thiol is derived from cysteine, and, in the absence of thiamine from the mix, no thiol was detected. The probable mechanism of formation from thiamine is shown in Scheme 5.16.[257]

The thiazole formed under the conditions used by Cerny and Davidek[250] was only singly labelled, the label appearing at C-2, suggesting that the compound had been formed from cysteamine and formaldehyde derived from ribose (cf. ref. 258).

Thiazoles and related compounds have been reviewed by Maga.[259] Vernin and Párkányi[216] have tabulated 25 thiazoles from seven model systems.

The low threshold of 2,4-dimethyl-5-ethylthiazole is noteworthy (2 ppb), as is its nutty, roast, meaty, liver-like aroma.[260]

2-Acetyl-2-thiazoline has an even lower threshold (1 ppb). It is formed from cysteamine with glucose[258] in relatively small proportions; instead when cysteamine and 2-oxopropanal are used,[261] it is formed along with 2-acetylthiazolidine and 2-formyl-2-methylthiazolidine. The temperature has a great effect on the proportion of the three compounds extracted into CH_2Cl_2 from reaction at pH 6, as can be seen from Table 5.4. The determination of 2-acetyl-2-thiazolidine after reaction at 25 °C was used as an alternative to condensation with o-phenylenediamine to determine 2-oxopropanal in foods and beverages, 0.04–47 ppm being found in the 17 items analysed.

The formation of 2-acetyl-2-thiazoline by reaction of cysteamine with 2-oxopropanal was confirmed by Hofmann and Schieberle,[262] who showed that cupric ions and oxygen increase the yield. The thiazolidine and the isomeric 2-thiazoline give 2-acetyl-2-thiazoline in 34 and 7% yield, respectively, on heating in water at 100 °C for 20 min, longer times leading to degradation. The results are

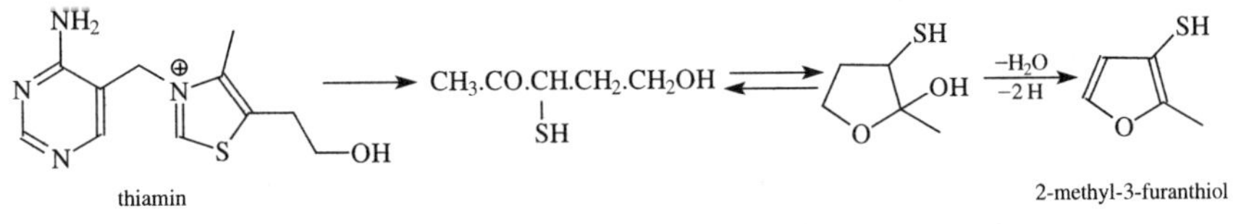

Scheme 5.16 *Proposed formation of 2-methyl-3-furanthiol from thiamine*[257]

Table 5.4 *Proportions formed of each of the three products of the reaction of 2-oxopropanal and cysteamine at three different temperatures*[261]

Temperature (°C)	2-Acetylthiazoline (%)	2-Acetylthiazolidine (%)	2-Formyl-2-methyl-thiazolidine (%)
0	2.0	40.0	58.0
25	8.5	89.0	2.5
100	81.0	19.0	0.0

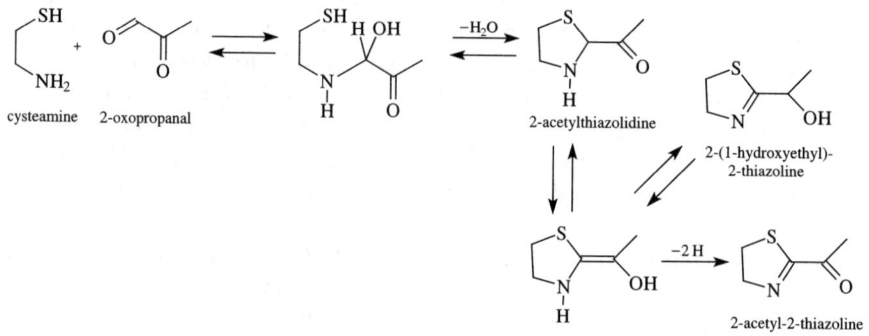

Scheme 5.17 *Formation of 2-acetylthiazolidine, 2-(1-hydroxyethyl)-2-thiazoline, and 2-acetyl-2-thiazoline from cysteamine and 2-oxopropanal*[262,263]

explained by Scheme 5.17. The E_a for the conversion of 2-(1-hydroxyethyl)-2-thiazoline to 2-acetyl-2-thiazoline was found to be 57.4 kJ mol^{-1} (13.7 kcal mol^{-1}).

Two alternatives were proposed for the final dehydrogenation; the enaminol could undergo a metal-catalysed peroxidation[263] with subsequent elimination of H_2O_2 or it could add to another molecule 2-oxopropanal to form the aldehyde-ammonia with subsequent elimination of hydroxyacetone.[262]

Rhlid et al.[264] have tried to exploit these results technologically by preparing 2-(1-hydroxyethyl)-2-thiazoline by microbial fermentation (cysteamine, ethyl L-lactate, and D-glucose with bakers' yeast) and incorporating it into a pizza recipe (5 mg per 50 g raw dough), leading to increases in the roasted, toasted, popcorn-like aromas.

Engel and Schieberle[265] showed that cysteamine and fructose in 0.1 M phosphate buffer, taken from 20 to 145 °C in 20 min, gave rise not only to 2-acetyl-2-thiazoline, but also to three other key aroma compounds, one of which was shown to be the novel N-(2-mercaptoethyl)-1,3-thiazolidine. It was formed in the highest amount at pH 8.0, near the isoelectric point (IEP) of cysteamine, which was explained by Scheme 5.18, the IEP being favourable for the formation of the thiirane. The structure of N-(2-mercaptoethyl)-1,3-thiazolidine had already been determined and its odour characterised as roast and popcorn-like.[266] It has a threshold of 0.005 ng L^{-1} in air, which is one-fourth of that of 2-acetyl-2-thiazoline. It is amphoteric and relatively unstable to heat.

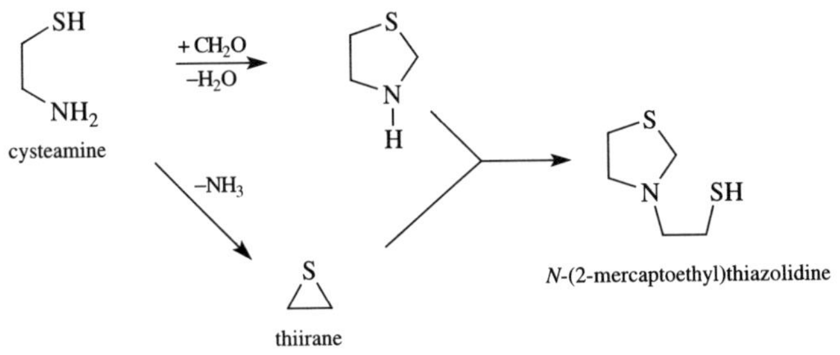

Scheme 5.18 *Formation of N-(2-mercaptoethyl)thiazolidine from cysteamine and formaldehyde*[265]

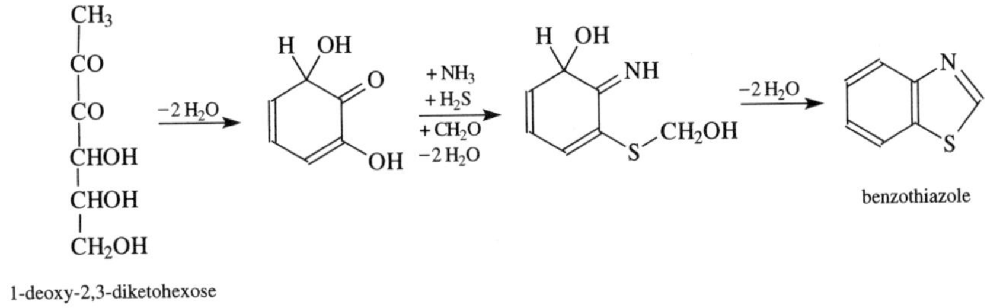

Scheme 5.19 *Proposed formation of benzothiazole from 1-deoxy-2,3-diketohexose*[216]

Much earlier, Mulders[267] had treated cysteine (0.01 mol) and cystine (0.1 mol) with ribose (0.025 mol) in phosphate buffer, pH 5.6 (35 ml), and diethyleneglycol (200 ml) under reflux at 125 °C for 24 h, identifying 45 components, including seven thiazoles, one of which was the 2-acetyl derivative.

Benzothiazole is a compound quite frequently encountered in food volatiles. Vernin and Párkányi[216] postulate its formation from 1-DH as shown in Scheme 5.19.

Shibamoto and Yeo[268] have compared microwave (700 W, high setting, 15 min) and thermal treatment (reflux, 100 °C, 40 h) for a glucose–cysteine system. The conditions used were determined by the onset of browning and aroma formation. The two sets of conditions gave samples with similar popcorn and nutty flavours, but the microwaved samples also gave pungent, raw, and burnt aromas, absent from the conventionally heated ones. The sample prepared conventionally at pH 9 contained much higher amounts of methylpyrazine, 2,6-dimethylpyrazine, and HDMF, whereas the microwaved one gave a much higher amount of 4,5-dimethyloxazole and was the only one to produce 2,3-dihydro-3,5-dihydroxy-6-methyl-4H-pyran-4-one. Such data, to some extent, explain the differences in acceptability of the two types of heating. For browning to occur in microwaving, a minimum of 10% moisture is required. Surprisingly, microwaving at pH 2 gave about twice the absorption at 420 nm than at pH 9 (about 1 AU), the absorption for pH 5 and 7 samples being nearly 0 (< 0.1 AU).

Another less common heterocyclic sulfur volatile is 1-[H]-pyrrolo-[2,1c]-1,4-thiazine, identified in bread crust by Baltes and Song [249] and previously obtained by Guntert et al.[269] from a ribose–cysteine system.

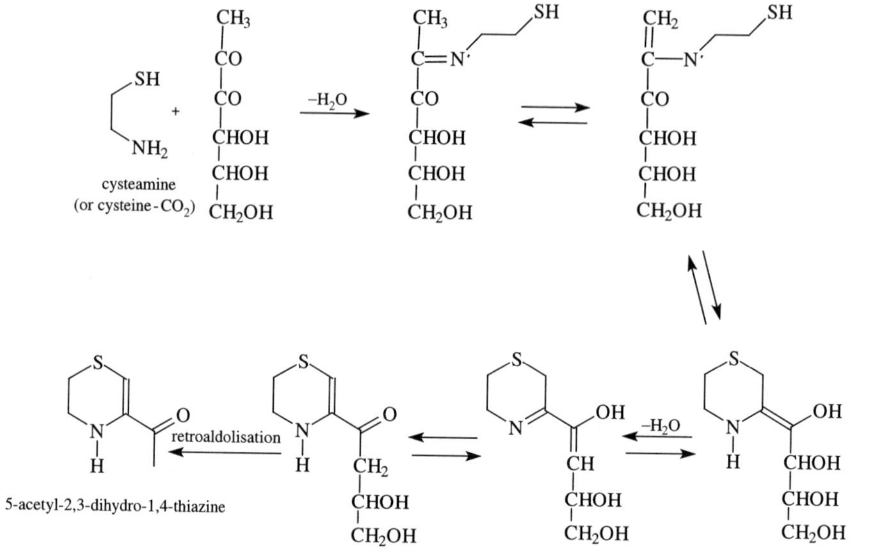

Scheme 5.20 *Proposed formation of 5-acetyl-2,3-dihydro-1,4-thiazine from cysteamine (or cystine) and 1-deoxy-2,3-diketohexose*[270]

Another popcorn-like odorant, 5-acetyl-2,3-dihydro-1,4-thiazine ($T = 1.25$ ppb), is formed from diacetyl and cysteamine, optimally at pH 7, whereas formation from cysteine and carbohydrate has its optimum at pH 5.[270] The mechanism was depicted as in Scheme 5.20.

Canning of liver sausage encounters a problem from the formation of burnt off-odours,[271] the intensity of which was found to be linearly related to the loss of free Gly, Glu, and Thr in model experiments. Addition prior to canning of reduced glutathione left flavour and colour unaffected, but the addition of N-acetylcysteine inhibited burnt flavour formation, optimally at 0.15%.[272] However, an acid taste developed, but was prevented by the further addition of 0.25% diphosphate.

2.1.8 Cyclopentanes

Cyclotene (3-methylcyclopent-2-en-2-olone) is an important caramel odorant produced in the Maillard reaction (see ref. 273). It has 12 tautomeric forms (see Scheme 5.21). Its formation had been proposed to involve the condensation of two molecules of hydroxyacetone (acetol),[274] but, in comparing pyrolysis (250 °C, 20 s) with reaction in aqueous solution (120 °C, 3 h, sealed tube) for glucose–glycine mixtures, the glucose being singly labelled at C-1 to C-6 in a series of experiments, Wnorowski and Yaylayan[212] found that the glucose skeleton was incorporated intact 100% into the cyclotene formed, with C-1 in the methyl group and C-4 in the carbonyl C, under both sets of conditions.

2.2 Olfactory Threshold

The first attempt to apply odour threshold values to rationalise the extensive information obtained by the analysis of the volatile products of Maillard reactions

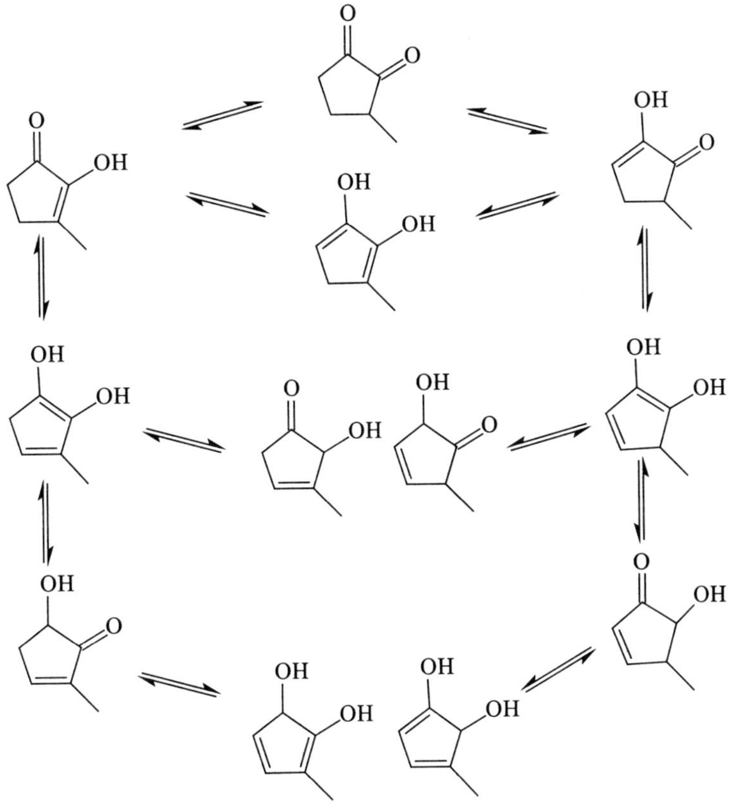

Flavour and Off-Flavour Formation in Nonenzymic Browning 83

Scheme 5.21 *The 12 tautomeric forms of 3-methylcyclopent-2-en-2-olones*[274]

was that of Arnoldi,[275] but was hampered by the lack of threshold data. This still holds, even if not to the same extent.

Strecker aldehydes, the main volatiles of Group 2, tend to have low threshold values (phenylethanal, 4 μg L^{-1} in water – thresholds will be given in these units unless stated otherwise and are taken from Rychlik *et al.*[204] and Maarse[276] –; 3-methylbutanal, 0.35; methional, 0.2). Some Group 1 volatiles have comparable olfactory power (HDMF, 0.6; butanedione, 4), but others do not (norfuraneol, 2100; maltol, 9000; acetic acid, 22000). The threshold value of acetaldehyde (15), although a Strecker aldehyde, is only moderately low.

Group 3 volatiles, formed by further interactions, can reach even lower threshold values than those mentioned so far, particularly when sulfur is involved, usually derived by the break down of cysteine (bis(2-methyl-3-furyl) disulfide, 0.00002; 2-methyl-3-furanthiol, 0.0004; 2-furylmethanethiol, 0.005; dimethyl trisulfide, 0.01; dimethyl disulfide, 0.16; dimethyl sulfide, 0.3; hydrogen sulfide, 10). Into some volatile sulfur compounds other atoms as well as the sulfur of cysteine are incorporated (2-acetylthiazoline, 1; 5-acetyl-2,3-dihydro-1,4-thiazine, 1.25; 2-acetylthiazole, 10). Other volatiles are more likely to be derived from methionine (methanethiol, 0.2).

For some compounds, threshold values are only available in air, but 2-acetyl-2-thiazoline (0.05 ng L^{-1} in air) provides a link with the data given above: 2-propionyl-2-thiazoline (0.02 ng L^{-1} in air) and N-(2-mercaptoethyl)-1,3-thiazolidine (0.005 ng L^{-1} in air).[265]

Proline also has special significance in terms of leading to specific volatiles with low threshold values. Proline, not being a primary amine, cannot undergo the Strecker degradation, but becomes channelled into forming volatiles in which the pyrrolidine ring is preserved (2-acetylpyrroline, 0.1) or expanded (2-acetyltetrahydropyridine, 1.6). The aroma produced in an aqueous glucose–proline sealed system in 1 min at 200 °C has been shown by Roberts and Acree[277] by Charm analysis to be due to 2-acetyl-3,4,5,6-tetrahydro-1H-pyridine (63%), 2-acetyl-1-pyrroline (19%), 2-acetyl-1,4,5,6-tetrahydro-1H-pyridine (12%), and HDMF (4%). The role of butanedione (0.5%) was understated by a factor of about 8, because its yield was only 12%. Despite being the main volatile produced, 5-acetyl-2,3-dihydro-1H-pyrrolizine contributed only 0.3%. Another major product, maltoxazine, exerted no effect on the aroma.

The Strecker reactant, *i.e.*, the α-dicarbonyl, picks up ammonia during the Strecker reaction and can then form heterocyclic volatiles, particularly pyrazines. The most powerful odorants among these are the 2-ethyl-3,5-dimethyl- (0.04) and the 2,3-diethyl-5-methyl-derivatives (0.09). It is worth noting that most other alkylpyrazines have much higher thresholds and thus make relatively little contribution to the aroma of heated foods.

2.3 Odour Quality[278]

Odour is not just a question of intensity, but also of quality. One can try to pick the odour qualities most relevant to Maillard reaction products from Harper's list of 44:[279]

Almond-like	003	Warm	021
Burnt, smoky	004	Cooked vegetables	034
Meaty(cooked)	013	Sweet	035

The numbers attached are those from Dravniek's scheme of 146 odour qualities,[280] from which the following can usefully be added:

Malty	024	Nutty (walnut, *etc.*)	115
Popcorn	026	Coffee-like	118
Like burnt paper	062	Burnt milk	121
Caramel	094	Bakery (fresh bread)	126
Chocolate	097	Maple (as in syrup)	133
Molasses	098	Grainy (as grain)	137

Some years ago, Lane and Nursten[176] described the odours obtained from more than 400 model Maillard systems and sorted the descriptors used into 14 groups, as follows:

1. Sweet, boiled sugar, caramel, toffee
2. Chocolate, cocoa

Flavour and Off-Flavour Formation in Nonenzymic Browning 85

3. Bread, crusty, biscuits, cakes, toast
4. Meaty, beefy
5. Potato, potato skins, potato crisps
6. Fruity, aromatic ester
7. Celery, chicory, leeks, Brussels sprouts, turnips
8. Puffed wheat, sugar puffs
9. Nutty
10. Floral
11. Ammoniacal
12. Unpleasant, 'caused coughing'
13. Aldehydic
14. Burnt, charred, scorched, acrid, potato crisps, toast, smoky

Some close links were apparent. Thus, Group 10 was associated only with the amino acid phenylalanine, whereas Group 2 odours could be derived from 13 amino acids and Group 3 from 12.

In relation to the aromas produced on exposure to microwave radiation, Yaylayan *et al.*[281] examined many combinations of sugars and amino acids, grouping the latter into aliphatic, hydroxylated, aromatic, secondary, basic, amide, acid, and sulfur-containing ones. The odours observed were grouped into eight and they have been assigned to the above groups, as far as possible, below:

Caramel	1
Meaty	4
Nutty	9
Meaty + vegetable	
Fragrant	6
Roasted vegetable	
Baked potato	5
Baked	3

For caramel aroma, aliphatic amino acids were required; for meaty, sulfur-containing ones; for nutty, amides or hydroxy ones; for meaty and vegetable, sulfur-containing and aromatic ones; for fragrant, aromatic ones; for roasted vegetable, aromatic, sulfur-containing, acid, aliphatic and basic or amide ones; for baked potato, sulfur-containing ones; and for baked, secondary ones. In each case, additional amino acids were or were not present in the mixtures heated.

The odour descriptors for quite a number of individual Maillard products are given by Kerler and Winkel.[282]

With regard to a glucose–proline system, Roberts and Acree[277] have examined the sensory aspects in much more detail by applying Charm analysis (see the Olfactory Threshold section above). Four compounds provided most of the aroma: 2-acetyl-3,4,5,6-tetrahydro-1H-pyridine (burnt, caramel; 63%), 2-acetyl-1-pyrroline (popcorn; 19%), 2-acetyl-1,4,5,6-tetrahydro-1H-pyridine (burnt, caramel; 12%), and HDMF (cotton candy; 4%). All Maillard systems of interest need to be submitted to similarly detailed analyses.

3 Sapid Compounds

3.1 Bitter Compounds

By now, quite a range of bitter-tasting compounds has been isolated and characterised in Maillard-type reactions.

Thus, Tressl et al.[219] characterised eight 2-(1-pyrrolidinyl)-2-cyclopentenones and 11 cyclopent(b)azepin-8(1H)-ones from proline–monosaccharide and proline–cyclic enolone systems. The compounds possessed bitter tastes, with the former exhibiting concomitant astringency. The bitter thresholds in water of Structure **33** and **34** were 50 and 10 ppm, respectively.

The mechanism of formation was postulated to be as in Scheme 5.22.

The effectiveness of various precursors of these compounds was investigated quantitatively by Ottinger and Hofmann.[283] Hexose-derived cyclotene was the common precursor for both **33** and **35**, as well as **34** and **36**. The formation of each compound is very much determined by the nature of the N-containing precursor. Thus, for example, pyrrolidine (e.g., formed by thermal decarboxylation of proline) plus cyclotene produced **33** and **35** only, whereas 1-pyrroline (derived from proline

	R	R'	R''	T
37	H	H	H	$2.5 \cdot 10^{-4}$
38	H	H	CH₃	$2.5 \cdot 10^{-4}$
39	H	CH₃	H	$5.0 \cdot 10^{-4}$
40	H	CH₃	CH₃	$1.0 \cdot 10^{-3}$
41	CH₂OH	H	H	$1.0 \cdot 10^{-3}$
42	S for O in both furyl rings			$6.3 \cdot 10^{-5}$

Structures 5.2 *Bitter compounds*

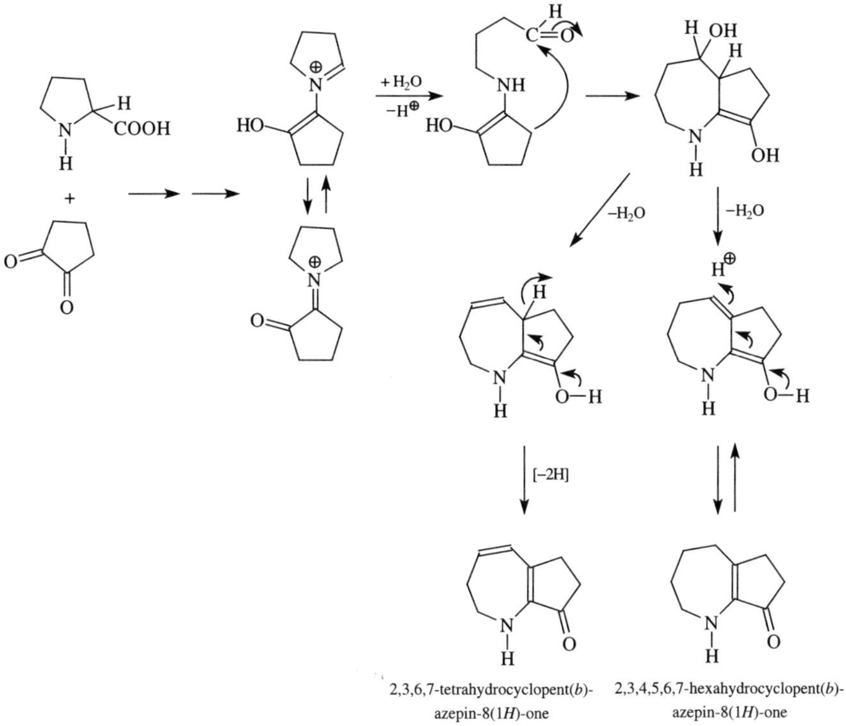

Scheme 5.22 *Formation of cyclopent(b)azepinones*[219]

by Strecker degradation) plus cyclotene produced **36** exclusively. Neither N-source gave rise to **34**, which was by far the dominant product of the interaction of proline and cyclotene. Even in the presence of Cu^{2+}, it remained the main product, thus casting doubt on the significance of the oxidation in Scheme 5.22.

Bitter compounds are also formed in solutions of alanine with xylose and rhamnose.[284] Twenty-six HPLC fractions were obtained, seven of which were shown to have high impact on taste dilution analysis. Structures **37-41** accounted for 57% of overall bitterness. The compounds have low threshold values; introduction of methyl groups into the furyl rings increase the threshold value. On the contrary, substituting the furyl ring-O by S (**42**) lowered the threshold value to almost 10^4 times lower than that of caffeine on a molar basis.

When xylose and alanine were heated, the bitter-tasting compounds corresponding to Structures **43** and **44** (see Scheme 5.23), were obtained,[285] their yield increasing to about six times in the presence of furfural. Ribose was about as effective as xylose in yielding **43** and **44**, which were unable to be generated from glucose or fructose. Valine and leucine gave 3–4 times the level of **43** and **44** as glycine or alanine in a phosphate buffer, pH 5.0. Changing the pH from pH 3.0 to 9.0, reduced the level of **43** and **44** about 4 times. The use of $[^{13}C_5]$ribose led to the pathway proposed in Scheme 5.23.

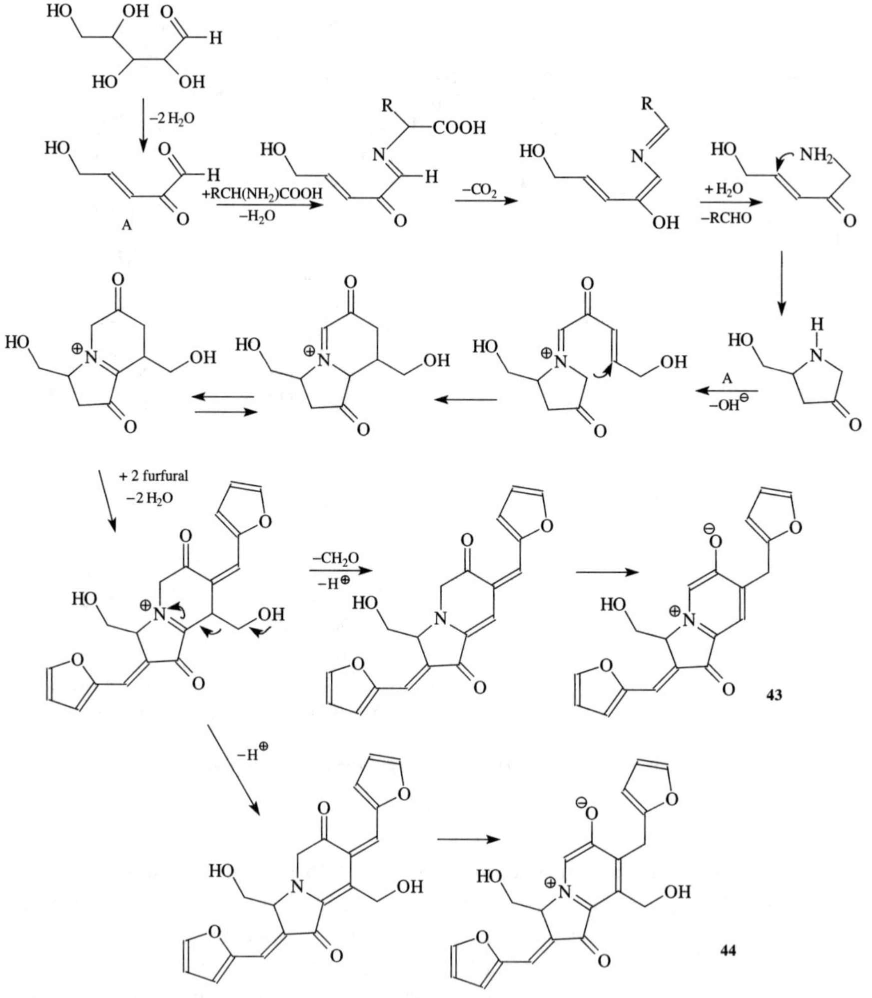

7-(2-furylmethyl)-2-(2-furylmethylidene)-3,8-bis(hydroxymethyl)-
1-oxo-2,3-dihydro-1*H*-indolizinium-6-olate

Scheme 5.23 *Formation of bitter-tasting indolizinium-6-olates*[285]

3.2 Sweetness Enhancers

Ottinger *et al.*[286] have applied their comparative taste dilution analysis (cTDA) to examine the extractable products from heated aqueous D-glucose and L-alanine that were not solvent-extractable. One HPLC fraction proved to be a strong sweetness enhancer. It was isolated and submitted to LC-MS and NMR, both 1D and 2D; the results, together with its synthesis from HMF and alanine, unequivocally identified it as the inner salt of *N*-(1-carboxyethyl)-6-(hydroxymethyl)pyridinium-3-ol (alapyridaine, Structure **45**). It has no taste on its own, which in many applications would be an advantage. Depending on the pH, it lowers the detection threshold of sweet sugars, amino acids, and aspartame, the

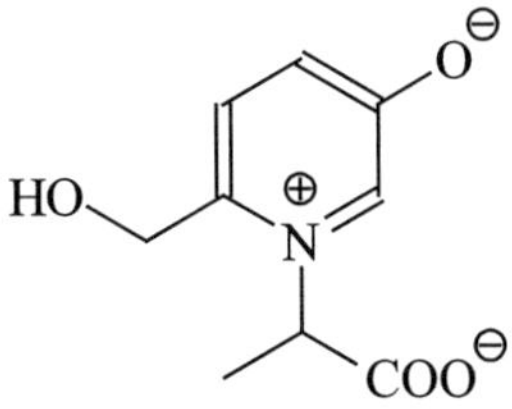

45
alapyridaine

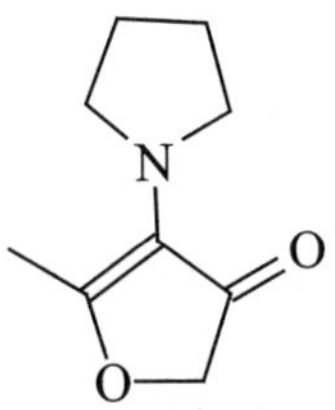

46 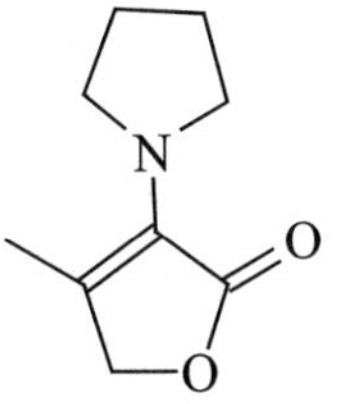　　　**47**

Structures 5.3

best example being an equimolar mixture of glucose and alapyridaine with a threshold only 1/16th of that of glucose alone. The physiologically active enantiomer is the (+)-S, the (−)-R one being without any effect on sweetness. Alapyridaine is relatively stable, >84% being recovered after 5 h at 80 °C in aqueous solution, pH 3–7.

3.3 Cooling Compounds

Although cooling compounds, of which menthol is probably the best known, have been recognised for many years and are of special significance, in relation particularly to toothpaste and to tobacco smoking, their production in Maillard reactions has only been recently demonstrated.[287] Compounds with cooling effects were isolated both from roasted glucose–proline systems and from dark malt, the strongest effects being exerted by **35** and **33**. This surprising finding was followed by a structure–activity investigation,[288] which led to **46** and **47**, both odourless with threshold values for cooling of 1.5–3.0 and 0.02–0.06 ppm, respectively, the latter lying at a level about 35 times lower than that of (−)-menthol ($T_{cooling}$ = 0.9–1.9 and $T_{mint-like}$ = 0.1–0.2 ppm). Whereas the cooling effect with (−)-menthol is inevitably accompanied by the odour of mint, the cooling effect of compounds such as **46** and **47** is odourless and therefore potentially has a much wider range of applications.

4 Summary

Overall, the contribution of the Maillard reaction to flavour and off-flavour is striking in its scope and amazing in its intricacies.

CHAPTER 6

Toxicological and Protective Aspects

1 Toxic Products

Toxicological aspects of the Maillard reaction and its products have been reviewed by Lee and Shibamoto[289] and by Friedman.[290]

In this chapter, carcinogenic products will be considered first, followed by mutagenic ones, which constitute the main focus. Smaller sections then deal with genotoxic products, allergenicity, other toxicological aspects, and protective effects.

2 Carcinogenic Products

Carcinogenicity is itself an exceedingly complex topic. Enzymes are important in relation to carcinogenicity. Xenobiotic enzymes play a major role in the detoxification of dietary carcinogens and mutagens, and intestinal Caco-2 cells provide useful models for assessing the effects of xenobiotics on enzyme activity. Thus, Faist et al.[291] have studied the effect of melanoidins both on a Phase-I enzyme, NADPH cytochrome c reductase (CCR), and a Phase-II enzyme, glutathione S-transferase (GST). Where the results achieved a significant level, activity was reduced, thus implying that detoxification would be less efficient in the presence of the melanoidins tested (equimolar glucose–glycine dry-heated for 2 h at 125 °C; equimolar glucose–casein dry-heated for 144 h at 90 °C). The only exception was the latter system in relation to the Phase-I enzyme. The differences between the effects of the two heating regimes on glucose alone are worth noting. They imply the formation of different products.

Faist et al.[292] have also examined the effect of dietary CML on the expression of GST in the rat kidney. In their Study 1, casein-linked CML was administered at two pharmacological doses [110 and 300 mg CML per kg body weight (b.w.) per day] for 10 d. In Study 2, supplementary breadcrust was used to give a daily intake of 11 mg (kg b.w.)$^{-1}$ d^{-1} for 42 d. In Study 1, the glutathione level in the kidneys increased 43 and 65%, respectively, and Phase II GST activity increased 12 and 96%, respectively, over the control. In Study 2, the protein content of the isoenzyme GST 1π-1 increased, but that of 1μ-1 and 1α-1 remained the same as in the control. Both studies were supported by experiments on Caco-2 cells. Exposure to purified CML

(5 μM, 96 h) resulted in a 46% increase in GST activity and exposure to 0.5 mg breadcrust ml^{-1} cell culture medium in a 38% increase with GST 1π-1 protein increasing by 40%, implying that CML is having a protective effect.

Low-molecular-mass browning products (< 1 kDa) from ultrafiltered Japanese soy sauce inhibited CCR, but increased GST activity at low concentrations (25 mg per 100 ml), thus showing evidence of a chemopreventive effect in intestinal cells *in vitro*.[293] More than 90% of soy sauce components have molecular masses < 1 kDa.

A dichloromethane extract of the reaction product of glucose and glycine in the presence of sodium nitrite, after purification, had low cytotoxicity and no genotoxicity, but did act as a tumour promoter in a two-stage oncogenesis with C3H10T1/2 cells initiated with benzo[*a*]pyrene.[294]

β-Lactoglobulin, glycated with ribose, arabinose, galactose, glucose, rhamnose, or lactose, showed no increase in cytotoxicity in the methylthiazoletetrazolium assay against COS-7 and HL-60 cells.[163]

As early as 1939, Widmark[295] observed that an extract of roasted horse muscle produced a carcinogenic effect when applied to mouse skin. Once the Ames test had been developed, the focus moved onto the mutagenicity of highly heat-treated foods (see the next section).

Currently, attention with regard to carcinogenicity has focused on acrylamide. In 1994, the International Agency for Research on Cancer (IARC) classified acrylamide in Group 2A as a 'probable carcinogen in humans' (see ref. 296). When a tunnel was being constructed in Sweden, acrylamide monomer was released due to the incomplete polymerisation of the component used in the grouting agent. During a study of people and wildlife exposed to acrylamide as a consequence, levels of Hb-acrylamide adduct, significantly higher than expected, were found in blood samples from the *control* group. This led to the suggestion that acrylamide in food might be responsible and, accordingly, analysis of food samples followed, first in Sweden and then in the UK and elsewhere (see Table 6.1).

It seems that significant amounts of acrylamide are formed in foods only when they are starchy and are fried, baked, or otherwise processed at a relatively high temperature.

According to Tareke *et al.*,[297] adducts of acrylamide with the N-termini of Hb have been regularly observed in people with no known environmental exposure to acrylamide. The mean level of the adduct in Swedish adults was estimated to correspond to a daily intake of about 100 μg acrylamide.

Table 6.1 *Acrylamide levels in some foodstuffs (from www.slv.se and www.food.gov.uk)*

Sample	Acrylamide (ng g^{-1})
Potatoes, raw	n.d.
Potatoes, boiled	n.d.
Potatoes, chipped and fried	Up to 3500
Potatoes, frying chips overcooked	Up to 13000
Potatoes, crisps	Up to 2300
Crispbread	Up to 4000
Breakfast cereal, rice-based	Up to 250
Breakfast cereal	Up to 1400

n.d., not detected.

The main uses of polyacrylamide are as a flocculant in the treatment of the water supply and in paper and pulp processing. Water is thus also a potential source of acrylamide found in human beings. Polyacrylamide has further uses in the removal of suspended solids from industrial waste water as well as in many other areas, such as cosmetics and soil conditioning. Smoking is known to involve exposure to acrylamide.

One important question concerns the origin of acrylamide in food. The Maillard reaction is an obvious candidate and asparagine, which already possesses an amide group, becomes a more specific target. Accordingly, Mottram et al.[296] treated an equimolar mixture of the amino acid and glucose at 185 °C in phosphate buffer in a sealed tube and obtained 3100 μmol mol^{-1} amino acid acrylamide. Formation did not occur below about 120 °C and exhibited a maximum at about 170 °C. Acrylamide was not detected (<7 μmol mol^{-1}) in parallel experiments with glycine, cysteine, and methionine at 185 °C. Glutamine and aspartic acid gave only trace amounts (about 10 μmol mol^{-1}). Heating a dry mixture of asparagine and glucose at 185 °C (i.e., in the absence of buffer solution) gave only 350 μmol mol^{-1}. Under such dry conditions, glutamine and aspartic acid gave traces of acrylamide and the other amino acids gave none, except for methionine (70 μmol mol^{-1}). When butanedione was used as a Strecker dicarbonyl in place of glucose, asparagine led to 890 μmol mol^{-1} acrylamide in buffer and to 560 μmol mol^{-1} dry. Asparagine alone at 185 °C did not produce acrylamide, confirming the requirement for a dicarbonyl reagent in a Strecker reaction context. Butanedione did not produce acrylamide in significant amounts with any of the other amino acids, except methionine (84 μmol mol^{-1} dry). Methionine would be converted into methional, which can break down into acrolein, ammonia, and methanethiol. Acrolein is readily oxidised to acrylic acid, which with ammonia can be converted into its amide. Acrolein can also be derived from lipids.

In parallel work, Stadler et al.[298] heated 20 amino acids individually at 180 °C for 30 min and obtained acrylamide (3.6 and 0.6 μmol mol^{-1}) from methionine and asparagine, respectively. When mixed with an equimolar amount of glucose, the yield increased spectacularly, especially with asparagine (370 μmol mol^{-1}). Replacing asparagine by its monohydrate or adding a small amount of water increased the yield even further (960 μmol mol^{-1}). Varying the carbohydrate (fructose, galactose, lactose, sucrose) gave comparable yields of acrylamide. Pyrolysing the glucosylamines of asparagine, glutamine, and methionine at 180 °C for 20 min gave 1305, 14, and 8 μmol mol^{-1} acrylamide, respectively. Fructosylasparagine similarly gave 1420 μmol mol^{-1} acrylamide.

Results with $^{13}C_6$-labelled glucose implied that asparagine is the source of the carbon atoms of acrylamide. Similarly, there was 98.6% incorporation of ^{15}N-amide-labelled asparagine into acrylamide and no incorporation of ^{15}N-α-amine-labelled asparagine.

It is worth noting that, in being converted into acrylamide, the Strecker aldehyde from asparagine needs to have a hydroxy group reduced as well as to lose a molecule of water.

The free asparagine levels of cereals and potato are relatively high; in wheat flour 167 mg kg^{-1} have been found, equivalent to 14% of the total free amino acids,[296] whereas the corresponding figures for potato are 940 mg kg^{-1} and 40%.[299]

Overall, it seems that acrylamide in food can be derived in multiple pathways, the Maillard reaction of asparagine representing the principal one. An industry perspective on acrylamide in food has recently been presented.[641]

3 Mutagenic Products

Japanese work[300,301] led to the discovery of a new group of highly mutagenic compounds, now classified as heterocyclic aromatic amines (HAAs). Many of these have been shown to be multi-site tumour inducers in long-term animal studies on rodents and monkeys (*cf.* ref. 302). Studies on human cells *in vitro* have demonstrated that HAAs are metabolised into bioactive compounds and form DNA adducts. These results, in combination with epidemiological data, suggest a carcinogenic effect of HAAs for humans who are genetically susceptible and/or moderately to highly exposed.[303]

By now the number of HAAs has grown to over 20.[304] They are divided into two main classes: the aminocarbolines and the amino-imidazo-azaarenes (AIAs). A variable number of methyl groups is present in various positions. The aminocarbolines are subdivided into α-, β-, and γ-carbolines and dipyridoimidazole (a δ-azacarboline), represented by, *e.g.*, AαC and MeAαC, harman and norharman, Trp-P-1 and Trp-P-2, and Glu-P-1 and Glu-P-2, respectively. All the AIAs possess a 2-aminoimidazo group with a methyl group attached to one of the ring nitrogens. This part of the structure is assumed to originate from creatine and is fused onto either a quinoline (IQ, MeIQ), quinoxaline (IQx, MeIQx, DiMeIQx, TriMeIQx), or pyridine (PhIP, 4'-OH-PhIP, DMIP, TMIP) system (see Structures **6.1**).[290,302]

The IARC has classified some HAAs as possible (MeIQ, MeIQx, and PhIP) and one (IQ) as a probable human carcinogen (*cf.* ref. 305).

By use of the Salmonella mutation assay (the Ames test), mutagenic activity was demonstrated in the outer layer of meat after frying, and a high proportion of this activity in fried meat and also in broiled fish has been traced to imidazoquinolines and an imidazoquinoxaline, the so-called IQ compounds. Indeed, the mutagenic potency of HAAs in the Ames test is greater than for any other class of chemical.[304] Some results with *Salmonella typhimurium* TA98 with S9 activation are given in Table 6.2. In general, TA100 produces fewer revertants per μg of HAAs; this strain detects base-pair substitution rather than frame-shift mutation.

Mutagen formation in a variety of foods has been quantified by Barnes and Weisburger;[306] the results are shown in Table 6.3.

IQ compounds seem to be formed in a Maillard reaction, via 2-methylpyridine or 2,5-dimethylpyrazine, together with creatinine, an amino acid (glycine or alanine), and a reducing sugar, at relatively high temperature and under relatively anhydrous conditions.[307] Using such mixtures as a model system in diethyleneglycol–water [6:1 (v/v), b.p. *ca* 128 °C] under reflux for 4 h, up to 100×10^3 revertants per ml were obtained.

Similar model experiments led to 7,8-DiMeIQx.[308] When using fructose and alanine in such experiments, 4,8-DiMeIQx was identified as the main mutagen present.[309]

Phenylacetaldehyde and its aldol condensation product with creatinine are very important intermediates in the formation of PhIP. The corresponding Schiff base could not be found in model systems or in fried meat.[310]

The incorporation of [^{14}C]glucose into IQx, MeIQx, and DiMeIQx has been confirmed.[311]

When 20 amino acids were individually heated with glucose and creatinine in an aqueous model system, the presence of IQx, MeIQx, and DiMeIQx was established in most of the samples, PhIP, Trp-P-1, and Trp-P-2 being present in some in addition. The last two compounds have not received much attention hitherto, as

Structures 6.1 *Structures of some heterocyclic aromatic amines*

they had been thought to be formed exclusively under extreme cooking conditions and, consequently, were not expected to be present in the Western diet. However, by now, these compounds have been identified in foods cooked at moderate temperatures.[311]

In order to test whether heat is an essential requirement, Kinae et al.[312] have shown that heterocyclic amines can also be formed under mild conditions. They kept a mixture of glucose (175 mM), glycine (350 mM), and creatinine (350 mM) in phosphate buffer (0.2 M, pH 7.4) at 37 °C for 84 d and were able to show that, not only had mutagenic activity in the Ames test been produced, but HPLC and LC-MS confirmed the presence of MeIQx. Similarly, a mixture of glucose (45

Toxicological and Protective Aspects

Table 6.2 *Revertants obtained with heterocyclic amines in the Ames test with S. typhimurium TA98 with S9 activation (adapted from Finot[31] and Wong and Shibamoto[638]*

Compound	Revertants
IQ	433k
MeIQ	661k
IQx	75k
MeIQx	145k
4,8-DiMeIQx	183–206k
7,8-DiMeIQx	163–189k
Trp-P-1	39k
Trp-P-2	104.2k
Glu-P-1	49k
Glu-P-2	1.9k
PhIP	1.8–2k
Aflatoxin B1	6k
Benzo[*a*]pyrene	0.32k

Table 6.3 *Mutagen formation in various foods[306]*

Food	Sample	Cooking procedure	Cooking time (min)	Revertants per sample
White bread	Slice	Broiling	6	205
Pumpernickel	Slice	Broiling	12	945
Biscuit	Each	Baking	20	735
Pancake	Each	Frying	4	2500
Potato	Small slice	Frying	30	200
Beef	Patty	Frying	14	21700

mM), phenylalanine (91 mM), and creatinine (91 mM) led to mutagenic activity due to PhIP.

Jägerstad *et al.*[302] summarised the situation as follows. HAAs have been reported in all kinds of cooked meat and fish products, particularly those that have been fried, grilled, broiled, barbecued, roasted, or smoked. They have been primarily found in the outer layers of heat-treated meat and fish, with minor amounts in the inner parts of fried meat. In the few cases where β-carbolines have been determined, they were present in the highest amounts. HAAs are generally found in ng g^{-1}, their formation being affected, as expected, by temperature, time, a_w, pH, and the amounts and proportions of precursors, such as creatine and amino acids, peptides, and proteins. The Maillard reaction is important for the formation of certain HAAs in liquid systems, *e.g.*, the AIAs. Dry heating also produces aminocarbolines, which seem to be favoured by free-radical reactions/fragmentations and low a_w.

Interestingly, Felton and Knize[304] report that, of the industrially cooked samples they examined, only one contained measurable amounts of HAAs, namely, MeIQx

(0.42 ng g^{-1}) and DiMeIQx (0.1 ng g^{-1}), whereas restaurant-cooked samples contained MeIQx (0.48–0.89 ng g^{-1}) and PhIP (1.0–13 ng g^{-1}). Traditional restaurant food preparation tends to produce much higher levels of HAAs than fast-food outlets. Tikkanen et al.[313] had found that the majority of Finnish flame-broiled fish, chicken, and pork contained HAAs (0.03–5.5 ng g^{-1}). Zimmerli et al.[314] analysed 86 Swiss cooked-meat (including poultry and fish) samples and 16 related commercial samples, finding the highest level of PhIP (13.1 ng g^{-1}) in fried pork with bacon (pork Saltim Bocca) (total HAAs, 18.4 ng g^{-1}). Grilled lamb fillet came second with 9.7 and 11.9 ng g^{-1}, respectively, and fried cubed beef (ragout) third with 6.0 and 6.0 ng g^{-1}, respectively. HAAs were not detected in half their samples. There was no significant difference between restaurant and domestically cooked samples. Klassen et al.[315] analysed 28 samples of cooked hamburgers and six samples of chicken preparations from the Ottawa region. The former contained PhIP (0.2–6, mean 1.38 ng g^{-1}), IQ (0.1–3.5, 0.58 ng g^{-1}), MeIQx (0.3–6.9, 1.01 ng g^{-1}), and 7,8-diMeIQx (0.1–2.9, 0.50 ng g^{-1}), as well as traces of 4,7,8-Me3IQx (< 0.1 ng g^{-1}), Trp-P-1 (< 0.1–0.3 ng g^{-1}), and Trp-P-2 (< 0.1–0.8 ng g^{-1}). The latter registered only PhIP (0.1–2.1 ng g^{-1}) and MeIQx (0.1–1.8 ng g^{-1}). Klassen et al.[315] have summarised previous results in a useful table. Skog et al.[316] assayed hamburgers and chicken fillets cooked in different ways: convection oven, deep-fat fryer, and contact fryer. For the hamburgers, mutagenic activity was detected only after deep-fat frying, but for the chicken fillets, mutagenic activity was produced in each method of cooking, deep-fat frying giving the highest level. For chicken, increased mutagenic activity correlated with increased weight loss during cooking. In convection-oven cooking, high temperature and high air velocity increased mutagenic activity, whereas steam decreased it, except at high temperatures plus high air velocity. HPLC led to the identification of MeIQx, PhIP, and co-mutagenic norharman (up to 4, 12, and 20 ng g^{-1} cooked meat, respectively). HPLC fractions were assayed for mutagenic activity, which was detected in several fractions in addition to those corresponding to MeIQx and PhIP. These fractions did not correspond to known heterocyclic amines. Other studies have also indicated the presence of additional mutagenic components in cooked and barbecued beef (cf. ref. 316).

The HAA content is subject to manipulation during food preparation. Felton and Knize[304] give a good example of this through the effects of a sucrose-containing marinade, which increased MeIQx, but decreased PhIP in grilled chicken. Despite the increase in MeIQx, marinading greatly reduced the total amount of HAA.[317] Another example is provided by beef patties, which, after receiving a microwave pretreatment for up to 2 min, exhibited a 3- to 9-fold reduction in mutagenicity and HAAs after being 'well done'. The reason found is that the drip taking place during the microwave treatment removes HAA precursors by up to 30%. Since the levels of creatinine in bovine kidney and liver are low, the mutagenicity produced on frying is lower. Addition of 10% or more of soy protein to meat, prior to frying, completely abolishes the formation of HAAs, as assessed by mutagenicity testing (see ref. 318). Antioxidants, such as BHA, also decrease the formation of HAAs, as does the addition of relatively small amounts of tryptophan and/or proline.[318] Both black tea and green tea lower the mutagenicity of IQ and of PhIP in TA98, and they also inhibit the development of DNA repair in freshly explanted liver cells exposed to IQ and

PhIP. The inhibition by tea of these two tests, which are indicators of genotoxicity, suggests that naturally occurring plant extracts can modify metabolic activation of HAAs. Tea polyphenols, epigallocatechin gallate from green tea and theaflavin gallate from black tea, display similar inhibiting action in both tests (see ref. 318).

Brittebo et al.[319] reported that melanin can strongly bind PhIP both *in vivo* and *in vitro*. In mice, high and selective binding of radioactivity in the pigment epithelium of the eye and the fur was found following a single dose (0.3–4 mg kg^{-1}) of ^{14}C-labelled PhIP in pigmented mice, whereas no such localisation was observed in albinos. The binding of PhIP to commercial melanin *in vitro* was resistant to most organic solvents and extreme pH values. Only methanol saturated with ammonia was effective. These results suggest that HAAs could also be strongly bound to melanoidins and, if such binding were resistant to digestion, they might well be excreted intact without exerting any harmful physiological effects.[302] Accordingly, Solyakov et al.[305] prepared three different melanoidin model systems, one from glucose–glycine (1:1, 125 °C, 120 min), a 'meat' one from glucose + 13 common amino acids (1:1, 125 and 180 °C, 120 min), and commercial ground roasted coffee beans. Aqueous suspensions were incubated with a reference solution containing seven HAAs at room temperature for 60 min. After incubation, the systems were centrifuged and analysed by HPLC. The first system bound 24% of PhIP, the second did not bind any HAAs, but the coffee system bound 68% PhIP, 58% harman, 62% norharman, 75% Trp-P-1, 77% Trp-P-2, 89% AαC, and 82% MeAαC. There was no binding with the dialysate (< 10 kDa) from any of the model systems.

In model systems reflecting normal cooking conditions, a relatively weak mutagenic effect is exerted compared with that of HAAs. The type of mutagenic effect of these systems reflects base-pair in contrast to frameshift mutation, as caused by HAAs. Moreover, their action is effectively inactivated by liver chromosomal enzymes (S9 fraction), in contrast to HAAs, which require S9 activation to mutagenic/carcinogenic species.[31]

Based on their data, Zimmerli et al.[314] estimated the mean exposure of Swiss adults to HAAs to be 5 ng (kg b.m.)$^{-1}$ d^{-1}, comparing this with values of about 2 ng (kg b.m.)$^{-1}$ d^{-1} obtained for elderly Swedes and for a US subgroup. They went on to estimate the theoretical cancer risk due to such an intake by extrapolation of the results of long-term animal experiments as 10^{-4}, *i.e.*, one additional tumour induced in a population of 10k individuals exposed during their entire lifespan. For comparison, the comparable risk due to naturally occurring radioactive potassium is 10^{-3} and the risk of dying by lightening in Switzerland is 10^{-4} to 10^{-5}.

Friedman[290] concluded in his review that the doses of IQ and MeIQ required to produce tumours in laboratory animals were 10–20 mg (kg b.m.)$^{-1}$, much higher than the amounts found in human diets. Even combining the TD50 for PhIP of 2.2 mg (kg b.m.)$^{-1}$ d^{-1} for rats with its concentration of 13 ng g^{-1} in some meat products, equivalent to about 3 μg in a 250 g portion, does not give rise to too much concern for a 70 kg man.

Even though it is difficult to assess the actual risk to humans presented by current exposure to HAAs and related compounds, it is sensible to consider the scope for reducing lifetime exposure to these substances through changes in food preparation, such as those already mentioned.

Some attention has been paid to the mutagenicity of heated sugar–casein systems.[320] In the Ames test, the mutagenicity of both glucose–casein and fructose–casein increased with heating time at 120 °C (in phosphate buffer, pH 6.8), the former system being designated mutagenic (number of revertants per plate greater than twice the number of spontaneous mutants under the same conditions) after heating for 60 min, whereas the latter was already mutagenic after 20 min. Mutagenicity was absent when the systems were incubated with S9 mix. Galactose, tagatose, lactose, and lactulose were tested similarly. Tagatose–casein was mutagenic already after heating for 20 min, galactose–casein only after 40 min, lactulose–casein after 60 min, and lactose–casein not even after 60 min.

The effect of some MRPs (xyl-Lys, glu-Lys, fru-Lys) on induced DNA damage in human lymphocytes and HepG2 cells has been examined by comet assay by Yen and Liao.[321] Undialysed MRPs caused significant DNA damage in lymphocytes at 0.05–0.1 mg ml^{-1}, whereas the undialysable (cut-off 7 kDa) required > 0.1 mg ml^{-1}. Both undialysed and undialysable MRPs caused DNA damage in HepG2 cells, but to a lesser degree. The undialysable MRPs did not affect the activity of glutathione peroxidase or lipid peroxidation in lymphocytes (TBARS), but decreased glutathione content and activity of glutathione reductase and catalase. The three undialysable MRPs gave weak free-radical signals at 0.1 mg ml^{-1}, the signals becoming more intense at 0.8 mg ml^{-1}. At 0.8 mg ml^{-1}, they gave 15.6, 11.9, and 11.6 8-hydroxy-2'-deoxyguanosine/10^5 2'-deoxyguanosine and 384, 242, and 375 μM H$_2$O$_2$, respectively, indicating that radicals were inducing the damage.

Since Amadori compounds are secondary amines, the question of the formation and reactivity of *N*-nitroso derivatives arises. Accordingly, Pool *et al.*[322] tested those derived from glucose with Ala, Asp, Phe, Gly, Ser, and Try in five strains of *S. typhimurium* with and without S9. The first three compounds were not mutagenic, the next two showed low but reproducible increases in the number of his$^+$ revertants in TA1535 (without S9), but the last, containing indolyl-nitrosamine-D-fructose-L-tryptophan, was mutagenic in all five strains, with and without S9. When separated, the Try compound was mutagenic in three strains, without S9.

4 Allergenicity

There is little to say about the allergenicity of Maillard-derived products, but HDMF enhances some types of allergen-sensitised allergy responses in mice[323] and, using a skin reactivity test, good correlation was found between the degree of browning and the allergenic response of a modified bovine β-lactoglobulin.[31]

5 Other Toxicological Aspects

The relationship between the Maillard reaction and Alzheimer's disease has recently been considered by Gasic-Milenkovic *et al.*[324] AGE concentrations in the serum and urine of healthy individuals increase after the ingestion of an AGE-rich protein meal, in a manner directly proportional to the amount consumed. Only one-third of the AGEs appearing in the serum had been detected in the urine after 48 h, the fate of

Toxicological and Protective Aspects

the other two-thirds remaining to be determined.[325] The tissues retaining them could well serve as a source of glycotoxins, which might even cross the blood–brain barrier and contribute to protein crosslinking in the brain.

There is some evidence that HAAs, especially IQ, play a role in the incidence of coronary heart disease.[318]

Some toxicological testing has been carried out on caramel food colours, including feeding studies on rats. Attention has been focused on 4-methylimidazole and 2-acetyl-4(5)-tetrahydroxybutylimidazole, which are only present in Class III and IV caramels, as potentially exerting toxic effects. Abnormalities, including lymphocytopenia, were attributed to the 4-methyl compound and JECFA stipulated a maximum level of 200 mg kg^{-1} caramel. By now, toxicological effects at the levels likely to be ingested by humans are considered negligible, but this maximum level continues to be observed, more as an indicator of good manufacturing practice.

6 Protective Effects

Apart from antioxidant activity (see Chapter 9), Maillard products have been found to have antimutagenic, antibiotic, and antiallergenic effects. Chemopreventive effects were mentioned at the beginning of this chapter.

AGEs exhibit a biphasic effect on the proliferation of cells, proliferative at low and antiproliferative at high concentrations, *e.g.*, at 1–10 and >20 μg ml^{-1} BSA-AGE, respectively.[326] Proliferation also depends on the type of AGE, *e.g.*, early-glycation products of BSA reduced the proliferation of pericytes by 40%, whereas AGE-BSA increased it 156%.[327]

Yen and Hsieh[328] have shown that the mutagenicity of IQ on *S. typhimurium* TA98 and TA100 was strongly inhibited in the presence of Maillard products from xylose–lysine (1:2 mol, 100 °C, pH 9, 1 h). They demonstrated that the effect was due to the formation of inactive adducts with IQ metabolites and not to the direct inhibition of the hepatic microsomal activation required to transform inactive IQ into a DNA-alkylating agent.

Scavenging free radicals may also play a role in the antimutagenicity of melanoidins.

According to Kato *et al.*,[329] nondialysable melanoidins from glucose–glycine degrade nitrite and prevent nitrosamine formation, both maximally at pH 1.2. Treatment with nitrite greatly enhances the antimutagenicity of the melanoidins.

Friedman[290] reports earlier work by his group, which showed that the presence of glucose during alkaline treatment of soybean proteins significally lowered the amount of lysinoalanine formed.

The susceptibility of bacterial growth to inhibition by Maillard products from xylose–arginine or glucose–histidine varied widely, as shown by Einarsson *et al.*[330–332] Gram-negative *Salmonella* were virtually not inhibited, whereas Gram-positive *Bacillus subtilis*, *Lactobacillus*, and *Staphylococcus* were strongly inhibited. *B. cereus* was not inhibited. The mechanisms thought to be operative were chelation of iron, inhibition of oxygen uptake, and inhibition of uptake of glucose and serine. Protein precipitation may be an additional factor.

Antigenicity can be modified by heating food proteins with reducing carbohydrates. For example, Oste et al.[333] heated a solid mixture of soybean trypsin inhibitor (KTI) and carbohydrate at 120 °C and analysed the dialysed product by ELISA. Glucose, lactose, and maltose lowered the antigenicity of KTI by 60–80%, starch being less effective. The decrease was rapid, occurring within 10 min with glucose, and 60% of lysine remained chemically available.

CHAPTER 7

Nutritional Aspects

Many of the nutritional, that is, mainly antinutritional, aspects of the Maillard reaction have been included in reviews by Friedman[290] and O'Brien and Morrissey.[334]

1 Effects on Availability of Essential Amino Acids

All amino acids, peptides, and proteins can participate in the Maillard reaction. Reactivity of essential and nonessential amino acids is of importance in relation to, for example, intravenous (parenteral) feeding, but, in general, by far the greatest attention is paid to lysine, because it is not only an essential amino acid, which may be limiting in the diet, but also because its ε-amino group remains reactive, even in peptides and proteins.

The determination of nutritionally available lysine was dealt in Chapter 2.

In an extensive study, van Barneveld et al.[335–338] examined the effect of heating field peas for feeding to growing pigs on the digestibility, availability, and utilisation of lysine. Dry heating was applied for 15 min at 110, 135, 150, and 165 °C. Ileal digestibility was little affected, but availability determined by the slope-ratio assay decreased from 0.96 for the unheated to 0.71, 0.77, 0.56, and 0.47, respectively. The results indicate that, unlike ileal digestibility values, estimates of lysine availability are sensitive to heat treatment and are a close reflection of lysine utilisation in heated protein concentrates. Thus, lysine availability values would be more suitable for use in diet formulation with heat-treated protein concentrates than ileal digestibility.

2 Effects on Ascorbic Acid and Related Compounds

Although ascorbic acid and particularly dehydroascorbic acid can undergo Maillard-type reactions, the loss of vitamin C due to the Maillard reaction is rarely an issue. The involvement of ascorbic acid and related compounds in the Maillard reaction is considered in Chapter 11.

3 Effects on Enzyme Activity

The Maillard reaction can affect the activity of enzymes in at least two ways, through glycation or through interaction with melanoidins.

Intestinal Caco-2 cells provide useful models for assessing the effect of xenobiotics on enzyme activity. Thus, Faist *et al.*[291] have studied the effect of melanoidins both on a Phase-I enzyme, NADPH cytochrome *c* reductase, and a Phase-II enzyme, glutathione S-transferase. Where the results achieved a significant level, activity was reduced, thus implying that detoxification would be less efficient in the presence of the melanoidins tested. However, some other results pointed the other way (see Chapter 6).

4 Interactions with Metals

From a nutritional point of view, many metals (in ionic form) are of interest. Sodium and potassium, and their relative amounts, are of great importance; however, they are unlikely to be affected by Maillard reaction products, but calcium and magnesium are, as well as iron and zinc. Trace metals, such as copper, cobalt, chromium, manganese, molybdenum, and vanadium, have nutritional significance, but little is known of the effects of the Maillard reaction on their absorption/utilisation. In excess, all metals prove toxic, but for some, such as aluminium, lead, mercury, and cadmium, there is no nutritional significance, only potential toxicity. Maillard products may be beneficial with these, but again few facts are available. Because of its suitability for protecting steel to be used in canning food, the concentration of tin in food can be relatively high, but the tolerance level is rarely exceeded.

Complexing, particularly of copper and iron, has great significance in terms of antioxidant activity (see Chapter 9).

Chelating activity in relation to copper was investigated by Terasawa *et al.*[339] by separating nondialysable melanoidin from glucose–glycine by pH gradient elution from a copper-chelated Sepharose 6B column. Six components were obtained. Their UV-vis spectra were almost the same and their electrofocusing profiles were similar. The electrofocusing bands were categorised into four major groups, pI 2.5–4.0, the main chelating activity being due to the band of pI 2.7.

Wijewickreme *et al.*[340] prepared nondialysable MRPs from glucose–Lys and fructose–Lys mixtures (0.8 M each) under 14 different conditions of time, temperature, and initial pH and a_w, monitoring browning by UV-vis spectrophotometry, A_{420}, yield of MRP, and its elementary analysis. The MRPs were fractionated by means of immobilised metal affinity chromatography. Although there were considerable differences in the relative concentration and number of unidentified intermediate compounds present in the MRPs, two principal components of *ca* 5.7 and 12.4 kDa were identified by MALDI–MS. Elementary composition and Cu-chelating affinity of crude and fractionated MRPs were greatly influenced by the type of sugar and reaction conditions. Glucose in general produced MRPs with relatively higher chelating affinity, possessing a higher proportion of C and less N than the fructose counterpart.

Not unlike Tarasawa *et al.*, Fogliano *et al.*[72] separated a glucose–glycine melanoidin by passing it down a column of washed Chelating Sepharose Fast Flow gel equilibrated with 0.2 M cupric chloride; only two fractions were obtained, one (26%) eluting immediately with 0.05 M phosphate/0.5 M NaCl buffer (pH 7.2), whereas the other (32%) would not elute even on lowering the pH to 3.5, but required addition of 0.1 M EDTA to the buffer at pH 7.2; > 40% thus remained irreversibly bound to the column. With Pb^{2+} and Fe^{2+}, only 10 and 5%, respectively, of the melanoidin was bound to the column. Melanoidins clearly differ in their affinity for metals.

The effect of diets containing MRP on copper metabolism in rats has been studied by Delgado-Andrade et al.[341] Four types of MRP were used, being produced by heating equimolar mixtures of glucose–lysine or glucose–methionine (GM) at 150 °C for 30 or 90 min. Apart from the control, each diet contained 3% of one of these MRP. Copper absorption and retention increased, as well as the efficiency of both processes, for all groups except GM30. The copper concentration increased significantly in the spleen, kidney, and skin only in rats fed with GM90.

5 Absorption/Elimination of Amadori Compounds

Experiments with ^{14}C-labelled fructosyllysine (FL) in rats showed that this Amadori compound is not actively transported out of the intestines. However, experiments with ligated gut segments indicated that uptake by passive diffusion is feasible, at least for ε-FL, whereas α-FL and fructosylmethionine are poorly absorbed.[342] The presence of FL affects the active transport of other compounds; thus the active transport of threonine, proline, and glycine was lowered, but that of lysine, methionine, and galactose remained unaffected.[343]

^{14}C-Labelled FL was also injected into the veins of rats, when it was almost completely excreted in the urine. Evidence such as this and that reported in the preceding paragraph show that Amadori compounds, and especially FL, do not have an appropriate transport system and so, if digested and absorbed, are excreted in the urine, largely as they are.

Accordingly, in human beings, ingestion of single doses of about 3 g of FL led to very rapid excretion in the urine, but faecal excretion continued for up to 72 h.[344] However, only 4.1% of the administered FL was eliminated in the urine, with a further 1.0% in the faeces, leaving 94.9% unaccounted for. For rats, the corresponding figures were 9.7, 1.5, and 88.8%. Only human infants, given casein-linked FL, have so far recorded higher excretion rates (urine, 16%; faeces, 55%), in keeping with their known enhanced absorption efficiency for molecules of higher mass.[345] The only plausible suggestion for the fate of the rest of the FL is decomposition by the microflora of the hindgut. This is in accord with the hindgut of rats having been shown to decompose FL, as well as pig faeces. It is also in accord with finding high amounts of FL in the droppings of germ-free chicken fed on heated protein.[343]

6 Digestion of Melanoidins

The fermentation of melanoidin by human gut bacteria has been evaluated using a glucose–gluten model system (150 °C, 2 h) that had been subjected to a simulated upper gut (peptic and pancreatic) digestion.[346] Both mixed culture chemostats and pure cultures were used. In the former, the melanoidins were well fermented, giving an increase in Gram-positive cocci, clostridia, and bacteroides, whereas, in monoculture, only the Gram-positive cocci isolated were able to break down the melanoidins. Animal studies have indicated that melanoidins escape digestion in the upper gastrointestinal tract, but they are susceptible to metabolism by the vast numbers and types of microbes present in the hindgut (in human beings, say, 500 species and 10^{14} cells). These microorganisms rely for growth on substances that have escaped digestion by the indigenous enzymes of the host. Depending on the type of metabolism taking place, there may be consequences for health or disease.[347] For example, the main fermentation products are short-chain fatty acids, which lower the pH and may be systematically metabolised to provide some energy for the host. On the other hand, many toxic compounds are also produced by the gut microflora and these can lead to diseases such as gastroenteritis and colon cancer.

CHAPTER 8

Other Physiological Aspects

1 Introduction

In Chapter 7, the nutritional aspects of the Maillard reaction were considered in some detail, but the Maillard reaction involves more than nutrition, and it will now be dealt with in the wider setting of health in general. In that context, four aspects of the Maillard reaction have come to the fore:

1. Complications in diabetes (cataract formation and nephropathy in particular).
2. Dialysis-associated amyloidosis.
3. Alzheimer's disease and ageing in general.
4. Atherosclerosis.

However, first both the formation of Amadori compounds and of 'advanced glycation endproducts' *in vivo* must be considered.

2 Formation of Amadori Compounds *in Vivo*

It was realised only relatively recently that the Maillard reaction also occurs *in Vivo*. When human haemoglobin is chromatographed on weakly acid ion exchangers, three minor haemoglobin fractions (normally representing 5–8% of the total) elute in front of the main fraction, HbA_0.[20,348] The third of these, HbA_{1c}, which is present in higher amounts in the blood of diabetic subjects, was shown to be identical with HbA_0, except that a 1-deoxyfructos 1 yl group was bound to the nitrogen of the N-terminal valine of each of the two β-chains. Such 'fructosylation' also occurs when HbA_0 is incubated with glucose; in principle, every protein in the body is open to such fructosylation both *in vitro* and *in vivo*. In principle, too, there is no reason why the Maillard reaction should stop at the Amadori product *in vivo* and, indeed, by now it is known that much can happen subsequently.

The functional characteristics of HbA_{1c} have been compared with those of normal HbA_0,[349] when the low-affinity conformation (T-state) of the former was found to be destabilised and its affinity for diphosphoglycerate was *ca* 2.6 times lower than that of the latter. Computer modelling showed that the two sugar residues are asymmetrically disposed within the diphosphoglycerate-binding site. Molecular mechanics and dynamics calculations concerning the interaction with diphosphoglycerate indicated that normally this substrate can adopt two stable orientations, whereas in HbA$_{1c}$ only one is possible. Thus, glycation does not prevent the binding of diphosphoglycerate, but it imposes a different mode of binding by changing the internal geometry of the complex and the surface distribution of the positive electrostatic potential within the binding pocket. The work has far-reaching implications for other molecules liable to glycation.

The types of modifications that have been described to date (**62–101**) are arranged systematically in Structures 8.1.

The modifications can be classified into six groups,[350] depending on the amino acid involved and whether the modification affects it alone or leads to crosslinking to another amino acid.

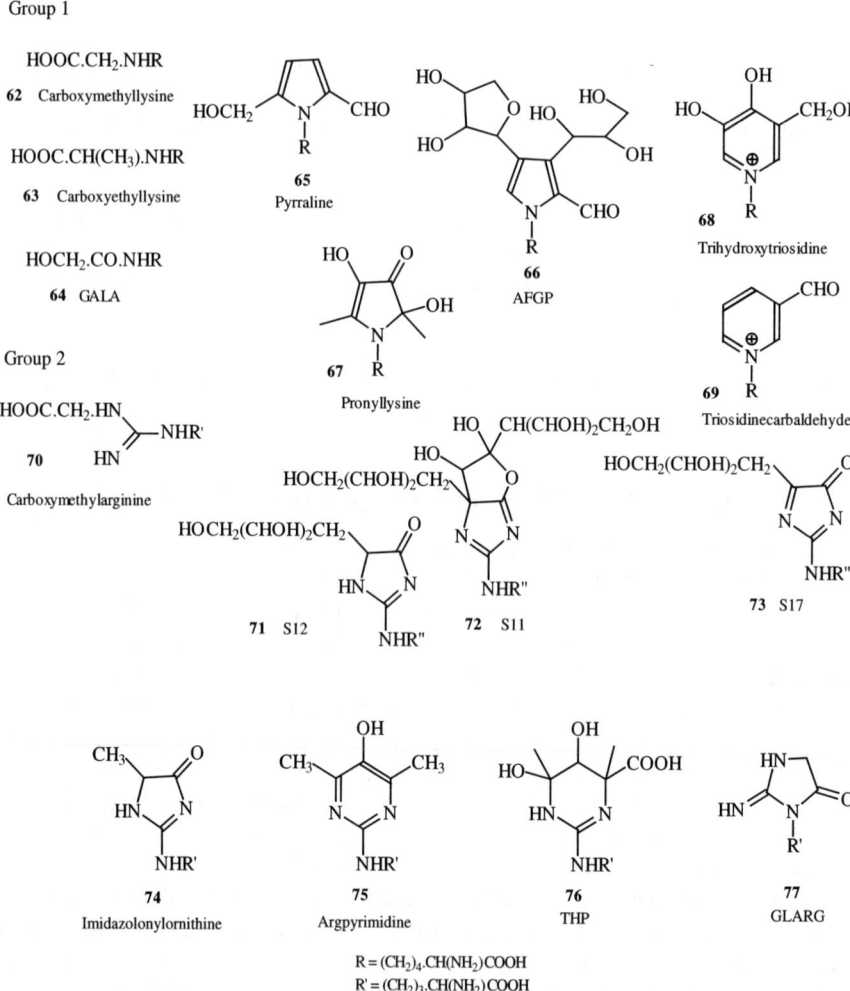

Structures 8.1 *Physiological modifications of amino acids observed*

Other Physiological Aspects

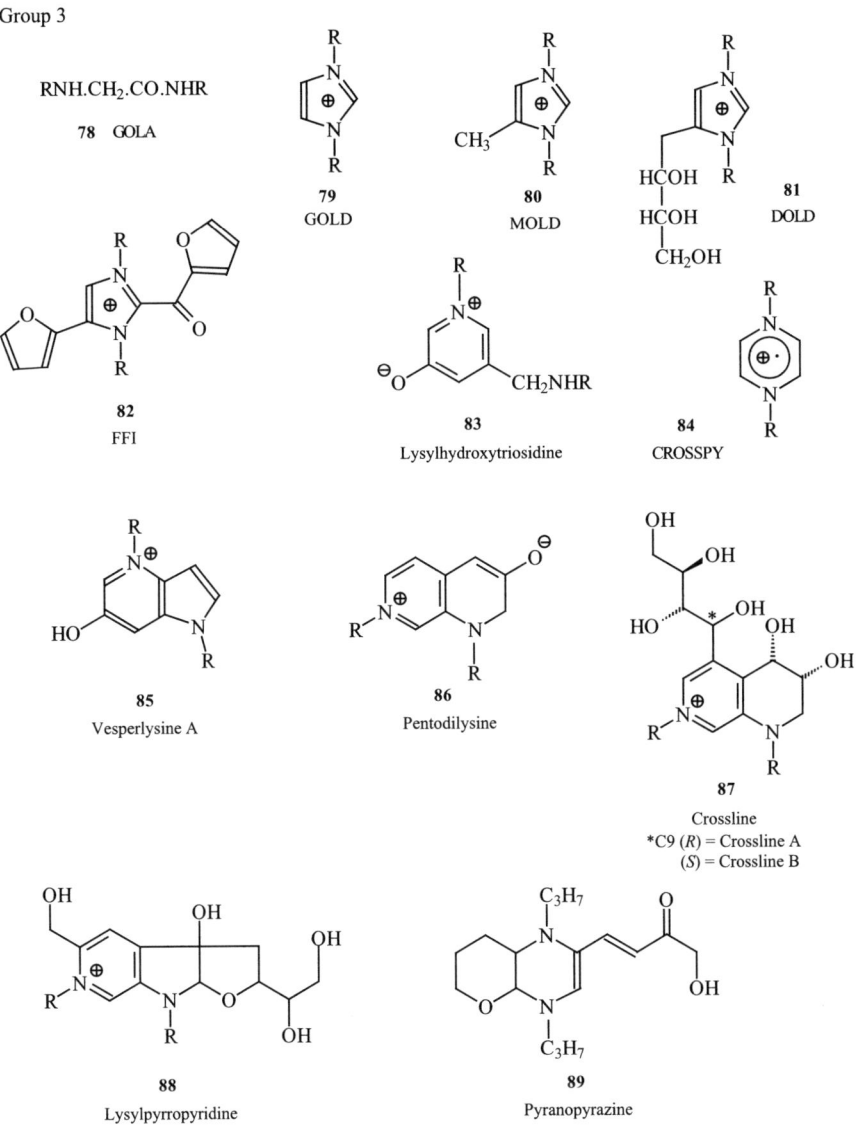

Structures 8.1 (cont.)

Starting with Group 1, lysine is the most frequently affected amino acid, since its side-chain amino group continues to be potentially available when it itself is bound into peptides and proteins. It gives rise to eight derivatives: N^ε-carboxymethyllysine (CML),[351,352] N^ε-(1-carboxyethyl)lysine (CEL),[351,352] glycolloyllysine (GALA),[353] pyrraline,[354] AFGP,[355] pronyl-lysine,[356] trihydroxy-triosidine, and triosidine-carbaldehyde.[357]

Thorpe and Baynes[352] have pointed out that, even in the absence of hyperglycaemia and hyperlipidaemia, CML is found in high concentrations in lipofuscin granules in

Group 4

No examples, as yet

Group 5

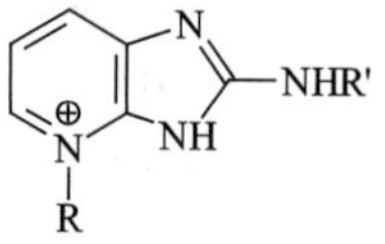

90 Pentosidine 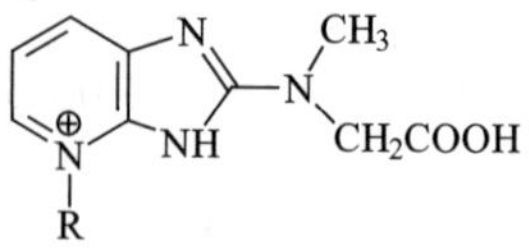
91 C-Pentosidine 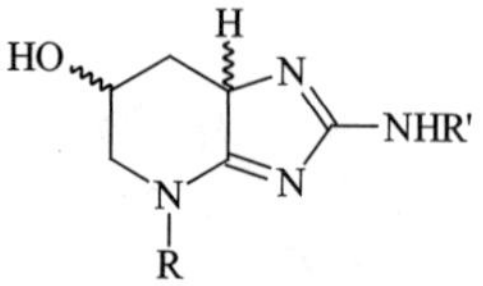
92 Pentosinane

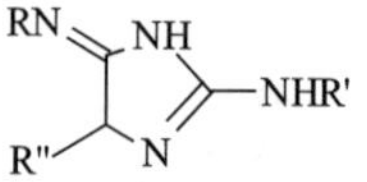

93 R″ = H   GODIC
94 R″ = CH$_3$ MODIC
95 R″ = CH$_2$.CHOH.CHOH.CH$_2$OH   DOGDIC
96 R″ = CH$_2$.CHOH.CH$_2$OH   DOPDIC

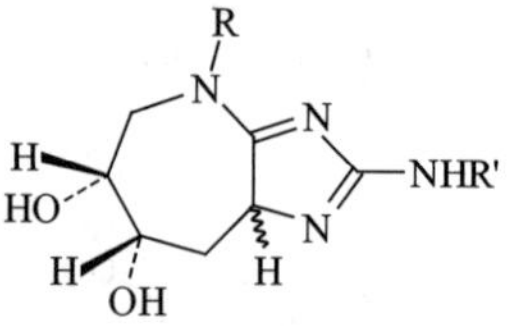

97 Glucosepan

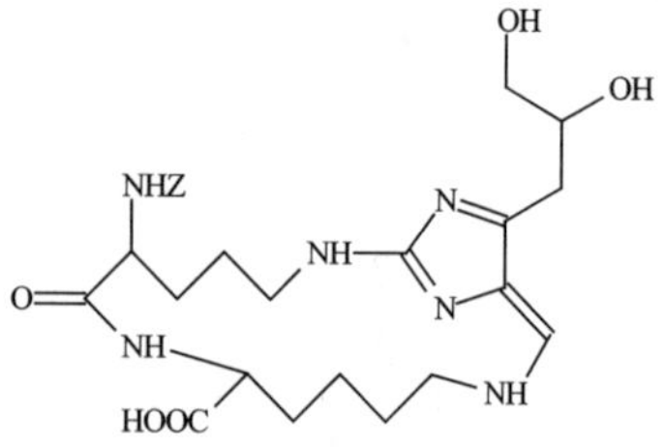

98 ALI

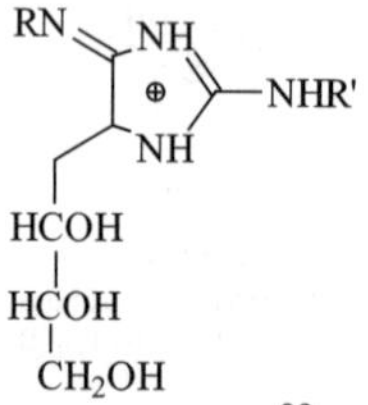

99 2-Ornithyldihydroimidazolium-4-imide
100 Isomeric ornithyldihydroimidazolium-4-imide 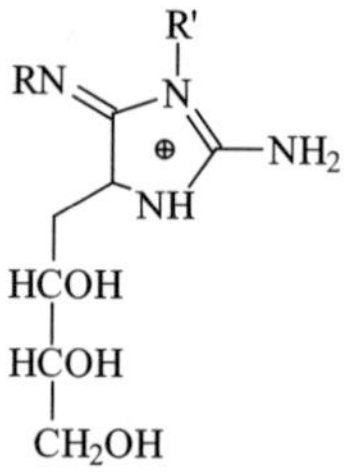
101 Arg-hydroxy-triosidine 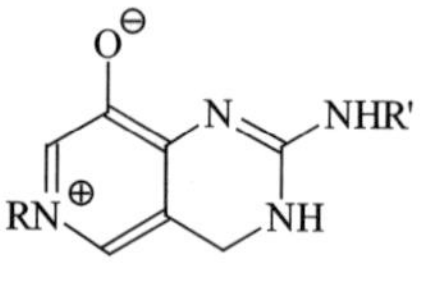

Group 6

No examples, as yet

Structures 8.1 (*cont.*)

adrenal glands, in aged and Alzheimer brain tissue, in protein deposits in dialysis-related amyloidosis, in collagen and plasma proteins in uraemia, and in liver proteins of rodents exposed to oxidative stress by iron overload. It is present also in foetal tissue, maybe as a result of the high concentrations of ribose metabolites used for DNA and RNA synthesis. It is also possible that increased CML at sites of tissue injury and inflammation results, in part, from stimulation of cell growth and development.

Group 2 is based on arginine, which can give rise to N^{ω}-carboxymethylarginine in parallel to the conversion of lysine into N^{ε}-carboxymethyllysine.[358] The reactive side-chain guanidino group of arginine can give rise to S12, S11, S17,[359] as well as to S16′, S16, S6, S7, S10,[360] imidazolonylornithine,[361,362] argpyrimidine,[363] THP,[364] and GLARG.[362]

Padayatti et al.[365] used a monoclonal antibody to measure argpyrimidine by competitive ELISA in water-soluble and water-insoluble fractions from young, aged, nuclear-cataractous, and brunescent-cataractous lenses. The water-insoluble proteins contained two to three times as much argpyrimidine as the water-soluble. Although there was no clear increase with the age of the donor in either the water-soluble or water-insoluble fraction, the argpyrimidine levels in brunescent-cataractous lenses were significantly higher (254 ± 155 pmol mg^{-1} protein, $P < 0.005$) than in age-matched, aged (16 ± 8) or nuclear-cataractous lenses (49 ± 26 pmol mg^{-1} protein). Lenses from diabetics at 50 ± 24 pmol mg^{-1} protein showed a modest increase when compared with age-matched normal ones. The levels observed correspond roughly to a modification of 0.002 and 0.04% of the arginine residues on α-crystallin in aged and brunescent lenses, respectively, but this considerably exceeds pentosidine (see below) levels. GOLA, which belongs to Group 3 below, has been found at 66 pmol mg^{-1} brunescent lens protein.[353]

The formation of argpyrimidine from ascorbate was considered a possibility, at least in the nuclear region. Glomb and Lang[60] found that an imidazolidine is the only product of the initial reaction of arginine with glyoxal at pH 4–5 and 20–50 °C, but it slowly degrades to N^{ω}-carboxymethylarginine **70**. At equilibrium, its *vic*-diol grouping is 86% in the *trans*-form. Its formation is reversible. Both it and **70** form **77** (GLARG) under strongly acidic conditions. THP is clearly related to argpyrimidine.[364]

Group 3 is based on two lysine residues, with crosslinking between them, and comprises GOLA,[353] the imidazolium crosslinks (glyoxal-lysine dimer, GOLD, and methylglyoxal-lysine dimer, MOLD,[366] and DOLD), lysylhydroxytriosidine,[357] CROSSPY,[367] vesperlysine A (Fluorophore LM-1),[368] pentodilysine,[369] the crosslines,[370] and lysylpyrropyridine.[359] The pyranopyrazine has only been produced synthetically, using propylamine as a model for lysine.[371] FFI also has an imidazolium crosslink, but is known to be produced by acid treatment during sample processing.[355]

Group 4 should then be based on two arginine residues, with crosslinking between them, but, as yet, there is no instance of such elaboration to report.

Group 5 is based on a lysine crosslinked with an arginine residue and comprises the important pentosidine,[372–375] as well as C-pentosidine,[376] pentosinane,[377] GODIC and MODIC (gl$\underline{y}$oxal or $\underline{m}$ethylglyoxal $\underline{d}$erived $\underline{i}$midazoline $\underline{c}$rosslink),[378] DOGDIC and DOPDIC ($\underline{de}$oxyglucosone- and $\underline{de}$oxypentosone-$\underline{d}$erived $\underline{i}$midazoline $\underline{c}$rosslink,[377] glucosepane,[379] ALI (arginine-lysine-imidazole),[380] arg-hydroxy-triosidine,[357] and potentially 2-(N^{δ}-ornithyl)-imidazolium-4-imide and its isomer.[381]

GO reacts much more extensively with BSA than MGO, giving about four times the crosslinking.[378] (It is worth noting that physiological concentrations in blood are GO 150 and MGO 80 nM, increased by 1 and 5–6 times in diabetes, respectively. Such concentrations are about 1000 times lower than those used experimentally.) GO derivatises about 13 mmol mol^{-1} Arg, demonstrating high potential for protein crosslinking. A glucose–BSA system (in air and without metal-complexing agent) gave rise to about 3.7 GODIC, 1.3 glucosepan, and 0.5 mmol mol^{-1} Arg MODIC. The amount of glucosepan corresponds to 32 mmol mol^{-1} BSA.[379] Lederer and Bühler[379] have put forward the hypothetical pathway leading to glucosepan and pentosidine shown in Scheme 8.1.

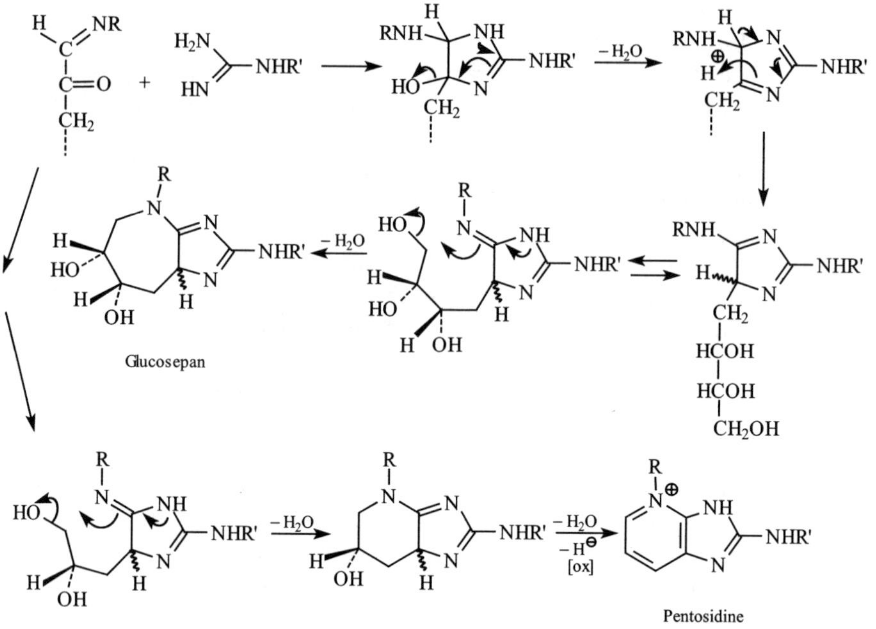

Scheme 8.1 *Hypothetical pathway to glucosepan and pentosidine*[379]

Clearly, there also needs to be a Group 6, 'Others'. Tressl et al.[382] have shown by means of model experiments that crosslinking lysine with cysteine through maleimides is possible in principle. The pathway illustrated in Scheme 8.2 was elucidated by using 4-aminobutyric acid/1-[^{13}C]arabinose. The product readily adds thiols, such as cysteine, across the double bond under very mild conditions (pH 7, 32 °C).

MRX, involving arginine and cysteine, could well represent another type of crosslink.[383] Histidine can be involved in crosslinks, such as N^{π}-HAL,[384] derived via dehydroalanine. It has been synthesised along with a number of other crosslinks.[385]

It should be noted that N^{ω}-carboxymethylarginine (CMA) was not found until it was realised that, unlike CML, it is unstable to acid hydrolysis. By means of enzymic hydrolysis, using peptidase and pronase E, Odani et al.[358] detected 124 and 85 ng CMA ml^{-1} plasma ($P = 0.017$) in Type 2 diabetics and normal subjects, respectively. An antibody to CMA has been raised.[386]

An amino acid analyser has been adapted for the determination of CML and CEL.[387]

A monoclonal antibody to CML (6D12) has been used quite widely to demonstrate CML in tissues, such as in the kidneys of patients with diabetic nephropathy and atherosclerotic lesions of arterial walls. However, it has been found to react not only with CML, but also with CEL. Therefore Nagai et al.[388] have set out to prepare polyclonal CML-specific antibody (PCMS) and its CEL counterpart (PCES). They appear to have been successful.

S12 has been identified in 3-DG–insulin and 3-DG–angiotensin II systems, when treated under physiological conditions at 37 °C and pH 7.6.[360] It has also been detected in the kidneys and aorta of diabetic patients using an anti-S12 monoclonal antibody.

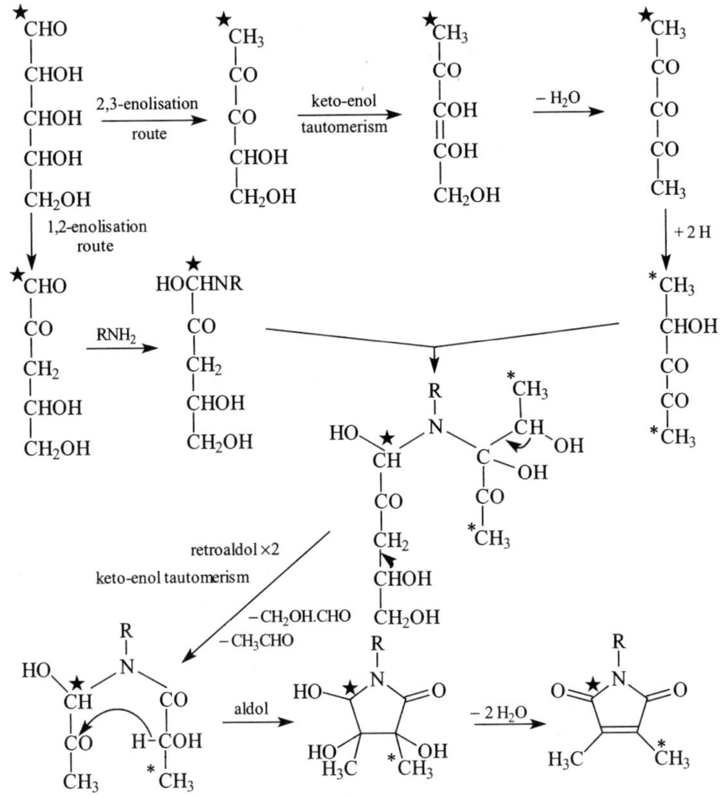

Scheme 8.2 *Mechanism of formation of pentose-specific dimethylmaleimides with protein-crosslinking potential.*[382] ^{13}*C-label:* (★)=100%, (*)=50%

Antibodies raised against MGO-Arg adducts (mAb3C and mAb6B) specifically recognised argpyrimidine as the principal epitope and allowed the demonstration of the accumulation of argpyrimidine in the arterial cell wall of rat brain after occlusion of the middle cerebral artery, followed by reperfusion.[389] The implication is that argpyrimidine contributes to the progression not only of diabetic complications, but also to tissue injury caused by ischaemia/reperfusion.

A monoclonal antibody (mAb5A3) has also been developed for protein-bound THP.[364] Using it, THP adduct has been found in macrophage-derived foam cells in human atherosclerotic lesions.

A method for determining lysylpyrropyridine *in vivo* (rat-tail collagen, plasma), after complete hydrolysis with proteases, has been developed.[390]

Sekine *et al.*[391] have determined urinary pentosidine and pyrraline in 75 diabetic (retinopathy: none 21, simple 26, proliferative 28), 50 nephropathic (diabetic: 24), and 22 control volunteers. Urinary pentosidine was increased in diabetic patients, the proliferative retinopathy group giving the highest levels. However, urinary pyrraline did not increase. There was good correlation between the pentosidine levels in urine and plasma of 22 control and 28 nephropathy volunteers with r^2 being 0.79. Urinary pentosidine was higher in microalbuminuria than in the control group and higher in macroalbuminuria than in microalbuminuria, pyrraline levels again being left unaffected. Since oxidation is involved in the formation of pentosidine, but not in the formation of pyrraline, oxidative stress rather than the Maillard reaction may well be the driving force behind diabetic complications.

Plasma levels of creatine as well as of pentosidine increase in diabetic nephropathy, which leads to the possibility of creatine replacing arginine in the formation of pentosidine.[376] Accordingly, lysine was incubated at 60 °C for 48 h with ribose and creatine, when on HPLC a novel fluorescent peak was detected. It reacted with an anti-pentosidine antibody and was identified by NMR and FAB-MS as C-pentosidine.

The pathways of the formation of a number of Group 5 products have been elucidated further by Biemel *et al.*[377] (see Scheme 8.3). That the bicyclic compounds, glucosepane, pentosinane, and pentosidine, are formed at most in negligible amounts when starting with 3-DG or 3-deoxypentosone was completely unexpected, since, on paper, they seemed reasonable precursors. The bicyclic compounds were formed in substantial amounts only from the corresponding sugar or Amadori compound, the latter yielding particularly high amounts with only minor proportions of DOGDIC and DOPDIC. The Amadori compounds are thus pivotal to the reaction scheme.

The pathways for hexose and pentose are differentiated mainly through the relative chemical stability of the homologues, glucosepane and pentosinane. The former is a proper AGE under physiological conditions, but the latter is smoothly oxidised to an intermediate, which is subsequently dehydrated to the advanced glycoxidation product, pentosidine. At this stage, it is not known whether oxidation or dehydration is rate determining; the role of metal ions needs to be clarified as well.

Experiments with [$^{13}C_6$]-D-glucose showed that the sugar's carbon chain remains intact in the course of crosslink formation. Thus, the formation of bicyclic compounds from carbohydrate fragments is likely to be at most only a minor pathway.[392]

Since the human plasma levels of glucose are more than 100 times those of pentoses, it is glucosepane and DOGDIC that are expected to be of prime physiological significance.[377]

Looking back, the variety of substituents, including crosslinks, encountered in physiological research, is really remarkable and raises the important question of their implications in the area of foods. In principle, any of the above types of substituents could be present on proteins, peptides, and amino acids found in foods (see Section 3.1 below). Under appropriate conditions, the crosslinks would be expected to be able to affect texture, in particular, but also other functional properties.

3 Formation of 'Advanced Glycation Endproducts' *in Vivo*

Nonenzymic glycosylation has for some years been considered an important contributor to diabetic complications. Increased glycation of haemoglobin, albumin, and immunoglobulins has been invoked, at one time or another, to explain the haematological abnormalities, altered pharmacokinetics, and increased susceptibility to infection among diabetic patients.[393] The glycation hypothesis gradually evolved into

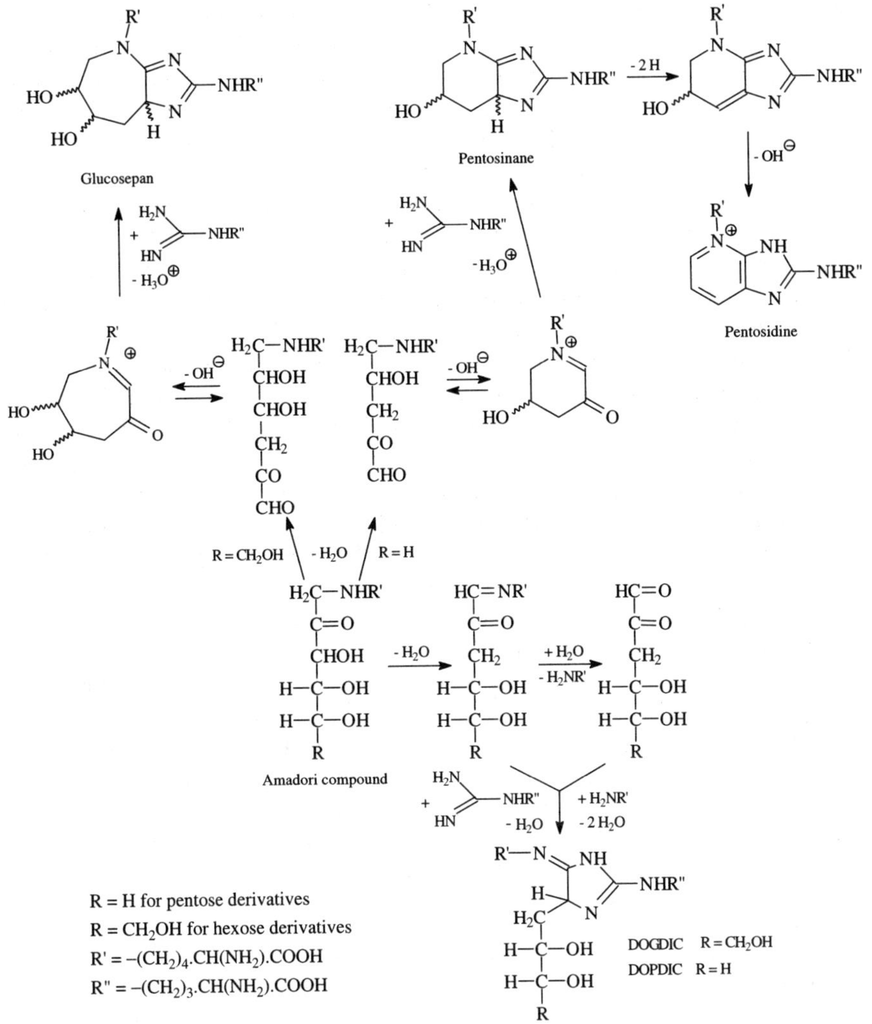

Scheme 8.3 *Pathway for the formation of lysine-arginine crosslinks, glucosepan, pentosinane, pentosidine, DOPDIC, and DOGPIC*[115,377]

the AGE hypothesis, as attention shifted to the chemical modifications, especially the crosslinking of proteins that occur subsequent to glycation.[394] The AGE hypothesis offers a reasonable explanation for the gradual development of chronic complications in diabetes. Long-lived proteins, such as collagen and crystallins, would naturally have a much greater chance of AGE accumulation. Carbohydrates other than glucose, such as ascorbate, fructose, and ribose, were drawn into the picture as possible precursors for AGEs in diabetes and inhibition of AGE formation, rather than glycation, became the target for the pharmacological management of diabetes.[395]

The modification by AGE of even less long-lived proteins is also possible. Makita et al.[396] were able to show by means of AGE-specific antibodies that circulating haemoglobin is modified to the extent of 0.42% in normal individuals, increasing to 0.75% in patients with diabetes-induced hyperglycaemia. The correlation between Hb-AGE and HbA$_{1c}$ gave $r = 0.9$ ($P < 0.001$), consistent with its being derived from the latter. The antibodies did not react with HbA$_{1c}$. Aminoguanidine, administered over 28 d, decreased Hb-AGE significantly ($P < 0.001$), preventing its formation, but not affecting existing Hb-AGE or HbA$_{1c}$.

Szwergold et al.[397] proposed that intracellular nonenzymic glycation is controlled by enzymic deglycation, catalysed by fructosamine-3-kinase (FN3K). This enzyme, which has been purified, sequenced, and cloned, phosphorylates fructoselysine to the 3-phosphate, which is unstable, decomposing to lysine, 3-DG, and P$_i$. Szwergold et al.[398] have found that enzymic deglycation activity in vivo involves both FN3K-dependent and -independent mechanisms, which has led them to propose a nonenzymic glycation/enzymic deglycation hypothesis. FN3K is an entirely novel enzyme, 35 kDa with 309 amino acid residues and no significant homology to any mammalian protein. It appears to be expressed in man in all the tissues examined. Using human erythrocytes owing to their convenience, lysates consistently gave more glycated haemoglobin than intact cells on 7-d incubation. This shows that the capacity to deglycate is lost with the integrity of the cells, as expected. The rate of glycation in intact cells with L-glucose was about five times that with the D-isomer, because only the latter is deglycated enzymically. 6-Deoxy-D-glucose behaves virtually indistinguishably from the parent sugar, but 3-DG does not act as a substrate for FN3K; nevertheless, it gives a higher degree of glycation in haemolysates than intact cells, providing evidence of a deglycation pathway independent of FN3K.

The browning of proteins by glucose is catalysed by transition metal ions and by air. The AGEs in tissue proteins best characterised are CML and pentosidine, products of a combination of glycation and oxidation of hexoses or ascorbate. In view of facts such as these, the AGE hypothesis evolved to accommodate a role for oxidative stress.[399] Multiple autoxidative mechanisms are involved in the formation of AGEs, as illustrated in Scheme 8.4 for the formation of CML from glucose, including:

1. The Wolff pathway of oxidative glycosylation,[400] initiated by the autoxidation of glucose to form reactive intermediates, such as arabinose and glyoxal.[401]
2. The Namiki pathway, involving cleavage of the Schiff base, forming protein-bound aldehydes and glycolaldehyde and glyoxal.[402]
3. The Hodge pathway, proceeding by oxidative cleavage of the Amadori compound.[6]

There is evidence that the Wolff and Namiki pathways prevail in phosphate buffer in vitro, but the Hodge pathway may be important in vivo.[403,404]

Other carbohydrates, such as ascorbate, fructose, and glycolytic intermediates, may give rise to CML and other AGEs in vivo. As for the formation of CML, other oxidative pathways may also be involved, such as lipid peroxidation[405] and oxidation of serine.[406] Thus CML can be an AGE, but that is not necessarily always the case.

Sometimes AGEs have been localised using formalin-fixed paraffin-embedded tissue sections after heat-induced epitope retrieval, a process that could be responsible for the conversion of Amadori products to AGEs. Hayashi et al.[407] therefore compared the reactivity of a monoclonal antibody to CML as an important AGE in

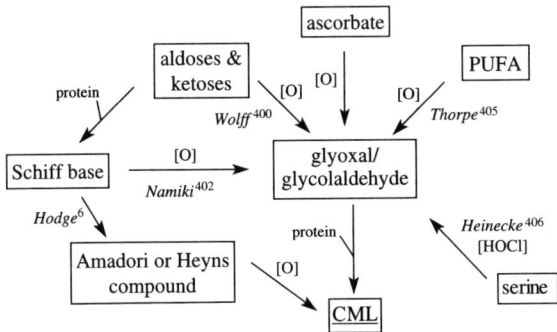

Scheme 8.4 *Pathways of formation carboxymethyllysine (CML) in vitro*[351]

tissue sections before and after heat treatment. In heat-treated sections, the nuclei of human and rat epidermis and the cytoplasm of rat liver were strongly stained with the antibody (6D12), whereas, in frozen sections, staining of the same sites was negligible. To gain further insight, glycated HSA, as a model Amadori product, was heat-treated and CML was determined by antibody and by HPLC. CML was generated by heat treatment above 80 °C and increased with time. Its formation could be inhibited in the presence of $NaBH_4$ (132 mM), diethylenetriaminepentaacetic acid (1 mM), or aminoguanidine (90 mM). Reactive intermediates, such as glucosone, 3-deoxyglucosone, 2-oxopropanal, and glyoxal were detected by HPLC after heat treatment at 100 °C. Heat clearly brings about important changes.

While much attention has been focused on the role of oxidative stress in the development of diabetic complications, the Maillard reaction can proceed effectively via dicarbonyl intermediates formed by several routes in the total absence of metal ions and air. Dicarbonyl compounds, such as dehydroascorbate (DHA), 3-deoxyglucosone (3-DG),[408] and 2-oxopropanal (MGO),[409] are present in tissues at micromolar concentrations, which are increased in the blood of diabetic patients. These dicarbonyl compounds brown proteins efficiently under anaerobic conditions. Pentoses and tetroses also brown proteins rapidly and crosslink them, even under rigorously anaerobic, antioxidative conditions,[410] apparently by the rearrangement and hydrolysis of Amadori compounds to form deoxyglycosones. While DHA is produced by the oxidation of ascorbate, 3-DG and MGO are formed from glucose and its metabolites by non-oxidative routes. 3-DG, and probably its isomer, 1-deoxy-2,3-glucodiulose, are formed by rearrangement and hydrolysis of the Amadori compound of glucose with protein[6] or by mixed enzymic and nonenzymic processes, involving phosphorylation of fructose and β-elimination of phosphate from fructose-3-phosphate.[411] MGO is formed by the elimination of phosphate from the metabolites, glyceraldehyde-3-phosphate or dihydroxyacetone phosphate, but it is also a product of amino acid metabolism.[409]

The efficient browning of proteins by 3-DG and MGO under anaerobic conditions emphasises that oxidation is not essential for Maillard reactions *in vivo*, although oxygen and oxidation appear to be rate-limiting for browning and crosslinking of proteins by aldo- and ketohexoses, as well as ascorbate.

3.1 Parallels *in Vivo* in Foods

Several papers have already linked these discoveries with foods. In the first, CML was found in rat urine, with excretion rates of 4–19%.[412] The excretion of CML was definitely related to the intake of CML, rather than to the intake of fructoselysine or lysinoalanine, although adding fructoselysine free of CML did increase the CML excreted, so fructoselysine must be being metabolised to CML or converted into it in the diet before consumption. CML had previously been found in the urine, in quite a high proportion, of premature infants and of hospitalised youngsters, without a clear indication as to its source.[413]

In another paper, pyrraline (ε-pyrrole-lysine), determined by reversed-phase HPLC with electrochemical detection, was shown to increase in nonfat dried milk progressively over 6 h at 80 °C from 2 to 133 ppm.[414] It has been found in a range of processed foods, the highest value of 227 ppm being recorded for a gravy stored for 6 h at 80 °C. Although pyrraline degrades on storage under moist conditions, it increased progressively at 110 °C in a powdered meal replacer, whereas furosine levels peaked at 1–2 h and then declined rapidly to about one-third after 5 h.

More recently, pyrraline was found to increase almost linearly with time during the storage of freeze-dried milk at 70 °C. The amount formed increased with moisture content, reaching more than 5000 mg kg^{-1} protein in 50 h with 9% moisture.[354] Values of up to 3100 mg kg^{-1} protein were found in some samples of milk or whey powder. Remarkably high amounts (200–3700 mg kg^{-1} protein) were also found in bakery products, indicating that up to 15% of the lysine residues may have been modified.

Pentosidine, too, has been found in foods.[354] Ion-exchange chromatography with direct fluorescence detection gave a detection limit lower than 50 μg kg^{-1} protein. The levels in food ranged from 'not detectable' to 2–5 mg kg^{-1} protein for sterilized and evaporated milk and up to 35 mg kg^{-1} protein for some bakery products and coffee, a range of concentrations comparable with those in plasma and urine. Pentosidine also increased with storage, but, compared with the crosslinks due to lysinoalanine and histidinoalanine (up to 3000 mg kg^{-1} protein), it does not play a major part in crosslinking food proteins.

Imidazolonylornithine has also been detected in coffee and in bakery products,[354] in amounts ranging from 4 to 13 g kg^{-1} protein, indicating that 20–50% of arginyl residues may react with 2-oxopropanal.

Soy sauce also contains some of these reaction products: pyrraline (3.6), imidazolones (53.8), lysylpyrropyridine (2.1), and pentosidine (0.7 nmol g^{-1} soy sauce powder). The first two were determined in the total soy sauce by immunochemical reaction, whereas the last two were determined in the nondialysable fraction (>0.5 kDa) by acid hydrolysis and HPLC.[360]

3.2 Glycoxidation

Oxidative browning *in vitro* can be distinguished from nonoxidative browning by the formation of CML. Although ribose and 3-DG brown proteins at essentially the same rate under both types of system, CML is formed only under oxidative conditions.[20,51] This requirement for oxygen in the formation of CML and pentosidine should be considered to be a good approximation, rather than absolute. These AGEs could

Other Physiological Aspects

be formed from precursors during the work-up of samples for analysis or by intermolecular redox reactions, which could generate oxidised precursors, even under antioxidative conditions. The role of oxidation in the browning of proteins via the Maillard reaction is clearly dependent on both precursor and pathway. For C_6 aldoses and ketoses, oxidation increases the rate of browning. For ascorbate, oxygen is essential for browning, but, for smaller sugars and dicarbonyl compounds, it is not.

While it is relatively easy to assess the role of oxidation in the formation of specific AGEs *in vitro*, the distinction between glycoxidation products and nonoxidative AGEs *in vivo* is not always obvious. Consider GO as an example. It is a product of the oxidation of glucose[415] and its carbon atoms are in an oxidation state of +1 relative to the carbon atoms in glucose, oxidation state 0. In contrast, MGO, with its carbon atoms at an average oxidation state of 0, is not a product of the oxidation of glucose and is formed during anaerobic glycolysis from triose phosphates.[348] Would CML, a product of the reaction of GO with protein, be considered as a glycoxidation product, while its homologue CEL, which is derived from MGO, be considered as a nonoxidative Maillard product?[351] CML and CEL are present at similar concentrations in proteins (Table 8.1), both increase with age in lens proteins and skin collagen, and their concentrations in the lens correlate strongly with one another.[354] They also increase to similar extents in skin collagen of diabetic patients, compared with age-matched control animals.[351] A similar situation exists for the imidazolium crosslinks, GOLD and MOLD (see Table 8.1).

Crosslinking by glutaraldehyde (Gu), formaldehyde (F), and glyceraldehyde (Gy) has been compared using ribonuclease (RNase) as the model protein.[416] Gu reacted almost instantaneously, producing a yellow coloration and complete aggregation (SDS–PAGE). Gy reacted considerably more slowly, forming increasing amounts of dimers; trimers appeared after 1 h, and some large aggregates were present after 48 h. Loss of lysine was about 20% immediately and almost 60% after 48 h. F reacts similar to Gy, but slightly more slowly.

Irrespective of their origin, dicarbonyl compounds are detoxified by common antioxidant-related mechanisms, such as the glyoxalase pathway, aldehyde reductases, or aldehyde hydrogenases, yielding, for example, D-lactate from MGO or 3-deoxyfructose from 3-DG. The efficiency of these pathways depends on cellular levels of reduced glutathione and/or NADPH. High levels of dicarbonyls in tissues may therefore result from overloaded or compromised antioxidant defence systems. Thus, increases in MGO and 3-DG may result from oxidative stress, even though

Table 8.1 *Estimated levels of AGEs in human tissue proteins about 80-yr old*[351]

Marker	Lens protein	Skin collagen	Reference
CML (mmol mol^{-1} Lys)	3.9	1.7	Dyer et al.[633]
			Lyons et al.[634]
CEL (mmol mol^{-1} Lys)	4.6	0.52	Ahmed et al.[421]
GOLD (mmol mol^{-1} Lys)	0.15	0.04	Frye et al.[420]
MOLD (mmol mol^{-1} Lys)	0.77	0.4	Frye et al.[420]
Pentosidine (mmol mol^{-1} Lys)	0.006	0.03	Dyer et al.[633]
			Lyons et al.[634]
o-Tyrosine (mmol mol^{-1} Phe)	—	0.023	Wells-Knecht et al.[635]
MetSO (mmol mol^{-1} Met)	—	185	Wells-Knecht et al.[635]

these compounds are not formed from glucose by oxidative reactions. Rather than focus on the classification of Maillard reaction damage *in vivo* as glycoxidative or non-oxidative, it is more relevant to appreciate that increases in the rate of formation or steady-state concentration of dicarbonyl compounds, regardless of their origin, is consistent with lowered antioxidant defences and a resultant increase in levels of AGEs in tissue proteins.[351] This has led Baynes and Thorpe[417] to propose that the increase in reactive carbonyl compounds in diabetes results in part from the stress of excessive substrate, and also from a failure/overload of detoxification pathways.

Although CML and CEL are formed from carbohydrates and were originally classified as AGEs, they have since also been detected as products of lipid peroxidation. They may be formed directly from GO and MGO, or from other precursors, during the metal-catalysed oxidation of PUFA in the presence of protein. GO and MGO are formed alongside other carbonyl compounds, more characteristic of lipid oxidation, such as malondialdehyde (MDA) and 4-hydroxynonenal (HNE). The formation of CML and CEL, as well as adducts of MDA and HNE to protein, during lipid peroxidation is inhibited by AGE inhibitors,[351] such as aminoguanidine[395] and pyridoxamine.[418] Under these circumstances, it is more appropriate to use the term advanced Maillard reaction product (AMRP) to describe the broad range of products formed by carbonyl-amine chemistry during the Maillard reaction *in vivo* (**62–101**). AMRPs may be divided into AGEs and advanced lipoxidation end-products (ALEs), as well as into a mixed group. While the formation of AGEs may or may not require oxygen, the formation of ALEs is clearly an oxygen-dependent process.

It may be more appropriate in individual cases, where the origin is not clear, to redesignate an AMRP as an EAGLE, either an advanced glycation or an advanced lipoxidation endproduct.[419]

Diabetes has played an important role in the above considerations, but, in many ways, diabetes is a form of accelerated aging. Baynes[419] wrote 'Aging, the gradual, time-dependent decline in survivability of a biological system, results from the decline in the functional capacity of its physiological sub-systems, as measured in humans, for example, by changes in lung expansion volume, creatinine clearance, cardiac output or immunological responsiveness.' Most of the current biomarkers of aging are AMRPs. Aging is most apparent in long-lived proteins, such as the crystallins of the lens and collagens. The crystallins develop brown colour and fluorescence, associated with aggregation, crosslinking, and insolubilisation. The aging of collagens leads to stiffening of tendons and decreased aortic compliance.

The glyoxal- and methylglyoxal-mediated crosslinks, GOLD and MOLD (**79** and **80**) increase in lens protein with age ($r^2 = 0.69$, 0.75, respectively, both $P < 0.001$), but account only for 0.2% of the lysine residues,[420] whereas CML and CEL account for about 0.4% each (with age, $r^2 = 0.85$, 0.90, respectively).[421] Even so, they are present at higher concentrations than pentosidine and dityrosine, constituting, for tissue at age 80, about 200, 800, 4, and 3 μmol mol^{-1} Lys, respectively. In pooled skin collagen of 85-yr old subjects, GOLD, MOLD, and pentosidine constitute 40, 400, and 40 μmol mol^{-1} Lys, respectively.[422,423] At 100 mol Lys mol^{-1} triple-stranded collagen, MOLD is thus present at 0.04 mol mol^{-1} collagen, which is to be compared with 1–5 mol mol^{-1} collagen for enzymatically formed crosslinks,[424] making it clear that its contribution is very limited.

When RNase is treated with MGO, both inter- and intramolecular crosslinks are formed, but MOLD accounts only for about 5% of the crosslinks in the dimer.[420]

There are differences in the rate of accumulation of AGEs between collagens of different tissues. Indeed, Verzijl et al.[425] showed that CML, CEL, and pentosidine not only increased linearly with age ($P < 0.0001$), but so did general measures of AGE formation, such as browning and fluorescence ($P < 0.0001$), and the values for all were higher for cartilage collagen than skin collagen. The digestibility of cartilage collagen by bacterial collagenase decreased linearly with age (ca 0.3% yr^{-1}, from about 30% at 20 yr to about 10% at 80 yr), proportionally to the extent of glycation. The level of arginine and of the sum of lysine and hydroxylysine decreased with age ($P < 0.0001$, 0.01, respectively), readily attributed to modification by the Maillard reaction. The sum of the three AGEs determined and the level of AGE fluorescence correlated with the mean degree of modification of the three amino acids ($P < 0.001$, 0.05, respectively). However, the sum of the three AGEs even in cartilage collagen of an 80-yr old reached only 5.6 mmol mol^{-1} Lys. The ratio CML/CEL was 6.1, 3.4, and 1.3 for collagen from cartilage, skin, and lens protein, respectively. The value 1.3 is close to the ratio obtained by Ahmed et al. in vitro from collagen and 3-deoxyglucosone or ascorbate, whereas the other two are closer to that similarly obtained, but with ribose, glucosone, or glucose.[421]

Turnover rates were thought to be an important factor in explaining differences in the rate of accumulation of AGEs between collagens of different tissues. This hypothesis was tested by comparing the accumulation of CML, CEL, and pentosidine with the degree of racemisation of the protein's L-aspartic acid, as a measure of residence time.[426] CML levels in human articular cartilage collagen increased linearly with age ($r = 0.89$, $P < 0.0001$), as in skin collagen ($r = 0.80$, $P < 0.0001$). The rate of accumulation in cartilage was higher than in skin ($P < 0.0001$). In parallel, levels of CEL and pentosidine also increased with age ($P < 0.0001$ in all cases), those in cartilage increasing more rapidly ($P < 0.0001$ for both). The rate constants for aspartic acid racemisation, determined for dry cartilage and skin collagen from Arrhenius plots for 120–160 °C, were shown to be identical (1.78×10^{-6} and 1.71×10^{-6}, respectively), validating the use of % D-Asp levels as a measure of residence time. In both cartilage and skin collagens, % D-Asp levels increased linearly with age ($r = 0.95$, $P < 0.0001$ and $r = 0.78$, $P < 0.0001$, respectively) and the % D-Asp accumulation was higher in cartilage than in skin ($P < 0.0001$), confirming that collagen had a longer residence time, i.e., slower turnover, in cartilage than in skin. In both human cartilage and skin collagens, CML, CEL, and pentosidine levels increased linearly with % D-Asp ($P < 0.0001$ for each), demonstrating that protein turnover is indeed a key factor in the rate of AGE accumulation. Furthermore, the relationship between AGE level and % D-Asp is identical for cartilage and skin, supporting protein turnover as an important determinant of AGE accumulation. The rate constants for D-Asp accumulation allowed half-lives for collagen to be calculated: 15 and 117 yr for skin and cartilage, respectively. These values can be compared with ~200 and >500 yr determined for cartilage and dentin previously.[427,428]

Age biomarkers would be expected to accumulate more quickly in short-lived animals. Comparative data for two AGEs in skin collagen are available[419] for 30-yr-old humans and 9-month-old rats (each at about 25% of their respective maximum life-spans) for CML (0.4 and 0.057 µmol mol^{-1} Lys, respectively) and pentosidine (8 and 0.58 µmol mol^{-1} Lys, respectively). They show greater accumulation in the *longer-lived* species, contrary to expectation. Facts like these make it difficult to argue that the total burden of Maillard and oxidation products in tissue proteins is a determining factor in the aging process.

If protein modification is not the key to aging, could DNA modification be more significant? After all, lifespan is likely to be programmed genetically. But the genome is protected by constant repair processes and damage is due only to the cumulative 'errors' made in these processes. The damage is attributed to reactive oxygen species, which also initiate chain reactions via carbohydrates and lipids. The cell nucleus is relatively deficient in lipids, but carbohydrates are important components and are integral to DNA. The Maillard reaction can thus be thought to amplify oxidative stress by the propagation of reactive oxygen species and by the generation of reactive carbonyl compounds.[419] Such considerations underline the importance of obtaining detailed, quantitative information on these issues and their far-reaching consequences.

Using four monoclonal antibodies raised against AGEs, namely 6D12 for CML, KNH-30 for CEL, 1F6 for fluorolink, and 2A2 for an as yet unknown epitope, Ling *et al.*[429] showed that all four structures were present in most tissues and cells in rats at birth and that AGEs accumulated and increased in distribution gradually as the rats aged. Except for fluorolink, they were even present in almost all tissues of the 10-d foetuses tested and tended to increase with foetal development. Highest concentrations of CEL were found in rats throughout their lifespan in cerebral neuronal cells and of the 2A2 epitope in cardiomyocytes, whereas in the foetus highest concentrations of CML were found in the gastrointestinal tract, of CEL in the encephalon and spinal chord, of fluorolink in the heart, and of the 2A2 epitope in the lungs. HPLC showed the 13-d foetus to have 725 ± 87 mmol CML per mol Lys. These results suggest that AGEs may well be generated endogenously by the foetuses.

Since 3-DG-imidazolone is generated only by reaction with 3-DG, it is potentially an important marker for 3-DG modification *in vivo*.[430] Using monoclonal antibodies, BSA incubated with 3-DG at 37 °C for four weeks was shown to have greatly increased reactivity to the probe for 3-DG-imidazolone and to that for CML, but only slightly increased reactivity to that for pyrraline. HPLC allowed the degree of the three modifications to be quantified as 1.10, 0.84, and 0.33 mol mol^{-1} BSA, respectively. CEL, GOLD, MOLD, DOLD, and argpyrimidine were below the HPLC detection limit.

Originally, glycohaemoglobins were determined using ion-exchange chromatography (see above) and HPLC with cation-exchange columns became the routine method for their analysis. The percentage of HbA_{1c} in Hb can also be determined immunologically. Most recently, a capillary electrophoretic method has been developed[431] using an electrolyte of 27 g L^{-1} malic acid (pH 4.6) with dynamically coated columns, first with polycations and then with polyanions. Samples from 105 patients were analysed (3.5–10.8% HbA_{1c}). The CE results tended to be 2–3% lower than those obtained by immunoassay, but the correlation between them was better ($r^2 = 0.962$) than between HPLC and immunoassay ($r^2 = 0.781$). The CE method separates HbA_{1c} not only from HbA_0, but also from HbA_{1a}, HbA_{1b}, HbA_{1d}, carbamylated Hb, foetal Hb, etc.

Other Physiological Aspects 121

4 Glycation of Phospholipids

Biologically, proteins have had most of the attention with regard to glycation, but, in principle, it has always been realised that any amino compound bearing at least one hydrogen atom on its nitrogen is able to participate in the Maillard reaction.

Aminophospholipids are therefore susceptible, but they have already been considered in Chapter 3.

5 Glycation of Nucleic Acids and their Components

3-Deoxyglucosone has been found to react readily with 2'-deoxyguanosine, two major products being isolated and identified as diastereoisomers of N-(1-oxo-2,4,5,6-tetrahydroxyhexyl)-2'-deoxyguanosine.[360] AGEs have been produced from DNA also by reaction with 3-DG or glucose.

6 The Role of the Maillard Reaction in the Lens

In the lenses of dogs that have been diabetic for 5 years, maintained with insulin but with poor glycaemic control, mean levels of crystallin glycation, fluorescence, vesperlysine A, and pentosidine increased significantly.[432] These changes correlated highly with those in dura mater collagen; however, whereas pentosidine crosslinks were significantly elevated in collagen from dogs with moderate levels of HbA_1 (*i.e.*, ~8.0%), lens pentosidine levels were normal in this group and were only significantly elevated in the animals with poor glycaemic control ($HbA_1 = $~9.7%). Thus, it seems that there is a tissue-specific glycaemic threshold for pentosidine formation, *i.e.*, glycoxidation in the lens. This threshold may in part be linked to a marked acceleration of crystallin glycation with values $> 8\%$ and/or changes in lens membrane permeability.

Since vesperlysine A was elevated in dogs with moderate glycaemic control, whereas pentosidine, a glycoxidation product, was elevated only in dogs with poor control, the former becomes a unique marker for mild hyperglycaemia.[368]

Hofmann *et al.*[367] have obtained evidence that glycolaldehyde can interact with ε-amino groups to crosslink proteins with generation of colour due to pyrazinium radical cations (*cf.* CROSSPY **84**) under physiological conditions. This has direct bearing on, for example, cataract formation.

Polyphenols appear to have an inhibitory effect on diabetic cataract formation in rat lenses.[433] Fresh rat lenses were incubated in 15 mM xylose to mimic hyperglycaemia. The lenses turned opaque, mimicking cataract, within 4 d. They were treated with anthocyanins and with the extracts of many vegetables and fruits, including Japanese radish, grape skins, and coloured rice. Grape skins were found to be potent inhibitors of lens opacity formation at 100 μg mL^{-1} in the medium and, of the anthocyanins present, delphinidin-3-glucoside and malvidin-3-glucoside proved to be the most active. Since rice is a staple, 48 strains of coloured rice were examined, extracts of nine of which showed inhibitory effects at 100 μg mL^{-1}. It was reported that the effect here was due not to the anthocyanins present, but to the phenolic acids, protocatechuic acid in particular.

In keratoconus, the cornea is weaker and thinner at its centre. This can be counteracted by tissue engineering, *e.g.*, by crosslinking.[434] Such a process has also been patented.

7 Role of the Maillard Reaction in Nephropathy

AGE formation is related to hyperglycaemia in diabetes, but not in uraemia, because AGE levels do not differ between diabetic and nondiabetic haemodialysis patients, which is a mystery.[435] Pentosidine and CML were elevated in uraemic plasma and their levels were not correlated with fructoselysine levels. The addition of aminoguanidine and OPB-9195 lowered the yield of AGEs in both uraemic and control plasma. Carbonyl stress was therefore thought to be a contributory factor in uraemia.

The collagen of the myocardium of autopsied haemodialysis patients and of age-matched controls was extracted first with 0.9% NaCl and second with collagenase, giving a soluble fraction (SF) and a collagenase-soluble fraction (CF). Glycoxidation and lipoperoxidation products in these fractions were assessed by determining fluorometrically pentosidine and malondialdehyde (MDA), respectively.[436] Both pentosidine and MDA were higher in CF from patients than in that from the controls ($P < 0.05$). MDA correlated strongly with pentosidine in CF ($P < 0.0001$), but not in SF, which does not contain matrix collagen. The results suggest that glycoxidation and lipoperoxidation contribute synergistically to cardiac damage in haemodialysis patients.

Aortic samples, free of atherosclerotic lesions, were obtained from haemodialysis patients with end-stage renal disease and from age-matched controls.[437] The media, but not the intima, proved positive immunohistochemically for pentosidine, which was observed along and between elastic fibres. In both groups, pentosidine-linked fluorescence was higher in elastin than in collagen, and the HPLC level of pentosidine was higher in the elastin from patients than in that from the controls.

Yoshimura *et al.*[438] have shown immunohistochemically with the 6D12 probe that reactive substances (CML) accumulate in the vascular walls of cardiac tissues in dialysis patients. A partial decrease in this accumulation followed renal transplantation.

Autopsy samples were obtained from diabetic and control subjects, and renal tissue sections were examined immunohistochemically for the expression of membrane attack complement (MAC), C3, and CML, for CD59 and single-strand DNA as signs of apoptosis, and smooth-muscle actin (SMA). CML co-localised with MAC in areas free of CD59.[439] CML was present in the area of loss of SMA. Only a few smooth-muscle cells positive for single-strand DNA were detected in diabetic patients. CD59 is weakly positive in media except in the area of loss of SMA. The degree of medial smooth-muscle cell loss correlated with the per cent medial deposition of CML and MAC. The degree of medial smooth-muscle loss was higher in diabetic patients with renal dysfunction than in others. It seems that glycation promotes smooth-muscle cell necrosis in renal arteries of diabetic patients, the process being mediated by MAC.

In human subjects with uraemia, the blood plasma levels of the α-oxoaldehydes, glyoxal, 2-oxopropanal, and ribosone increased greatly, but 3-erythrosone, 3-deoxyerythrosone, 3-deoxyribosone, and 3-deoxyglucosone did not.[440]

7.1 Peritoneal Dialysis

In peritoneal dialysis (PD), sterilisation of the fluid by heat leads to the presence of reactive carbonyl compounds, which have been incriminated in the progressive deterioration of the peritoneal membrane in long-term PD patients.[441] Glutathione in conjunction with glyoxalase I, as well as aminoguanidine, counteracted the effect.

Although the concentration of glucose far exceeds that of its degradation products (GDP) in PD fluid, it is much less reactive. This makes it difficult to determine which is more important in relation to AGE formation. Millar *et al.*[442] have therefore incubated lysozyme as a model protein with 3.86% PD fluid or 100 μM 2-oxopropanal, glyoxal, or 3-deoxyglucosone or 100 mM glucose in 100 mM phosphate (pH 7.4) at 37 °C. Sterilisation of PD fluid may be carried out in twin bags, where glucose is compartmentalised during sterilisation, lowering the formation of GDPs from that in single bags. The solutions were either ultrafiltered every 2–3 d and replenished with fresh fluid, or were left to incubate continuously. Protein-bound AGE fluorescence (λ_{ex} = 350 nm, λ_{em} = 430 nm) increased progressively in all PD fluids. It was always higher in fluid from single than in that from twin bags (> 65% at 30 d) and with replenishment than without (147% for twin, 125% for single bag at 30 d), except that, for glucose, replenishment gave a 25% reduction in AGE fluorescence (at 15 d). It seems that GDPs are more important in AGE formation in PD fluid than glucose.

8 Role of the Maillard Reaction in Cancer

The intensely coloured Maillard product **25** (see Chapter 4) has been demonstrated to be a potent inhibitor of the growth of human tumour cell lines A431, LXFL529L, GXF251L, and CXF97L with IC_{50} of 3–9 μmol.[178] In addition, it has a detrimental effect on microtubule integrity.

The level of pentosidine of nondiabetic pulmonary cancer patients in treatment with the antineoplastic drug, cisplatin, has been found to be higher than in healthy subjects,[443] when the protein from urine, plasma, and erythrocyte membranes was examined by HPLC. Urine pyrraline was unaffected. This work implies a whole field of study of the interactions of medication with the Maillard reaction.

As regards xenobiotics, metabolism is generally divided into Phase I and Phase II, the former being concerned *inter alia* with reduction, oxidation, and hydrolysis and the latter with conjugation of the xenobiotic or its Phase I metabolites. The main determinant of whether exposure to a xenobiotic will result in toxicity is the balance between Phase I and II enzymes. In particular, induction of the Phase II enzyme, glutathione S-transferase, by antioxidants has been proposed as a promising strategy for cancer prevention.[444] Hence, the emphasis placed on the antioxidant properties of MRPs from a food point of view (see Chapter 9) also has relevance physiologically.[356] Lindenmeier *et al.*[356] showed that bread crust and different extracts of it exhibited increased glutathione S-transferase activity, as well as inhibiting cytochrome *c* reductase activity (Phase I), both effects implying protection against cancer *in vivo*.

Consideration of intervention is deferred to Chapter 14.

CHAPTER 9

Other Consequences of Technological Significance

1 Introduction

Ten aspects of the Maillard reaction will be considered in this chapter: the effect on a_w, on pH, on redox potential, on solubility, on texture, on foamability and foam stability, on emulsifying power, on volatile formation on storage, on binding of volatiles, and on other functionalities. The third of these has received by far the most attention and this is reflected here.

2 Effect on a_w

Production of water is one of the symptoms of the Maillard reaction (see Chapter 1). In particular, sugar dehydration (Reaction C of the Intermediate Stage) produces a substantial proportion of water, which, in theory, has several potential consequences. If a food is of low a_w, the moisture may be sufficient to encourage microbial growth. At low a_w, an increase in moisture would also increase the rate of browning. It could contribute to a deterioration of texture as well. Although these aspects need to be borne in mind, they are by no means the primary considerations.

3 Effect on pH

Lowering of pH is another one of the symptoms of the Maillard reaction (see Chapter 1). The pH of food and physiological samples is critically dependent on the components present, particularly buffers. In aqueous media, in the absence of buffers, the Maillard reaction brings about a considerable lowering of the pH. Thus, for example,[445] when solutions (1 molal with respect to the sugar and the amino acid) were refluxed for 120 min, the pH dropped as follows: xylose-Gly, 5.2 to 3.9; glucose-Gly, 5.2 to 4.1; xylose-Lys, 5.5 to 2.9; and glucose-Lys, 5.5 to 3.5. There are two principal ways by which this drop in pH is brought about: formation of acids [mainly by sugar degradation (Reaction D of the Intermediate Stage; see Chapter 1)] and conversion of amino groups into less basic, mainly heterocyclic, forms (Reaction G of the Final Stage and loss of some nitrogenous compounds as volatiles). The drop in pH takes place even though there is a loss of carboxyl groups (production of CO_2 is another symptom of the Maillard reaction).

The possible consequences of a drop in pH are manifold. The pH greatly affects the mechanism of the Maillard reaction itself (see Chapter 2), it affects microbial activity, and it may cause polymeric material, particularly proteins, to lose solubility and thus alter texture.

4 Effect on Redox Potential

Increased reducing power, *i.e.*, increased antioxidant activity or more negative redox potential, is also one of the symptoms of the Maillard reaction (see Chapter 1). This has considerable significance, because, as far as foods are concerned, one of the main ways in which they deteriorate chemically is through oxidation, particularly of unsaturated fats and oils, leading to oxidative rancidity. The topic has considerable physiological significance as well (see Chapter 8).

In the food field, attention has been paid to this aspect of the Maillard reaction for over 50 years, as can be seen from Table 9.1. However, it must not be assumed that the Maillard reaction is the only factor involved in producing the effects listed.

Overall, there is no doubt that the Maillard reaction produces substances with an antioxidative effect, but it is by no means clear as yet how and to what extent that effect is exerted. Melanoidins not only have increased reducing power, but are able to complex metals and, in this way, lower their ability to act as catalysts in lipid and other oxidations.

Determining antioxidative activity is one of the complicating factors. There are several methods and each deals with an aspect more or less different from those addressed by the others. Most of the methods that have been applied to Maillard products are briefly described in Table 9.2.

Surprisingly little attention has been paid to the electrochemical properties of Maillard products, even though the problems of electrode stability and of standard states are substantial. Rizzi[463] has studied carbohydrate–β-alanine systems (0.067 M-0.20 M) under standardised conditions, which included refluxing for 1 h in 0.1 M bis-Tris buffer (initial pH 6.9). Using a Pt/Ag–AgCl redox electrode, he observed increased negative voltages, consistent with the formation of reductones during Maillard reactions. The carbohydrate affected the change in voltage as follows: ribose (-192mV) > arabinose (-157mV) ~ xylose (-155mV) > rhamnose (-105mV) ~ glucose (-102mV) ~ lactose (-99mV) > fructose (-83 mV). Carbohydrates, which do not participate in the Maillard reaction, hardly affected the voltage: 2-deoxyglucose (-22mV), xylitol (-8mV), and sorbitol (-4 mV). In the absence of amine, xylose and glucose gave -42 and -25 mV, respectively, showing the effect of buffer alone to be minimal. No colour was formed. However, ascorbic acid, at 0.001 M and without amine or heating, gave -102 mV. On increasing the heating time to 6 h, the voltage of the glucose system decreased to -204 mV, accompanied by increased colour. Under similar conditions, but using 28 d at 37 °C, xylose–glycine gave -20 mV and xylitol–glycine -35 mV, which contrasts with the results of Glomb and Tschirnich,[114] who detected reductones under such physiological conditions by means of derivatisation.

Table 9.1 *Instances of antioxidant activity of Maillard reaction products in foods*

Product	Modification	Effect	Reference
Cookies	Addition of 1.5% glucose plus glycine or corresponding melanoidin	Superior to commercial antioxidants	Iwainsky and Franzke[467]
Cookies	Addition of 5% glucose to dough to replace an equal amount of sucrose	Better stability against oxidative rancidity on storage; darker colour	Griffith and Johnson[512]
Cookies	Addition of amino acids or protein hydrolysates to dough	Improved storage stability	Yamaguchi et al.[513,514]
Cookies	Addition of glucose and His to dough	More effective than addition of corresponding MRP	Lingnert[515]
Cookies	Addition of glucose and His to dough	Inhibition of development of rancidity	Lingnert and Eriksson[469]
Cookies	Addition of sugar and amino acids to dough	Strong antioxidative effect, Arg and xylose giving much less colour	Lingnert and Hall[481]
Butter	Model containing glucose and Lys	On heating, antioxidant capacity of ~5 g Trolox equiv. per 100 g dried extract develops	Bressa et al.[468]
Wheat, corn, oat	Toasting	Improved oxidative stability	Anderson et al.[516]
Fried rice cake chips	Addition of glucose and Trp	Improved storage stability	Tomita[517]
Drum-dried cereals	Prior toasting of wheat flour	Improved storage stability	Hauri et al.[518]
Soy bean sauces	Fermented with *Aspergi oryzae* for 6 min	MRPs formed with various types of antioxidant activity	Cheigh et al.[519]
Tomato juice	Heating at 95 °C for > 4 h	Better chain-breaking and O_2-scavenging activities	Anese et al.[520]
Tomato powder			Nicoli et al.[492]
Coffee	Medium dark roast gave maximum antioxidant properties	~10 min at ~200 °C gave maximum O_2-uptake and inhibition of crocein bleaching	Nicoli et al.[493]
Milk powder	Preheating 88–93 °C for 20 s	Greatly improved oxidative stability	Findlay et al.[521]
Liquid milk	Sterilization	Increased oxidative stability	Vandewalle and Huyghebaert[522]
Spray-dried whole milk powder	High heat	Increased stability over low heat	Binder et al.[523]
Spray-dried whey–buttermilk powder	Heat treatment	Improved storage stability	Hansen and Hemphill[524]
Margarine	Addition of 0.5% melanoidin	Inhibition of oxidation greater than with 0.01% dodecyl gallate	Franzke and Iwainsky[525]

Butterfat	Heating butter for 10 min at 200 °C	Increased storage stability	Josephson and Dahle[526]
Lard	Addition of MRP from glucose–Gly	Improved oxidative stability	Griffith and Johnson[512] (cf. ref. 466)
Vegetable oils	Addition of amino-reductones prepared from hexoses and secondary amines	Marked decrease in peroxide formation	Evans et al.[527]
Corn oil	Addition of mixture of glucose and Gly, more effective than either	Inhibition of development of rancidity	Maleki[528]
Sesame oil	Addition of 0.02% MRP from glucose-Gly	Almost doubled the induction period	Cämmerer et al.[466]
Safflower and sardine oils	Wheat gliadin and hen's egg albumin	Improved oxidative stability	Taguchi et al.[529]
Sterilised beef	Over cooking	Improved stability	Zipser and Watts[530]
Pork patties	Addition of MRPs	Improved oxidative stability	Bedinghaus and Ockerman[476]
Refrigerated cooked beef	Addition of retorted solutions of sugars and amino acids	Inhibition of warmed-over flavour	Sato et al.[531]
Frozen hamburgers	Precooked by frying	Improved quality	Dagerskog et al.[532]
Retorted turkey	Retorting produces inhibitory substances	Inhibition of warmed-over flavour	Einerson and Reineccius[533]
Turkey	Addition of honey or honey-Lys MRP	Greatest inhibition of oxidation by honey	Antony et al.[498]
Frankfurters	Addition of MRP from glucose and His or haemoglobin	Inhibition of development of rancidity	Lingnert and Eriksson[469]
Sardine products	MRP from glucose and His	Inhibition of autoxidation during storage	Tanaka et al.[534]

Rizzi[463] also exposed the Amadori compound from glucose and piperidine to the standardised conditions, but at 0.01 M without amine, which gave a reading of −153 mV and a deep-yellow colour. Yet heating at 77 °C for 15 min gave only −26 mV, but increasing the concentration progressively lowered the voltage, with 0.10 M leading to −362 mV. After 15 min at 77 °C, xylose-β–alanine gave +4 mV, in accord with the formation of the Amadori compound being slow relative to its decomposition.

Thiazolium salt XTT has been used to evaluate the extent of the Maillard reaction in UHT (ultra-high-temperature)-treated milk, where it proved more rapid and convenient than determination of lactulose, HMF, or furosine.[451] To establish the nature of the Maillard reaction products (MRP) involved, the interaction of the aminoreductone from lactose and n-butylamine ($\lambda_{max} = 319.5$ nm) with MRP was studied. Excellent correlation ($r = 0.967$, n = 19) was found between the increased absorbance at 319.5 nm and XTT reducibility, as measured at 492 nm, near λ_{max} for the formazan, the reduction product of XTT.

Table 9.2 *Methods for determining antioxidative activity*

Method	Comments	Reference
Redox	Pt electrode and Ag/AgCl, Cl^-_{sat} reference electrode, N_2 flushing	Anese and Nicoli[446]
FRAP	Ferric-reducing ability of plasma; uses Fe^{3+}-TPTZ	Rice-Evans[447] and Benzie and Strain[448]
Cerium sulfate	Potentiometric titration, using Fe^{2+} 1,10-phenanthroline as indicator	Cioroi[449]
Indophenol	Titration with 2,6-dichloro-indophenol	Tanner and Barnett[450]
Tetrazolium salt XTT	See p. 127	Shimamura et al.[451]
Polarography	10 or 30% lipid emulsified in 0.01 M KCl and 0.1 M phosphate buffer (pH 7.2). Usually 1×10^{-7} M haemoglobin. Time (s) to 90% oxygen uptake	Hamilton and Tappel[452]
Peroxide value	Peroxides liberate I_2 from KI and this is titrated with sodium thiosulfate	Lea[453]
AOXP Leucomethylene blue	N-Benzoyl derivative oxidised by lipid hydroperoxides in the presence of haemoglobin	Brand and Eichner[454] and Bright et al.[455]
Rancimat induction	Lag in lipid oxidation in a stream of air at relatively high temperature (90–130 °C)	Anese and Nicoli[446]
TBA value	Formation of malonaldehyde on oxidation of lipids and its reaction with 2-thiobarbituric acid	Gray[456]
Methyl linoleate decrease	By HPLC	Brand and Eichner[454]
Crocin bleaching		Tubaro et al.[457]
Hexanal oxidation	To hexanoic acid. Decrease in hexanal monitored by GC	Yanagimoto et al.[458]
Radical scavenging	Use of spin traps, such as 2-methyl-2-nitrosopropane, to convert unstable radicals into stable ones for determination by ESR	Perkins[535]
DPPH trapping	α,α-Diphenyl-β-picrylhydrazyl. Useful under lipophilic conditions	Brand and Eichner[454]
DMPD	N,N-Dimethyl-p-phenylenediamine. Oxidised to radical cation (by $FeCl_3$) that is decolorised by the antioxidant	Fogliano et al.[459]
TEAC ABTS trapping	Trolox equivalent antioxidant capacity. 2,2′-Azinobis-3-ethyl-thiazoline-6-sulfonate. Useful under hydrophilic and lipophilic conditions	Rice-Evans[447]
VCEAC	Vitamin C equivalent antioxidant capacity	Kim et al.[460]
ORAC Peroxyl radical trapping	Oxygen radical antioxidant capacity. Radical generated with azo initiators (ABAP, ADIBA)	Rice-Evans[447]
Oxygen uptake	Haemin catalysed	Berner et al.[461]
TRAP	Total radical-trapping antioxidant activity of plasma. Uses phycoerythrin and AAPH	Rice-Evans[447] DeLange and Glazer[462]

In order to achieve high throughput with ORAC determinations, Huang et al.[464] have developed a robotic eight-channel liquid handling system coupled with a microplate fluorescence reader, which has improved efficiency at least 10-fold, compared with the discontinued COBAS FARA II analyser. The CV was 15% or less and the limits of detection and quantification were 5 and 6.25 μM, respectively.

Fogliano et al.[72] assessed melanoidins from glucose–glycine, lactose–lysine (LL), and lactose–N-acetyllysine (LLa) by both the ABAP and DMPD methods (see Table 9.2). In the former assay, the activities relative to Trolox were 1/5, much less, and 1/5, respectively, whereas in the latter, the concentrations with an effectiveness equivalent to that of 3.6 μg mL^{-1} ascorbic acid were 65, 65, and 13 mg mL^{-1}, respectively.

This work was followed up by Borrelli et al.,[74] who used Standard Melanoidin (see Chapter 2) produced from dry-heated glucose–glycine (GG), alongside the LL and LLa, produced under aqueous conditions. In the ABAP assay, GG again gave a value comparable with that of LLa and about 1/4 that of Trolox, whereas in the DMPD assay, at 1 mg mL^{-1}, LLa had an activity of 1.48 μg mL^{-1} ascorbic acid, that of GG and LL being 9 and 13 times lower, respectively. These data suggest that such synthetic melanoidins act mainly as metal chelators and radical scavengers, rather than as reducing agents.

The nature of the medium, hydrophilic or lipophilic, in which the antioxidant is to be effective is a further important question, which has been tackled with respect to human plasma by Yeum et al.[465] They used ABAP as a hydrophilic radical generator and 2,2′-azobis(4-methoxy-2,4-dimethylvaleronitrile) as a lipophilic radical generator. In the former case, the rates of consumption of antioxidants decreased in the order, ascorbic acid > α-tocopherol > uric acid > lycopene > lutein > cryptoxanthin > β-carotene, whereas, in the latter case, α-tocopherol and carotenoids were depleted at similar rates, ahead of ascorbic and uric acid. The behaviour of melanoidins of different types under such conditions would be of interest.

When melanoidins are formed, many factors come into play. Here, several aspects will be considered (cf. ref. 466): differences between sugars and between amino acids, the molar ratio of amino acid to sugar, use of peptides and of hydrolysates, sugar–amino acid mixtures versus MRPs, properties of MRPs, and the mechanisms of antioxidant activity.

4.1 Differences between Sugars

Products from caramelisation show only marginal antioxidative activity at best.[19,467,468]

Lingnert and Eriksson[469] studied the antioxidant effect of MRPs from amino acids (Glu, Cys, Val, His, Lys, Arg) with different sugars (fructose, glucose, and xylose). All the products from fructose were the least coloured; except for Cys, they were the least antioxidant too. Xylose gave the most colour (except when combined with Val) and exhibited by far the greatest antioxidant effect (as measured polarographically) when combined with Arg or Lys. In suppressing hexanal formation from a linoleic acid emulsion, all three sugars were very effective when combined with His, only xylose-Arg giving a comparable result.

Subsequent work has also concluded that pentoses generally produce more browning and greater antioxidative activity than glucose[19,470,471] and that glucose leads to greater antioxidative activity than fructose.[472]

Dicarbonyl compounds, such as dihydroxyacetone, glyceraldehyde, and glyoxal, give melanoidins with higher antioxidative activity than do hexoses or pentoses.[466]

On the other hand, as the degree of polymerisation increases, the oxygen-scavenging power of glucose and its oligomers tends to decrease.[81]

4.2 Differences between Amino Acids

Of the six amino acids used by Lingnert and Eriksson,[469] the MRP from Arg and Lys with xylose exhibited by far the greatest antioxidant effect polarographically, with MRP from Arg being the greater. With regard to suppressing hexanal formation from a linoleic acid emulsion, His was very effective with any of the three sugars used, Arg-xylose being next in order of effectiveness, and then Lys with any of the three sugars. Using the peroxide value of linoleic acid, Yamaguchi et al.[473] found MRPs from xylose to be increasingly effective in the order glycine, lysine, arginine, histidine; for glucose, van Chuyen et al.[474] found the order to be glycine, arginine, lysine, histidine. In rats in vivo, these MRPs also had an antioxidant effect, as shown by liver TBA values, which gave the order glycine, lysine, arginine, histidine.

Glucose and amino acid (10 mmol of each) were refluxed in water (10 mL) and maximum colour formation (A_{450}) was reached in 15 h, when reducing ability was determined by a potassium ferricyanide method,[475] with the following approximate results: GABA, 6.0; Val, 4.1; Ser, 2.9; Ala, 2.7; Trp, 1.6; Gln, 1.1; Hypro, 1.0; Pro, 0.8; Asp, 0.6; Glu, 0.3 μg of ascorbic acid per 50 μg MRP. As a consequence, the product from GABA was separated by ethyl-acetate extraction, fractionation on Sep-pak SIL, and HPLC, and a compound isolated in 0.35% yield. Its structure was shown to correspond to 4-(2-formyl-5-hydroxymethyl-1-pyrrolyl)butanoic acid. However, its reducing ability is low (2.0 μg of ascorbic acid per 50 μg MRP), the reducing power being attributed mainly to 2,3-dihydro-5,6-dihydroxypyran-4-one (49.6 μg of ascorbic acid per 50 μg MRP), which reached its maximum concentration in 30 min. HDMF (39.2 μg of ascorbic acid per 50 μg MRP) was present in much lower proportions.

Under water-free conditions (170 °C, 20 min, not dialysed), Arg produced a stronger oxygen-scavenging effect than Gly, both in the presence of glucose and xylose. The same xylose-Arg product at 0.02 and 1% gave induction factors in the 'Rancimat' with sesame oil of 1.9 and 2.29, respectively, whereas xylose-Gly similarly had given 1.56 and 1.79,[466] implying the possibility of doubling shelf life. In contrast, 0.5% melanoidin from glucose-Arg, glucose-Gly, and maltose-Gly with lard gave induction factors of 1.05, 1.26, and 1.33, respectively.[81]

For a number of carbohydrate–Gly melanoidins (170 °C, 20 min, dialysed), free-radical content (EPR) and oxygen scavenging ran parallel,[81] except for maltotriose, where oxygen scavenging was greater than expected on the basis of such a relationship. Yet, the overall implication is that the oxygen-consuming properties are based on a radical mechanism.

In summary, one has to agree with the conclusion of Cämmerer et al.[466] that it is not possible to establish an overall ranking of amino acids according to antioxidative

Other Consequences of Technological Significance 131

activity of the derived melanoidins, because combination with different carbonyl compounds leads to different effects.[476,477]

4.3 Molar Ratio of Sugar to Amino Acid

Lingnert and Eriksson[469] examined MRPs from glucose and histidine in more detail, using each reagent at three concentrations; 2.5, 5.0, and 10.0 mM. As glucose concentration was increased, both colour and the antioxidative effect increased. On the other hand, increasing histidine concentration had no effect on the colour, but the antioxidative effect increased more sharply.

Waller *et al.*[478] fixed the total concentration of xylose plus arginine at 3.0 M, but varied the ratio from 0.25:2.75 to 2.75:0.25 (reflux, 20 h). Assessed at a dilution of 1:100, the equimolar mixture clearly gave rise to the highest antioxidative activity, which decreased symmetrically, as the ratio was changed in either directions. Even changing the ratio to 1:2 more than halved the activity. The high concentration for preparing the MRP needs to be noted.

4.4 Other Reaction Conditions

Other reaction conditions also greatly affect the antioxidative activity produced and are dependent to a certain extent on the nature of the assay employed. Cämmerer *et al.*[466] have summarised the situation by stating that, in general, higher initial pH favours the formation of melanoidins with higher antioxidative activity, the strongest antioxidative activity being obtained for the melanoidins formed in neutral or slightly basic media. Antioxidative activity also usually increases with temperature and time up to an optimum, and lower a_w seems to favour the formation of melanoidins with higher antioxidative activity.

4.5 Use of Peptides

Lingnert and Eriksson[469] compared the effectiveness in lowering hexanal production from linoleic acid of MRPs prepared from (a) 1 mM glycylhistidine, (b) 1 mM histidylglycine, (c) 1 mM glycine plus 1mM histidine, (d) 2 mM glycine, and (e) 2 mM histidine, each with 2 mM xylose. Although (d) was somewhat pro-oxidative, (a) and (c) were very effective, but, even so, they were exceeded by (b) and (e). That (e) was best is not too surprising in view of what has been said above, but it is noteworthy that (b), containing only half the amount of histidine, was equally good. Two additional points need to be made: peptides have been shown to be capable of leading to MRPs with antioxidative power and the nature of the amino acid providing the N-terminus seems to be significant.

4.6 Use of Hydrolysates

Can protein hydrolysates produce useful MRPs? Protein hydrolysates themselves already have some antioxidative activity and Lingnert and Eriksson[469] found this too, but the antioxidative effect was improved considerably on reaction with glucose.

They tested hydrolysates derived from brewer's grains, malt, and haemoglobin; all formed antioxidative MRPs. Compared with the MRP from glucose and histidine, higher concentrations were required to achieve an equal effect; however, conditions of hydrolysis and reaction with glucose had not been optimised.

Obretenov et al.[479] went a step further and used MRPs produced from protein hydrolysates (blood, sunflower grits) and starch hydrolysates. The strongest antioxidative effect on the induction period of lard, as determined by peroxide value, was obtained with a proportion of hydrolysates of 2:1 (blood/starch). The hydrolysate from blood was about twice as effective as that from sunflower grits.

Soy sauce has long been known to have strong antioxidant and free-radical-scavenging activity. To elucidate the position, Moon et al.[480] examined 29 different commercial samples obtained in Southeast Asia, using colour, nitrogen content, 3-DG content, FRAP, and TEAC. Large differences were found. Usually, white or light-coloured samples had low activity regardless of high nitrogen content, whereas sweet, dark-coloured ones, containing low nitrogen levels, showed a higher effect. FRAP correlated with brown colour ($r^2 = 0.963$) and 3-DG content ($r^2 = 0.970$), but not with nitrogen level. FRAP and TEAC were also highly correlated ($r^2 = 0.962$). Viscosity ranged very widely, from 3 to 6400 cP.

4.7 Amino Acid plus Reducing Sugar Versus MRP

Lingnert and Eriksson[469] compared doughs containing 0.1% histidine monohydrochloride monohydrate and 1% glucose with others containing 0.1% (based on the reactants) of the corresponding MRP with regard to stability of the cookies baked from them. No significant antioxidant effect was observed from the MRP either sensorily or through hexanal determination, whereas the addition of the reactants produced considerable effects, exceeding those from 1.6 ppm BHA/BHT (1:1). In consequence, they raised the question of the concentration of *antioxidative* MRPs as distinct from total MRPs.

Lingnert and Hall[481] followed up these experiments by measuring the antioxidative effect polarographically. Addition to cookie dough of histidine plus xylose or MRP from histidine plus glucose or arginine plus xylose produced not only strong antioxidative effects, but also considerable colour formation, whereas arginine plus xylose as reactants gave almost as strong an antioxidative effect with much less colour formation. Histidine plus glucose as reactants had almost no effect.

4.8 Relationship between Colour and Antioxidant Activity

There does not seem to be a clear relationship between colour formation and antioxidant activity. In their review, Cämmerer et al.[466] concluded that, although a strong positive correlation has been found in some work, this is not so in other studies. It seems that too many different compounds are involved in the progression of the Maillard reaction to give rise to a simple relationship between colour and antioxidative activity.

Brightly coloured pigments are formed early in model systems, and Murakami et al.[482] have shown clearly that formation of the blue colour in xylose–glycine mixtures, heated at 30 °C for 48 h, has a higher molecular mass on gel filtration (0.68 kDa) than the concomitant radical-scavenging activity (0.39 kDa). When heated at 100 °C, the radical-scavenging activity follows browning at least for the first 8 h, with a correlation coefficient of 0.914, both types of activity being centred on 7.8 kDa.

4.9 Heterocyclic MRP

Yanagimoto et al.[458] examined the antioxidative activity of heterocyclic MRPs in inhibiting the oxidation of hexanal by air. Of the three pyrroles, three furans, two thiazoles, three thiophenes, and two pyrazines tested, pyrrole-2-carboxaldehyde was by far the most active, closely parallelling BHT in giving about 100% inhibition at 50 μg mL^{-1} and about 90% at 5 μg mL^{-1}.

4.10 Pro-oxidant Activity

Since the crocin-bleaching method is based on competition kinetics, it can also be used to detect pro-oxidant activity (POA), for example, of early MRPs; the DPPH method cannot. In fact, in the presence of antioxidants, the crocin-bleaching rate [reaction (1)] is slowed down, because the antioxidant reacts with the radical first and the antioxidant radical formed [reaction (2)] reacts only slowly with the crocin [reaction (3)]. On the contrary, the pro-oxidant competes with the radical for the crocin [reaction (4)], thus increasing the extent of crocin bleaching:[446]

$$\text{ROO}\bullet + \text{crocin} \xrightarrow{\text{fast}} \text{ROOH} + \text{crocin}\bullet \text{ (bleached)} \quad (1)$$

$$\text{ROO}\bullet + \text{antioxidant} \longrightarrow \text{ROOH} + \text{antioxidant}\bullet \quad (2)$$

$$\text{Antioxidant}\bullet + \text{crocin} \xrightarrow{\text{slow}} \text{antioxidant} + \text{crocin}\bullet \text{ (bleached)} \quad (3)$$

$$\text{Pro-oxidant}\bullet + \text{crocin} \longrightarrow \text{pro-oxidant} + \text{crocin}\bullet \text{ (bleached)} \quad (4)$$

Manzocco et al.[483] compared the redox potential and the POA of some potent oxidants with those of some foods (milk, bread). For the POA, hydroxy radicals generated from H_2O_2, peroxy radicals from ABAP, and DPPH$\bullet$, as well as milk and bread, were allowed to react with crocin in aqueous solution at 40 °C and the POA was taken as the ratio of the decrease in A_{crocin} at 5 min to oxidant concentration:

	redox potential (Ag/AgCl)	pH	POA
H_2O_2	320 ± 2	7	0.030 ± 0.002
DPPH$\bullet$	229 ± 3	6.9	35.167 ± 1.979
ABAP	216 ± 2	7.5	96.138 ± 0.644
milk	273 ± 6	6.8	0.044 ± 0.008
bread	150 ± 3	7	0.011 ± 0.001
crocin	120 ± 5	7	

The difference between milk and bread was attributed to the presence of early MRP in milk, but melanoidins in bread.

Pro-oxidant conditions are favoured in infant formulae by the presence of iron and of vitamin C and can lead to oxidative damage to tryptophan residues, which here is of particular importance, tryptophan often being the limiting amino acid. Using α-lactalbumin as a model compound, as it is high in tryptophan, Puscasu and Birlouez-Aragon[484] studied the loss of fluorescence due to tryptophan ($\lambda_{ex}=290/\lambda_{em}=340$ nm) on incubation with lactose, preformed early and advanced MRP (from proteose-peptone, because it is low in tryptophan), H_2O_2/Fe^{2+}, or ascorbate/Fe^{3+}. In each case, after 3 h, there was an appreciable loss of tryptophan from the pH 4.6-soluble protein of about 28%. The MRPs, both formed and preformed, exhibited fluorescence at $\lambda_{ex}=350/\lambda_{em}=435–440$ (major) and $\lambda_{ex}=330/\lambda_{em}=420$ nm.

4.11 Properties of the MRPs

4.11.1 Molecular Mass

Yamaguchi et al.[473] found the greatest antioxidative effect in a fraction of a xylose–glycine product of about 4.5 kDa. Its antioxidative effect on an equal weight basis was greater than that of BHA, but less than that of BHT. There was a strong synergistic effect between MRP and BHA. Lingnert et al.[485] showed that most of the activity of a glucose–histidine product remained in the retentate (cutoff about 1 kDa), which exhibited an antioxidative effect six times that of the crude reaction mixture. In the reaction mixture, maximum activity occurred after heating for 20 h, in parallel with maximum colour formation,[469] suggesting that activity decreased when molecular mass reaches too high a value. Cämmerer et al.[466] stated that the scavenging ability of low-molecular-mass (LMM) melanoidins (0.5–1 kDa) is rather weak and therefore antioxidative activity appears not to be related to reductone content.

Tressl et al.[76,77] have isolated some of the oligomers formed in their model systems, based mainly on pyrroles (N-methyl and N-methyl-2-formyl), and have subjected them to tests for antioxidant activity (DPPH and Fe^{3+} thiocyanate). Although activity was found, its occurrence seemed somewhat haphazard.

The formation of the pronyl derivative of lysine **67** has already been mentioned in Chapter 8 (see Formulae 8.1). Such compounds have strong reducing action, as shown by Lindenmeier et al.[356] On heating N^α-acetyllysine methyl ester at 100 °C for 25 min in phosphate buffer, (pH 5.5), with starch, glucose, 3-deoxyosone, 1-deoxyosone, or diacetylformoin, the antioxidant activity developed from approximately 0, 0, 0.5, 2.7, or 1.4 to 0, 0, 1.5, 5.4, or 6.3 relative Trolox equivalents (RTE), respectively. The last reaction mixture was resolved into 26 fractions, by RP-HPLC the highest antioxidant activity residing in Fraction 14 (4.5 RTE), all other fractions exhibiting 1.5 RTE or less. The compound responsible for the antioxidant activity in Fraction 14 was shown to be pronyl-N^α-acetyllysine methyl ester by NMR, double-quantum-filtered δ,δ-correlation spectroscopy (DQF-COSY), LC-MS and UV-vis spectroscopy. Pronylglycine methyl ester was prepared similarly and was obtained as yellow crystals with $\lambda_{max}=363$ nm, molecular mass (MM)=0.215 kDa. The relative antioxidant activity of the two pronyl derivatives and ascorbic acid was 0.53,

0.49, and ~0.1 RTE, respectively. Pronyl groups could not be liberated by acid or enzymic hydrolysis, but by cleaving and reaction with methylhydrazine were converted into 5-acetyl-4-hydroxy-1,3-dimethylpyrazole, which could be determined by GC-MS after relatively simple clean-up. In this way, wheat gluten heated with starch or glucose for 1 h at 220 °C or with diacetylformoin for 30 min at 150 °C was shown to contain 4.0, 18.9, or 7100 mg kg^{-1} pronyllysine, respectively. Bread crust contained about eight times the number of pronyl groups than bread crumb. Sequential extraction of bread crust with water, 60% ethanol, and 50% 2-propanol, followed by freeze drying each fraction, as well as the residue, gave yields of 20.7, 2.0, 0.2, and 77.2 g 100 g^{-1}, respectively, the pronyllysine content of each being 30.6, 169.3, 42.9, and 88.5 mg kg^{-1}, respectively, and the antioxidant power of the first three being about 0.44, 1.36, and 0.96 RTE, respectively. When the above 60% ethanolic extract of bread crumb was fractionated using sequential membranes with cut-offs of 100, 30, 10, and 1 kDa, the fractions contained about 1, 23, 25, 52.3, and 35.2 mg kg^{-1} pronyllysine and had antioxidant activity of about 0.22, 0.4, 0.7, 0.9, and 0.74 RTE, respectively, showing that the majority of the active components were of lower molecular mass.

4.11.2 Elementary Composition

For the glucose-histidine MRP, purified by electrophoresis, Lingnert et al.[485] obtained:

Found: C, 54.3; H, 5.4; N, 12.6; O, 24.5; ash, 2.9%.

For histidine plus glucose, calculated (-4 H$_2$O): C, 54.8; H, 4.9; N, 16.0; O, 24.3%.

For the xylose-arginine MRP obtained by Waller et al.,[478] the corresponding figures are:

Found: C, 37.1; H, 7.2; N, 20.7; O, 22.7%.
Calculated (-4 H$_2$O): C, 52.4; H, 6.3; N, 22.2; O, 19.0%.
(-3 H$_2$O):   C, 48.9; H, 6.7; N, 20.7; O, 23.7%.

Yamaguchi et al.[473] reported analyses for five purified xylose melanoidins, which are given below, alongside calculated values for appropriate dehydrated products from the starting materials:

For xylose-glycine:
Found: C, 50.2; H, 5.3; N, 7.0; O, 37.5%.
Calculated (-3 H$_2$O): C, 49.1; H, 5.3; N, 8.2; O, 37.4%.

For xylose-lysine:
Found: C, 47.4; H, 5.2; N, 6.4, O, 41.0%.
Calculated ($-$H$_2$O):   C, 47.5; H, 7.9; N, 10.1; O, 34.5%.

For xylose-NH$_3$:
Found: C, 46.8; H, 5.4; N, 11.6; O, 36.2%.
Calculated (-2 H$_2$O): C, 45.8; H, 6.9; N, 10.7; O, 36.6%.

For xylose-arginine:
Found: C, 44.4; H, 5.8; N, 12.8; O, 37.0%.
Calculated ($-$H$_2$O):   C, 43.1; H, 7.2; N, 18.3; O, 31.4%.

For xylose-histidine:
Found: C, 47.3; H, 4.6; N, 11.5; O, 36.6%.
Calculated ($-$H$_2$O):   C, 46.0; H, 5.9; N, 14.6; O, 33.5%.

The results fall into two groups, those with about 6% N and those with about twice that amount. The latter tend to be superior antioxidants, but the relationship is only approximate. The result with glycine corresponds to that of Benzing-Purdie et al.[69] (see Chapter 2), whereas the data of Lingnert et al.[485] for the glucose product correspond, as far as %C is concerned, more nearly to the loss of a further three molecules of water. The ammonia product approximated quite well to a loss of 2 H_2O and the composition of the remaining three melanoidins is nearest to that of the corresponding glycosylamine, but with lower nitrogen content.

There is still much to learn here.

4.11.3 Conditions during Formation

The higher the a_w, the more quickly does the reducing power increase for glucose–lysine systems,[486] although reducing power is produced under conditions where browning is not yet occurring, i.e., it is attributable to Amadori compounds. The highest reducing power was seen at an a_w of 0.52.

Lingnert and Eriksson[469] obtained the highest antioxidant effect at an initial pH of 7–9 for glucose–histidine systems, pH 5 not being far behind and colour being at a maximum at about pH 5. For xylose–arginine systems,[478] antioxidative activity was highest at initial pH 5, exceeding that at pH 7 to some extent.

All the organic additives investigated by Waller et al.[478] with the xylose-arginine system had detrimental effects on antioxidative activity, except for pyridine, which doubled it.

4.11.4 Standard Melanoidin

Brand and Eichner[487] prepared Solution A and Standard Melanoidin (see Chapter 2) under slightly modified conditions and studied their properties. The reducing power was determined by the ferricyanide method, radical scavenging by DPPH• decolorisation, and antioxidant activity in relation to the decrease in methyl linoleate. In relation to 1 mol of glucose applied, Solution A reduced about eight times the mol ferricyanide that Standard Melanoidin did. When the melanoidin was produced by dialysis at room temperature rather than 4 °C, the amount of ferrocyanide produced dropped to one-third, but dialysis at room temperature, under nitrogen, caused a drop to only two-thirds. Similar results were obtained with radical scavenging: Solution A scavenged six times the radicals that Standard Melanoidin did, and dialysis under different conditions had effects similar to those on ferrocyanide production. With regard to methyl linoleate oxidation, the results were rather different (the amounts applied are expressed in mol reducing equivalent (RE) per 100 mol methyl linoleate). Increased amounts of Solution A (28 as compared with 7.0 mol RE per 100 mol) gave increased protection, but, remarkably, Standard Melanoidin (3.5 RE per 100 mol) had a greater protective effect than Solution A (7.0 mol RE per 100 mol). In addition, more surprisingly still, a lower amount of Standard Melanoidin (0.87 mol RE per 100 mol) actually produced a pro-oxidative effect.

Wagner et al.[488] examined four fractions from the standard glu–Gly reaction: Solution A, LMM melanoidin and high-molecular-mass (HMM) melanoidin (12.4 kDa cut-off), and the insolubles (IS). In the ABTS test, % inhibition of A_{734} in terms of Trolox concentration, HMM melanoidin was most effective, followed by the IS and Solution A. LMM melanoidin was least effective, but even it was progressively more effective at 0.01 < 0.05 < 0.10%; 0.10% HMM was about four times as effective as 0.01%. These results agree with those obtained by Yoshimura et al.,[489] who found that HMM melanoidin gave the highest inhibition of OH• formation, not only due to direct scavenging, but also due to stronger chelating activity, which is relevant to the ABTS method. In orange juice, IS gave the greatest effect, followed by LMM melanoidin, Solution A and HMM melanoidin having the least, roughly equal effect. The effects in apple juice were similar, as they were in grape juice, but here HMM melanoidin showed increased activity.

4.11.5 Mechanisms of Antioxidant Activity

At present, little is clear, except that the antioxidative effect is related to reducing power. However, with respect to the termination of free-radical chain reactions, compounds capable of forming stable free radicals are significant. Indeed, Lingnert et al.[485] obtained a strong ESR signal from their purified histidine-glucose MRP at a g factor of 2.0035 ± 0.0003. The increasing, polarographically determined antioxidative effect in the sequence, crude reaction mixture, retentate, and precipitate, was accompanied by an increased ESR signal. The supernatant was less antioxidative and gave a less intense ESR signal. When the retentate was incubated for 75 h in air, both effects decreased; both increased under nitrogen. Although the two effects seemed to change in parallel, they were not directly proportional to each other.

Lessig and Baltes[185] reported stable free radicals to be present in an analogous product from glucose-4-chloroaniline. Reaction H of Scheme 1.1 (see also Chapter 2) is indeed based on free radicals.

Hydroxyl and hydrogen radicals can be quantitatively detected by ESR as spin adducts (DMPO or PBN).[490,491] The scavenging ability of melanoidins for hydroxyl radicals is much higher than that of known scavengers, such as fructose, mannitol, or BSA.[490] At concentrations of 0.3 and 0.03%, glucose–glycine melanoidins scavenge 86 and 47% of hydroxyl radicals and 85 and 58% of hydrogen radicals formed by γ-irradiation.[489]

Superoxide radicals are another factor in oxidative damage. They can be determined with nitrobluetctrazolium (NBT), which then forms the colourless formazan. When melanoidins scavenge the superoxide radicals, the colour of the NBT persists.[490,491] The activity of a glucose–glycine melanoidin on superoxide radicals is equivalent to the effect of 16 units of superoxide dismutase. The effect of the HMM and LMM fractions of this melanoidin is almost the same. The reaction rate constant of the melanoidin was markedly higher than that of ascorbic acid. If this were due to the reductone structures embedded in the melanodin, it is difficult to explain why the reducing power of the melanoidins is only 0.7 that of ascorbic acid.[490]

Peroxy radicals are known to be responsible for the bleaching of carotenoid pigments, such as crocin. Such bleaching can be inhibited by antioxidants.[468,492,493] In determining the residual radical scavenging activity of tomato juice, crocin bleaching and the DPPH method gave parallel results.[446]

Antiradical efficiency (AE) can be determined using N,N-dimethyl-p-phenylenediamine dihydrochloride (DMPD). If the amount of melanoidin (mg mL^{-1}) required to reduce the concentration of DMPD cation radicals by 50% is EC$_{50}$ and the theoretical time (min) required for EC$_{50}$ to reach the steady state is TEC$_{50}$, then

$$AE = 1/(EC_{50} \times TEC_{50})$$

Morales and Babbel[494] examined the melanoidins (cut-off 110 kDa) from 12 systems, containing either glucose (G) or lactose (L) with glycine, histidine, lysine, tryptophan, cysteine, or methionine. The EC$_{50}$ values ranged from 0.43 (G–Gly) to 1.74 mg mL^{-1} (G–Trp) and the AE from 0.010 (G–Trp) to 0.087 (L–His). Melanoidins from medium-roast coffee powder similarly gave 2.06 and 0.015 mg mL^{-1}, respectively, whereas ferulic acid and Trolox gave 0.0186 and 0.0781 mg mL^{-1} and 2.67 and 0.34 mg mL^{-1}, respectively. These results show that the radical-scavenging ability of melanoidins is about 10 to 100 times less than that of phenols, with that of coffee melanoidins tending to lie even lower.

Using TEAC (ABTS$^{\bullet+}$ radical cation scavenging), which can be applied in aqueous as well as lipid systems, Nicoli et al.[493] found that, depending on the degree of roast, coffee brews had an effect only about 1/20th of that of α-tocopherol. Maltose–glycine melanoidin managed only 1/50th of the effect of α-tocopherol.[466] The redox potential of ready-to-drink coffee brew[495] decreased progressively as the coffee was given a light, medium, or dark roast from +109 to −35 mV, the antioxidative power thus increasing. Chain-breaking activity as measured by DPPH$^{\bullet}$ bleaching decreased at first and then increased. It was affected less by the degree of roast than the redox potential. Whereas storage of the dark-roast brew for 17 d at 30 °C under nitrogen left the redox potential unchanged, storage in air raised it to different degrees.

Richelle et al.[496] compared the capacity to delay in vitro LDL oxidation of cups of beverages, as normally prepared. The lag times were as follows: 0.7–2.5% soluble coffee, 292–948; 1.5–3.5% cocoa, 217–444; 1 tea bag per 220 ml of green tea, 186–338; of black tea, 67–277; and of herbal tea, 6–78 min. Addition of milk did not alter the antioxidant effect.

The presence of metal ions can have a confounding effect. Bersuder et al.[497] investigated the antioxidant effect of a heated (105 °C, 10 h, initial pH 7.0) mixture of glucose and histidine (100:33.3 mM) on a sunflower seed oil emulsion, both by Rancimat and DPPH, in the absence and presence of Cu^{2+}. The addition of copper to the MRP increased the protective factor (PF) by Rancimat by 31%, but decreased the disappearance of the DPPH radical by 13%. Histidine, heated or unheated, and propyl gallate showed the opposite behaviour, decreasing the PF, but increasing the disappearance of DPPH. It could be that the MRP reduced the pro-oxidant activity of the copper, whereas, particularly for propyl gallate, chelating with copper may have resulted in an increased ability to donate hydrogen to the DPPH radical.

Eichner[486] reported the results of an interesting experiment. By interacting a sodium linoleate solution, already containing 10 mol% hydroperoxide, with a preheated glucose–lysine system, he showed that the proportion of hexanal formed was reduced from 0.35 to 0.09 mol mol^{-1} peroxide decomposed. Concomitantly, hydroxylinoleic acid was formed from the peroxide by reduction; it does not give rise to rancid products, the production of which is thus reduced.

It is not surprising that honey-lysine Maillard products have an antioxidant effect.[498] In turkey meat, increased levels of this MRP or of honey increase the antioxidant effect, but surprisingly the honey is the more effective, possibly because it gives a better dispersion or solubilisation of any MRP formed in the meat.

4.11.6 Effects on Other Food Components

Most food colours are sensitive to reduction to some degree. The effect of thermally degraded fructose and glucose on the azo dyes, FD & C Red Nos. 2 and 40 (**102** and **103**, respectively), has been investigated by Ross.[499] When treated with 0.1 M fructose in 1 M phosphate buffer (pH 7.16), the half-lives of 6.73×10^{-5} M Red 2 under nitrogen and in air and of Red 40 under nitrogen were 93, not determined, and 135 min at 61 °C; 60, 12, and 38 min at 77 °C; and, 54, 5, and 31 min at 100 °C, respectively. At room temperature, there was no loss of colour, even in 40 d. At 37 °C, Red 2 under nitrogen had $\tau_{1/2} = 19$ d. Storing the fructose solution for 2 d under refrigeration before the experiment, reduced $\tau_{1/2}$ by a factor of more than 2, but heating the fructose solution at 100 °C for 1 h in air had little effect on $\tau_{1/2}$. Parallel experiments with Red 2 and glucose produced no loss of colour in 28 d under nitrogen or in air, but at 100 °C under nitrogen, $\tau_{1/2} = 120$ min. Fructose is clearly the stronger reducing agent under these conditions than glucose. The effect of added amino acid would be expected to be considerable.

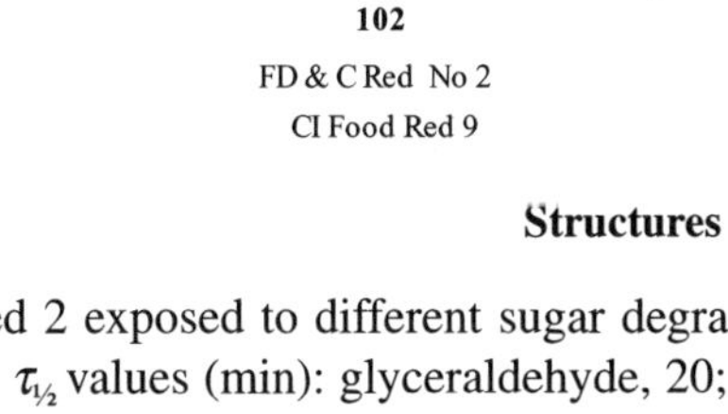

102
FD & C Red No 2
CI Food Red 9

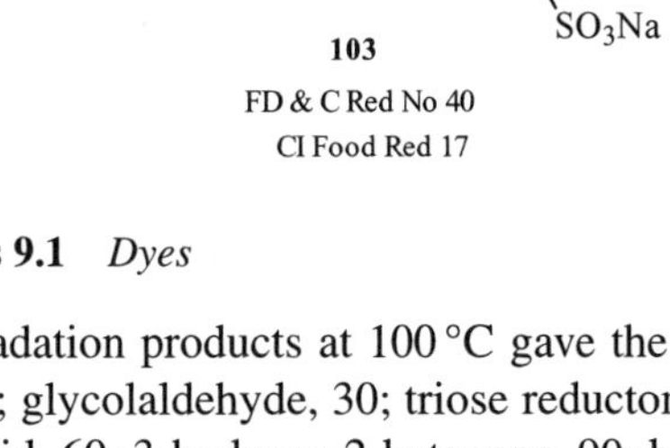

103
FD & C Red No 40
CI Food Red 17

Structures 9.1 *Dyes*

Red 2 exposed to different sugar degradation products at 100 °C gave the following $\tau_{1/2}$ values (min): glyceraldehyde, 20; glycolaldehyde, 30; triose reductone, 40; dihydroxyacetone, 50; 2-oxopropanoic acid, 60; 3-hydroxy-2-butanone, 90; butanedione, 300; HMF, 450; 2-oxopropanal, 1200; acetaldehyde, acrolein, and laevulinic acid, no reaction.

4.11.7 Volatiles from the Maillard Reaction

Heterocyclic volatiles have been examined for antioxidant activity through inhibition of the oxidation of hexanal to hexanoic acid.[500] Pyrroles had the most effect, all those studied acting as inhibitors (at 50 μg mL^{-1} over 40 d). Formyl and acetyl substituents enhanced the effect, *e.g.*, pyrrole-2-carboxaldehyde giving > 80% inhibition at 10 μg mL^{-1}. Of furans, the unsubstituted had the greatest effect (80% inhibition at 500 μg mL^{-1} over 40 d). With thiophenes, methyl and ethyl groups enhanced inhibition, but formyl and acetyl substituents lowered it. Thiazoles and pyrazines had no effect.

4.11.8 Voice of Practice

The final word on reducing power should perhaps remain with practice rather than theory. In Hungarian culinary practice, roux is prepared by frying flour in oil. Dworschák and Szabó[501] showed that considerable antioxidant activity is produced in this way, an activity which however drops greatly in 5 days. Greater activity was obtained on heating sucrose to make caramel or by frying potatoes and in both these cases the activity tended to increase on storage for 5 days. What else do the chefs know that cannot be interpreted as yet?

5 Effect on Solubility

The solubility of shellfish muscle protein can be improved by reaction with glucose, solubility in 0.1 M sodium chloride reaching 83% when > 60% of lysine residues have been modified, but decreasing progressively once > 80% of residues have reacted.[502]

6 Effect on Texture

The formation of Maillard products can improve the gel properties of dried egg white. Thus, heating dried egg white with a galactomannan (4:1, w/w) at 60 °C and 65% r.h. produced covalent bonds, as confirmed by SDS/PAGE. Gel strength and water-holding capacity increased compared with dried egg dry-heated without galactomannan, and reached a maximum after 3 days heating. The gels also became translucent. The galactomannan had been obtained from guar gum by hydrolysis with mannase.[503]

Model studies on ribonuclease showed glutaraldehyde to crosslink it almost instantaneously, lysine being no longer measurable, whereas both glyceraldehyde and formaldehyde crosslinked at a lower rate, only 40% of lysine becoming no longer available after 5 h at 37 °C.[504] These results were corroborated by experiments with wheat-protein fractions and the same aldehydes, glutaraldehyde reacting very quickly and the other two more slowly. Testing the reagents on bread and croissant dough, only glutaraldehyde at 200 mM and above was found to have a significant effect, modifying the dough properties and increasing the crumb strength of loaves through crosslinking the combined albumin and globulin fractions. There was no effect on croissant dough.

Pentosidine **70**, a lysine–lysine crosslink, has been found in foods, as already mentioned in Chapter 8.[354] Such crosslinks have potential significance for texture. The levels in food ranged from 'not detectable' to 2–5 mg kg^{-1} protein for sterilized and evaporated milk and up to 35 mg kg^{-1} protein for some bakery products and

7 Effect on Foamability and Foam Stability

Espresso coffee is visually characterised by a foamy layer on the top of the beverage. Petracco[505] has examined a hot-water extract of defatted roasted coffee powder after saturating it with ammonium sulfate, freeze-drying the precipitate, and exhaustively dialysing it (4.3% yield). When redissolved in water, it foamed well on shaking, giving a foam volume of 110%. A 0.4% (w/v) aqueous solution has a surface tension of 52.1 mN m^{-1}. Similar results were obtained on two samples of Brazilian Santos arabica and on robustas from Java and Uganda. Subfractionation by precipitation with isopropanol led to Subfractions A (about 60%, MM 34 kDa) and B (about 40%, MM 17 kDa) with foam volumes of 50 and 100% and surface tensions of 60.0 and 46.5 mN m^{-1}, respectively. A is designated as polysaccharide in nature, whereas B is melanoidin-like. The latter provides foamability, whereas the former stabilises the foam (the foam decrease in 24 h, is about 50 and 0%, respectively).

8 Effect on Emulsifying Power

Emulsion activity (EA) of ribonuclease A (RNase A, 10 mg mL^{-1}) has been shown to be enhanced by glycation with glucose-6-phosphate (G6P), using capillary electrophoresis to follow the changes in RNase A.[506] A cluster of about 20 peaks formed and migrated more slowly than RNase A monomer. The area of the cluster increased with temperature and with G6P concentration. EA decreased with the time of incubation in the absence of G6P. In its presence, EA reached a maximum with time of incubation, being highest for 60 mM G6P after 96 h at 30 °C and 18–24 h at 40 °C.

The emulsification properties of sodium caseinate can be improved by mixing it (1:1, w/w) in water with apple pectin (6% methoxyl) for 1 h and freeze drying.[507] The solid was heated for 48 h at 60 °C, r.h. 79%. SDS/PAGE showed that the two components had been linked covalently. The product had greater emulsifying properties and emulsion stability than either of its components in relation to corn oil in phosphate buffer (pH 7.4). The same applied when compared with gum arabic and glycerol monostearate. It maintained its superior performance over the other emulsifiers in the presence of 0.2 M NaCl and on heating to 90 °C for 10 min. The same was true for citrate buffer (pH 4.0), except that gum arabic came out slightly ahead.

9 Volatile Formation on Storage

Volatile formation has been used widely as a means of following deterioration on storage. A recent example is the following.

Buglione and Lozano[508] compared three types of grape juice: two red, Merlot and Criolla, and one white, Yellow Muscat. Whereas colour, as assessed at 420 nm, deteriorated fastest in Merlot (about 50% increase in 20 weeks), HMF became detectable only after 12 weeks at 10–30 °C, building up most quickly in Criolla.

10 Binding of Volatiles

Comparative aroma dilution analyses of the headspace of aqueous solutions, containing either the total volatiles isolated from a fresh coffee brew or these volatiles mixed with the melanoidins isolated from coffee brew, revealed drastic losses of odorous thiols, 2-furfurylthiol, 3-methyl-2-butenethiol, 3-mercapto-3-methylbutyl formate, 2-methyl-3-furanthiol, and methanethiol, in the presence of melanoidins.[509] The first compound was affected most, the reduction being 16-fold, and was accompanied by an overall reduction in roasty-sulfury aroma. The rapid loss of thiols was confirmed by stable-isotope dilution analysis. [^{2}H]-NMR and LC-MS gave strong evidence that the thiols become covalently bound via Maillard-derived pyrazinium compounds.

Hofmann and Schieberle[509] also carried out model experiments employing synthetic 1,4-diethyl diquaternary pyrazinium ions with, for example, 2-furfurylthiol, showing that the primary products were 2-(2-furyl)methylthio-1,4-dihydropyrazine, bis[2-(2-furyl)methylthio-1,4-dihydropyrazine], and 2-(2-furyl)methylthiohydroxy-1,4-dihydropyrazine. This supports the interpretation of the binding of the thiols as being covalent to pyrazinium intermediates.

In earlier work with unfractionated coffee melanoidins, Hofmann *et al.*[510] had already shown significant reductions of thiols in the headspace above aqueous model systems on addition of melanoidins, whereas aldehydes remained unaffected.

11 Other Losses of Functionality

From an examination of RNase A activity following incubation with 2-oxopropanal with or without 3,5-dimethylpyrazolecarboxamidine (a known Maillard inhibitor), Miller and Gerrard[511] concluded that loss in activity is related to glycation rather than to crosslinking.

CHAPTER 10

Implications for Other Fields

Three fields will be treated briefly here: soil science, textiles, and pharmacology. Other fields will no doubt become relevant in due course.

1 Soil Science: Humic Substances

The relationship between humic substances and melanoidins has been reviewed by Ikan *et al.*,[15] who provide an entry to the earlier literature. Both humic substances and melanoidins present enigmas, their structures still being largely undefined. This makes comparing and contrasting them doubly difficult.

Humic substances seem to be formed primarily by the degradation of lignin, complicated by interactions with other compounds, such as proteins and amino acids, produced from plant litter. Maillard[536,537] already recognised that the condensation of sugars with amino acids, peptides, and proteins also plays a role in the formation of humic substances.

Only 0.05% (about 4×10^{19} g) of the Earth's carbon is not locked up in sedimentary rocks and only about 9% of that is organic.[538] The organic carbon is divided roughly among seawater (45%), soil (40%), and terrestrial plants (15%). Humic substances are traditionally considered to comprise three main fractions: humic acids that are soluble in alkali, but insoluble in acid; fulvic acids soluble both in alkali and in acid; and humin insoluble both in alkali and in acid.

A modification of the lignin theory of the formation of humic substances is the polyphenol theory, which considers the key interactions to be those between quinones, derived either from polyphenols or lignin, and amino compounds.

Humic substances are very refractory; consequently, ^{14}C-dating suggests long average ages, such as 30 years for Suwannee River fulvic acid, 100–500 years for soil fulvic acid, 700–1600 years for soil humic acid, and 100–2400 years for soil humin.

A considerable amount of work has been done on comparing free radicals in humic substances and melanoidins by ESR,[539] as well as by ^{13}C-CP/MAS NMR spectra and δ^{13}C and δ^{15}N values. These studies have not been able to define the role played by the melanoidins, but do not exclude their participation either.

Birnessite, a manganese dioxide commonly present in soils, has been found to be a very effective catalyst of the browning of solutions of glucose and glycine in light, but also promotes the interaction in the dark.[540] Such reactions may have a role in the formation of humic substances in the soil.

The effect of montmorillonite and kaolin, saturated with calcium, aluminium, or cupric ions, as well as quartz, on humic-like substances formed from glucose-tyrosine was examined by Arfaioli et al.[541] All systems promoted their formation, the effectiveness being strictly related to the amount of added cation. Humification appeared to be due more to the cations than to the type of clay mineral. The clayey systems gave more complex (aromatic) substances than the quartz ones. The cations seemed more effective when free, *i.e.,* associated with quartz rather than with the clays. The nature of the cation was also important, cupric being the most active here. In the end, all systems took on a deep dark colour.

Burdon[542] has surveyed the current hypotheses for the structure of humic substances and has concluded that the various products from chemical degradations and NMR data are all consistent with their being mixtures of plant and microbial materials and their microbial degradation products. The examination of soil carbohydrates, proteins, lipids, and aromatics supported this view; the presence of colour, fluorescence, ESR signals, mellitic acid, and other features do not contradict it. Regarding the Maillard reaction, some free monosaccharides and the necessary amino species are present in soil, so it may proceed, but only to a small extent; it is not a major process. However, in marine environments, the relative abundance of carbohydrates and proteins makes them more probable precursors of humic substances than lignin or polyphenols.

The presence of Maillard reaction products has been detected in decayed plant material from an archaeological site dating back to ancient Egypt.[543]

2 Textiles

Yellowing is a serious concern for the textile industry. A high proportion of cases falls into the category of yellowing during storage or in transport. The problem is naturally most prevalent with textiles that are white or have only a pastel shade. It has been stated that there are more than 20 possible causes of such yellowing.[544]

Wool can be modified by means of the Maillard reaction with reducing sugars,[545] but, not surprisingly, this is accompanied by browning.

3 Pharmacology

Dihydroxyacetone is used widely to generate an artificial sun-tan, as it is the most effective of the five hydroxy-carbonyl sugar degradation products compared.[546] The browning occurs without sunlight, but the pigment produced does provide some protection against sunlight.

Sphagnum wound dressings can be 3–4 times as absorbent as cotton equivalents, and are also capable of reacting with proteins.[547] This reactivity gives them the potential of immobilising whole bacterial cells, as well as the enzymes, exotoxins, and lysins secreted by pathogens. Once immobilised, the enzymes, and therefore probably also the exotoxins and lysins, are rapidly inactivated by a Maillard reaction. Sphagnan, a pectin-like polysaccharide, has been identified in the cell walls of *Sphagnum* mosses and has been shown to contain highly reactive α-ketocarboxy groups, eminently suitable for interaction with amino groups. The proportion of

Implications for Other Fields

firmly bound protein is strongly correlated with the protein's basicity, but even a relatively acidic protein, such as pepsin, is bound to a significant extent.

The Maillard reaction also plays a role in the decomposition of some pharmaceutical preparations on prolonged storage. Thus, tablets containing vigabatrin, 4-amino-5-hexenoic acid, an anticonvulsant, undergo yellowing on ageing, attributed to the presence of Avicel, a microcrystalline cellulose.[548] Lactose, often used as an excipient, is an active participant in Maillard reactions and so is responsible for the lower stability of some preparations containing hydrochlorothiazide.[549]

CHAPTER 11

Nonenzymic Browning Mainly Due to Ascorbic Acid

Ascorbic acid will brown on its own in aqueous solution above 98 °C, producing furfural and carbon dioxide. Even in the presence of glycine, the carbon dioxide comes essentially from ascorbic acid (*cf.* Strecker degradation in Chapter 2). The browning with ascorbic acid also increases with pH and, above pH 7, autoxidation and browning occur even at 25 °C. Other reductones will react similarly. Glucose and fructose decrease the rate of browning and the same is observed with amino acids initially, although later they increase it.

Ascorbic acid degradation can be depicted as in Scheme 11.1.

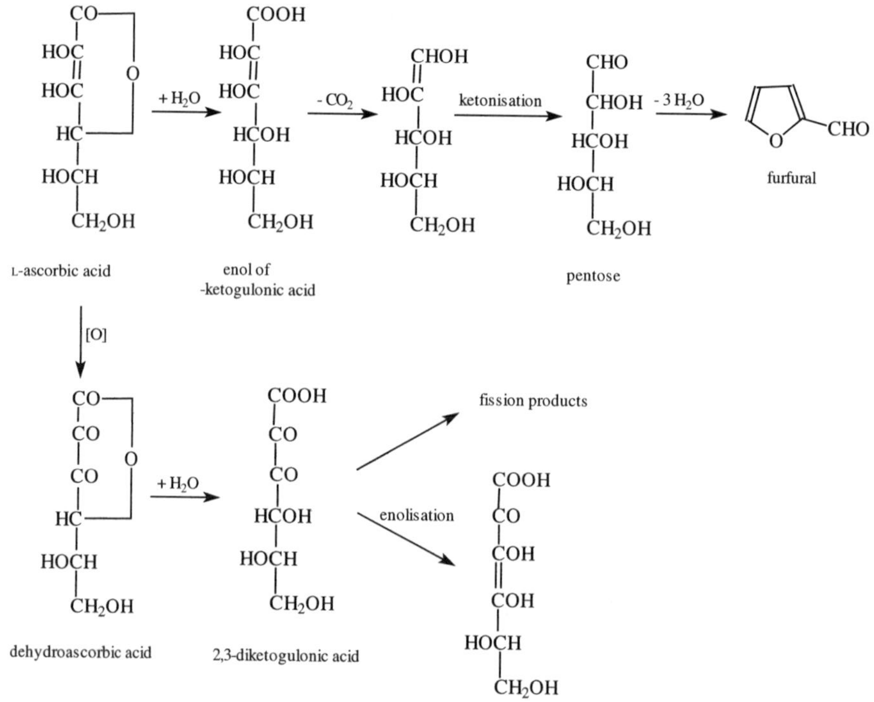

$$
\begin{array}{c}
\text{CO}\\
|\\
\text{HOC}\\
||\\
\text{HOC}\\
|\\
\text{HC}\\
|\\
\text{HOCH}\\
|\\
\text{CH}_2\text{OH}
\end{array}
\quad \xrightarrow{+\text{H}_2\text{O}}
\begin{array}{c}
\text{COOH}\\
|\\
\text{HOC}\\
||\\
\text{HOC}\\
|\\
\text{HCOH}\\
|\\
\text{HOCH}\\
|\\
\text{CH}_2\text{OH}
\end{array}
\quad \xrightarrow{-\text{CO}_2}
\begin{array}{c}
\text{CHOH}\\
||\\
\text{HOC}\\
|\\
\text{HCOH}\\
|\\
\text{HOCH}\\
|\\
\text{CH}_2\text{OH}
\end{array}
\quad \xrightarrow{\text{ketonisation}}
\begin{array}{c}
\text{CHO}\\
|\\
\text{CHOH}\\
|\\
\text{HCOH}\\
|\\
\text{HOCH}\\
|\\
\text{CH}_2\text{OH}
\end{array}
\quad \xrightarrow{-3\text{H}_2\text{O}}
\text{furfural-CHO}
$$

L-ascorbic acid → enol of -ketogulonic acid → → pentose → furfural

[O] ↓

dehydroascorbic acid → (+H₂O) → 2,3-diketogulonic acid → enolisation → fission products /

$$
\begin{array}{c}
\text{COOH}\\
|\\
\text{CO}\\
|\\
\text{COH}\\
||\\
\text{COH}\\
|\\
\text{HOCH}\\
|\\
\text{CH}_2\text{OH}
\end{array}
$$

Scheme 11.1

According to the scheme, ascorbic acid is readily converted into pentoses, but this does not explain the differences in behaviour between ascorbic acid and pentoses. 2,3-Diketogulonic acid (and the corresponding enol) is very unstable and readily develops brown coloration even at low temperatures.[550] It can also undergo oxidative fission.

At pH 7.0 and 37 °C, the degradation of ascorbic acid continues further, the main products being threose, glyceraldehyde, xylosone, and 3-deoxyxylosone.[551] Threose is more reactive compared with an aldopentose or an aldohexose. At pH 7.0 and 37 °C, it has a half-life of about 3.5 d. It seems probable that threose is a major factor in Maillard reactions involving ascorbic acid.

In very acid media, *e.g.*, lemon juice (pH 2.5), sugar–amine reactions are unlikely and browning occurs mainly by the degradation of ascorbic acid to highly reactive carbonyl compounds (*e.g.*, 3-deoxypentosone and 3,4-dideoxypentosulos-3-ene), which then interact with amines to give colour.[552,553] Clegg[554] found the browning of lemon juice to be proportional to the concentration of ascorbic acid, and to take place under aerobic conditions and not in a sealed vessel. This also applied to model systems. pH had a significant effect. Browning was maximal at pH 4.5 and increased further in the presence of citric acid. Glucose had no effect, but amino acids (hydrolysed casein) caused a further increase. The role played by organic acids, particularly citric, is important, but has not been explained, although the presence of citric acid was linked to the formation of a brown fluorescent pigment.[555] Furfural, although formed, was not active in colour development.

The thermal degradation and browning of ascorbic acid (AA) in orange juice at 20–45 °C under aerobic conditions have been modelled.[556] The formation of dehydroascorbic acid (DHAA) and pH were also monitored. Limited AA degradation could be described by first-order kinetics, but, when only small amounts of AA were retained, the kinetics became sigmoidal. The Weibull model was used to describe this pattern ($R^2_{adj} > 0.995$). The rate constant increased with temperature according to an Arrhenius-type relationship, with $E_A = 38.6$ kJ mol^{-1} (9.22 kcal mol^{-1}) and $k = 64.4 \times 10^{-3}$ h^{-1} at 32.5 °C. Prior to the maximum rate of degradation, pH, DHAA concentration, and browning remained fairly constant, but then increased. This was thought to be a result of the conversion of DHAA into AA following first-order kinetics in relation to DHAA and second-order in relation to AA, with different sensitivities to temperature.

On storage, orange juice gradually turns brown. The relationships between the main components involved were explored with model systems, consisting of sugars, AA, citric acid, and amino acids, stored at 50 °C for 2 months.[557] Browning was assessed at 420 nm. The solution turned brown gradually during storage, AA contributing most to the browning in the first 2 weeks, but sugars took over thereafter. AA had completely decomposed within 3 d. Arg and Pro promoted browning. Five products (3-hydroxy-2-pyrone (HP), HMF, furfural, 5-hydroxymaltol, and 2-furoic acid) were detected by HPLC. HP, furfural, and furoic acid were derived from AA and HMF from fructose. HP increased till day 3, but then decreases, while all the other compounds increased gradually during storage. A solution of HP stored on its own turned slightly brown. Furfural in the presence of amino acids turned yellow. Solutions of the other three substances did not brown, irrespective of the presence of amino acids.

The formation of HMF in blood-orange juice and model systems of fructose, glucose, and sucrose at different acid pHs and temperatures was studied by Arena et al.[558] Changes in the ultraviolet showed that the enediol was being formed and was slowly converted into HMF, giving pseudo first-order rate coefficients. Glucose degrades much more slowly than fructose. The reactivity of sucrose depends on pH and temperature, the activation energy being about 80 kJ mol^{-1} (20 kcal mol^{-1}) higher than that for fructose, the difference being ascribed to preliminary hydrolysis. The first-order rate coefficients for orange juice were similar to those for the model systems at comparable pH and temperature, indicating that sugar degradation occurs without the intervention of other compounds, thus excluding the Maillard reaction.

Obretenov et al.[559] prepared Standard Melanoidins, modelled on Ames,[85] using AA and glycine, lysine, and glutamic acid, respectively, and compared their behaviour in relation to the retention of isoamyl acetate with Standard Melanoidin prepared from glucose–glycine. Release was determined by assessing headspace concentration with 100 μm PDMS fibres. The last melanoidin tend to decrease release when isoamyl acetate was at 0.1 ppm, but increased release at 1 and 10 ppm. The melanoidin from AA and glycine behaved similarly, but that from AA and lysine did not. With 100 ppm melanoidin and 0.1 ppm isoamyl acetate, release was increased, whereas almost all runs with 1 ppm volatile caused a decrease and 10 ppm caused little difference from the control (water only). The melanoidin from AA and glutamic acid caused strong decreases in release at 0.1 and 1 ppm volatile, becoming even stronger with increased melanoidin concentration. At 10 ppm volatile, the melanoidin still tended to decrease release, but not very greatly. Overall, the amino acid used to produce the melanoidin seems to have had more effect than changing the reactant from glucose to AA. Nonpolar amino acids were thought to produce melanoidins of low polarity that have a solvating effect on volatiles, whereas lysine and glutamic acid produce polar melanoidins that exert dipole–dipole and dipole–ion interactions with isoamyl acetate, leading to higher retention.

AA and aspartame are often present simultaneously in soft drinks. When heated to 90 °C, the AA is oxidised to DHAA and condenses with aspartame to form the *N*-2-pyron-3-yl derivative (*cf.* ref. 557), which is strongly bitter and causes an off-flavour.[560] On storing solutions containing AA + aspartame and DHAA + aspartame (0.5% of each) at 5 and 37 °C for up to 150 d, the pyrone was formed in both solutions at 37 °C, but only in the latter at 5 °C. Intensity of bitterness correlated nearly linearly with browning. The highest amount of pyrone (>1 g L^{-1}) had been formed in the latter solution at 37 °C at 60 d.

Impairment of AA homeostasis in diabetic humans and animals seems to be linked to the pathogenesis of diabetic complications, but AA degradation is difficult to disentangle from that of sugars. Nishikawa et al.[561] have therefore developed a novel technique to follow AA catabolism, based on 6-deoxy-6-fluoroascorbic acid (FAA) and ^{19}F-NMR spectroscopy, usually without the need for chromatographic separation. FAA was injected into normal and STZ-diabetic rats, whose plasma levels of FAA subsequently reached 42 and 27 μM, respectively, implying accelerated cellular uptake in diabetics due to tissue depletion or accelerated oxidation and elimination in the urine. The urine contained 12–15 fluoro-substituted degradation products and the excretion ratio of FAA to total fluoro-degradation products was higher in the diabetics. In the kidneys, the ratio of fluoro-DHAA to total FAA in diabetics was more than double that of the normal, which suggests strongly that the kidney plays a major role in AA homeostasis and that its impairment is likely to be responsible for the depletion of the total body pool of AA in diabetes. The data confirm the need for AA supplementation in diabetes.

In STZ-diabetic rat livers, the levels of mRNA of l-gulonolactone oxidase, catalase, and glutathione peroxidase were decreased at 6 weeks, as well as that of plasma alpha 1 proteinase inhibitor 3.[562] AA synthesis enzyme and recycling enzyme mRNAs were also decreased, as was the level of AA itself. It seems that the antioxidative defence system had been severely damaged.

From a Maillard reaction point of view, one might have thought that the relatively elevated levels of glucose would act protectively of AA, but the real situation seems to be more complex.

CHAPTER 12

Caramelisation

Sugars, polysaccharides, polyhydroxycarboxylic acids, reductones, α-dicarbonyl compounds, and quinones will undergo browning in the absence of amino compounds.

Such reactions, even in the absence of catalysts, are important in the food industry, but they require high temperatures, not often encountered. Glucose, for example, decomposes only above 150 °C. Caramelisation is accelerated by carboxylic acids and their salts, phosphates, and metallic ions, but, even when catalysed, the energy requirements exceed those of sugar–amine reactions.

As with the Maillard reaction, odorous compounds are formed, water and carbon dioxide are liberated, the pH drops during the reaction, colour formation is markedly increased by increasing the pH, oxygen has only a slight enhancing effect on colour production, and the reaction is inhibited by sulfur dioxide.

The main reactions are 1,2-enolisation (Lobry de Bruyn-Alberda van Ekenstein rearrangement, *cf.* Amadori rearrangement), dehydration to furfurals, and fission (see ref. 563).

Freeze-dried systems, containing sucrose and organic acids, stored at 55 °C, can undergo rapid nonenzymic browning even at low r.h. Under these conditions, protein *lowers* the rate of browning, probably through its buffer action. The browning is attributed to reducing sugar formed by hydrolysis at moisture contents below 1% (r.h. = 0.1).[564]

Similarly, lysine decreases the loss of fructose, when equimolar solutions (0.05 M each) are heated at different pH values at 100 °C.[565] The effect is most marked at pH 8–11. Ajandouz *et al.*[565] estimated that caramelisation accounted for 40–62% of the total UV-absorbing products (A_{294}) obtained in lysine–fructose systems (pH 4–7) and 10–36% of the colour (A_{420}). When 0.05 M glucose is heated alone for 2 h at 100 °C in 0.05 M phosphate buffer (pH 7.5), 47% is not recovered, individual essential amino acids (0.05 M) decreasing the loss to 42–29%, thus also apparently protecting the glucose from caramelisation.[566]

The volatiles produced by sugar degradation can make an important contribution to flavour. Pyrolysis of glucose at 300 °C allowed 56 compounds to be identified, the main product being 1,4;3,6-dianhydroglucopyranose.[567] More than 100 volatiles were separated when glucose was heated at 250 °C for 30 min in air or in nitrogen; 4-hydroxy-2-pentenoic acid lactone, 1-(2-furyl)propane-1,2-dione, and 3-methylcyclopentane-1,2-dione were identified as new products.[568]

Caramel colours were dealt in Chapter 4.

Caramelisation

The effect of high pressure (400 MPa) on glucose caramelisation has been investigated by Moreno *et al.*[110] At pH < 10, neither intermediate nor advanced products increased under low or high pressure, but at pH 10 they increased greatly at atmospheric pressure. At high pressure they were completely suppressed.

The effect of thermally degraded fructose and glucose on two azo dyes was considered in Chapter 9.

CHAPTER 13

Inhibition of Nonenzymic Browning in Foods

1 Introduction

Inhibition of nonenzymic browning in foods is important mainly for two reasons:

(a) In cases where colour formation is undesirable, its results are unsightly; for example, scallops brown so much that canning is not successful.
(b) In cases where browning flavours are undesirable. It is often only necessary for the reaction to have proceeded to a very limited extent for off-flavours to have become noticeable.

2 Six Main Ways to Inhibit Nonenzymic Browning

There are six main ways in which nonenzymic browning can be reduced or prevented:

1. Refrigeration is effective, since browning has a high-temperature coefficient, the rate increasing 3–6 times with a 10 °C rise in temperature. Most foods will not brown below −10 °C during normal storage, say, for up to a year. Foods containing ribose brown much more readily than those containing glucose, the rates of browning of ribose:xylose:glucose being in the ratio >100:6:1.
2. Sulfur dioxide (see ref. 553) is frequently used, concentrations of up to 12,000 ppm (1.2%) being encountered, since it is more effective as the concentration is increased. Increased concentrations increase the time lag before browning starts, but, once it does, it proceeds at the normal rate. High concentrations should not be present in food products to be consumed, since even 30 ppm can at times be detected by taste. Dried fruits may contain up to 2000 ppm; however, much would be expected to be lost in subsequent cooking.

    There are legal limits to the concentrations of sulfur dioxide that are permissible in consumables. Thus the EU stipulates limits for sulfur dioxide (E220),[569] available from all sources (sulfites of various types, E221-224, 226-228), as illustrated by the following (mg kg^{-1} or mg L^{-1}):

beer	20
dried mushrooms	100

cider	200
lime and lemon juice	350
white vegetables, dried	400
dehydrated granulated potato	400
breakfast sausages and hamburgers	450
horseradish pulp	800
dried apricots, peaches, grapes, prunes, and figs	2000
dried banana	1000
dried apples and pears	600
concentrated grape juice for home wine-making	2000

Chemically, the obvious explanation for the use of sulfur dioxide is its reversible combinations with carbonyl intermediates of the browning reaction, but the substances that have been identified show that the situation is considerably more complex (see below).

3. Lowering the pH is useful up to a point. It may favour the initiation of alternative mechanisms, but these tend to be less rapid. If the system contains ascorbic acid, then any measures taken to conserve vitamin C will help to prevent initiation of nonenzymic browning by its oxidation. Even the addition of extra ascorbic acid will prove useful for some time, but, once oxidised, browning will be worse than without it.

4. Dehydration can be effective in preventing browning. However, the rate of browning often exhibits a maximum with a moisture content of 5–30% (about 50% e.r.h.), i.e., partial dehydration may make browning worse rather than better. Many fruit juices, for example, brown more when concentrated or even dehydrated and need to be extra dry before browning is really controlled. Orange crystals brown even at just above 1% moisture content. Whereas orange juice is normally concentrated to 6:1, lemon or grapefruit juice is concentrated only to 4:1. Tunnel driers may be operated in such a way as to prevent exposing fruit at intermediate levels of water content to the hottest air. With dehydrated potatoes, it has been found that it is more effective to reduce the moisture content from 7 to 4% at 570 ppm SO_2, than to increase the sulfur dioxide content from 400 to 1000 ppm, although, below 6% moisture, rancidity becomes a risk.

a_w values affect the formation of different volatiles differently. Some, such as methylpyrazine, exhibit a maximum at 0.65–0.75, whereas others, such as hexanedione, increase, and yet others, such as 2-ethylthiazole, decrease with increasing a_w.[125]

5. Since the browning reaction is essentially one between amino acids and carbonyl compounds, it is reasonable to attempt to control it by removing either one or other of these groups of substances. In a sense, sulfur dioxide operates by making carbonyl compounds not available.

An example of removal of carbonyl compounds is dried Chinese egg white, which used to be the best product available, because it had been allowed to ferment naturally for 48–72 h, the main carbonyl compound, glucose, being removed in this way. However, fermentation may allow dangerous bacteria to multiply. Yeast fermentation is therefore preferred, but gives yeasty flavours.

An enzyme mixture of glucose oxidase and catalase can be used instead:

$$C_6H_{12}O_6 \xrightarrow[O_2 + H_2O]{\text{oxidase}} \text{gluconic acid} + H_2O_2$$

$$H_2O_2 \xrightarrow{\text{catalase}} H_2O + \tfrac{1}{2} O_2$$

The same enzyme mixture has been recommended for the removal of head-space oxygen in packages.

Jiang and Ooraikul[570] have similarly reduced browning in potato chips and crisps by dipping them in 0.04–0.10% (v/v) glucose oxidase solution at 40 °C for 30 min before frying, claiming, thereby, to have increased the L value (*Lab* lightness) by approximately 5.

The removal of lactose before drying skimmilk is one way of reducing browning in this system.

The loss of glycogen by degradation to reducing sugar is higher in pork than in beef and hence the former browns more. Cooked meat is more stable than uncooked meat on dehydration, because the amylolytic activity is inactivated.

6. The counterpart to the removal of carbonyl compounds is the removal of amines; for example, heat can remove proteins by coagulation (and filtration), or they may be removed by adsorption on charcoal or by ion-exchange. Soluble protein is the important fraction in meat. Fish is leached before dehydration for the parallel reason. Sorghum starch makes a better glucose syrup than corn starch. At first sight, this is surprising, since corn starch contains less than the 1% total protein present in sorghum starch. However, sorghum starch only contains 0.01% soluble protein, but corn starch 0.07%, seven times as much.

There is not much to add about Methods 1, 3, 5, and 6; however, their relevance to the problems of inhibiting nonenzymic browning must always be kept in mind. Methods 2 and 4 are considered further below.

3 Chemistry of the Inhibition of Nonenzymic Browning by Sulfite

Sulfite is regarded as the most effective inhibitor of nonenzymic browning in dehydrated fruits and vegetables. It is also very effective in many other food products and has the advantages of also preventing enzymic browning and microbial growth, as well as being of low toxicity. It has a special function in modifying flour proteins for biscuit making, increasing extensibility.

The pK values of sulfurous acid are 1.81 and 6.91. SO_3^{2-}, therefore, plays no role below pH 5 and dissolved SO_2 disappears above pH 4. HSO_3^- is the main species at pH 5, a pH near that of many foods.

It is generally accepted that bisulfite forms addition compounds with one or more of the reactants, but the amount required is far less than that needed for equivalence with the amount of reducing sugar present:

$$NaHSO_3 + R.CHO \rightleftharpoons R.CH(OH).SO_3Na \quad \text{note reversibility}$$
$$\text{hydroxysulfonic acid}$$

It seems therefore that bisulfite reacts with intermediates, and this fits in with the fact that it usually prevents the reaction, rather than merely slowing it down. However, it could be that only the open-chain form of the sugar reacts.

At any rate, the bisulfite compound of acetaldehyde has been shown to be present in wine.

Many ketones react similarly, but only if the carbonyl C atom is part of a four- to seven-membered carbon ring or has a methyl group attached to it.

Aldoses behave like aldehydes, but react much more slowly. The bisulfite addition compound is much more unstable. When, say, acetaldehyde is added to a glucose–bisulfite solution, the bisulfite will be transferred to it.

Ketoses do not form bisulfite addition compounds. This fact shows also that formation of the bisulfite compound of the sugar is not essential, since SO_2 prevents browning by ketoses also.

Bisulfites react almost as readily with olefinic bonds as they do with aldehydes:

$$NaHSO_3 + R\text{-}CH{=}CH\text{-}R' \longrightarrow R\text{-}CH\text{-}CH_2\text{-}R'$$
$$|$$
$$SO_3Na$$

sulfonic acid

Consequently, unsaturated aldehydes or ketones can react at both types of groups (*e.g.*, 3,4-dideoxyhexosulos-3-ene).

4 Analysis

Hydroxysulfonic acids are readily decomposed in acid or alkali to regenerate the carbonyl compound:

$$\underset{\underset{SO_3Na}{|}}{\overset{\overset{OH}{|}}{R.CH}} + HCl \xrightarrow[\text{faster on heating}]{\text{slow at room temperature}} RCHO + H_2O + NaCl + SO_2$$

$$\underset{\underset{SO_3Na}{|}}{\overset{\overset{OH}{|}}{R.CH}} + NaOH \xrightarrow{\text{rapid at room temperature}} RCHO + Na_2SO_3 + H_2O$$

Aldose-bisulfite addition compounds are the least stable and decompose even in aqueous solution on heating.

Acid-induced decomposition is the basis of many methods of determining the presence of sulfite in foodstuffs. One of the best known methods is the Monier–Williams procedure, in which the sample is refluxed with HCl, and the SO_2 driven off is estimated iodometrically.

It is important to note, however, that sulfonic acids formed by the addition of bisulfite to a double bond do not decompose on heating with acid, i.e., the Monier–Williams procedure will not assess sulfite bound in this form.

The following is not a probable explanation of the action of SO_2:

$$\begin{array}{c} \text{OH} \\ | \\ \text{R-CH} \\ | \\ \text{SO}_3\text{Na} \end{array} + \text{R'NH}_2 \underset{\longleftarrow}{\overset{\text{can still react}}{\longrightarrow}} \begin{array}{c} \text{R'NH} \\ | \\ \text{R-CH} \\ | \\ \text{SO}_3\text{Na} \end{array} \xrightarrow{\text{can't}} \text{Schiff base}$$

Experiments with glucose solutions have shown that the amount of sulfite bound increases when the solutions are heated strongly beforehand. It seems as if the carbonyl compounds derived from glucose are responsible for this, having a greater affinity for sulfite than glucose, and competing successfully with it when only limited amounts of sulfite are available.

Burton et al.[571,572] have shown that, when sulfite is added to a model glucose–glycine system, it is bound initially in a form (probably as bisulfite addition compound) recoverable by the Monier–Williams method, but the recoverability then falls off. This is only due to a small extent to the oxidation of sulfite to sulfate and therefore points to addition to unsaturated carbonyls. Of the compounds examined, browning was most rapid with α,β-unsaturated carbonyls. Removal of the unsaturation, as with SO_2, virtually destroys the browning potential of these compounds. It is probable that the concentration of the unsaturated carbonyls governs the amount of sulfite required rather than the concentration of the initial reducing sugar. Even though furfurals can be regarded as unsaturated carbonyls, they were detected only when browning had reached an advanced stage and so, only under exceptional circumstances, can early browning be attributed to furfurals.

Dideoxyosulosenes, part of the 1,2-enolisation pathway, possess an olefinic bond that can react with bisulfite:

$$\begin{array}{c} \text{CHO} \\ | \\ \text{CO} \\ | \\ \text{CH} \\ \| \\ \text{CH} \\ | \\ \text{CH}_2\text{OH} \end{array} \quad \xrightarrow{\text{HSO}_3^-} \quad \begin{array}{c} \text{CHO} \\ | \\ \text{CO} \\ | \\ \text{CH}_2 \\ | \\ \text{CH.SO}_3\text{H} \\ | \\ \text{CH}_2\text{OH} \end{array}$$

3,4-dideoxypentosulos-3-ene 　　　3,4-dideoxy-4-sulfopentosulose

Inhibition of Nonenzymic Browning in Foods

Compounds of this type are much more stable than bisulfite compounds and therefore have low browning potential.

Osuloses can react with bisulfite twice (see Scheme 13.1). In the past, compounds of this type have been thought to play a significant role. Their formation is still reversible.

More recently, sulfonic acids have been isolated from model Maillard reactions and from food products treated with SO_2 (see Table 13.1).

Table 13.1

	SO_2 remaining (mg kg^{-1})	3,4-dideoxy-4-sulfoosuloses[a] (mg kg^{-1})
Dried cabbage (5 yr old)[b]	537	2509
Swedes (before drying)[b]	1941, 2114	0
Swedes (dehydrated)[b]	1148, 1120	80, 84
Swedes (cooked)[c]	0	222, 144
White wine	300	0
Sulfited lemon juice	280	<5

[a] As 3,4-dideoxy-4-sulfopentosulose.
[b] Dry weight basis.
[c] On dry weight before cooking.

From the above work on swedes,[573] using ^{35}S, it is clear that the sulfo-osuloses (about 60 ppm SO_2) by no means account for the quantity of SO_2 lost (about 2000 ppm). Physical loss does not occur to any extent > pH 4 (confirmed with dehydrated potato), but it does occur during jam-making, < pH 4. (Strawberry jam lost 25% in this way, but 95% reduction in measurable SO_2 had occurred. Under acidic conditions not much sulfo-osulose would be expected, *cf.* wine.)

The role of sulfite has also been examined by means of glucose/N^α-acetyllysine model systems.[367] Without sulfite, there was a lag phase in colour formation as measured at 420 nm of 10 min; with 2% sulfite, this was extended to over 30 min, and, with 20% sulfite, to 40 min. However, the rate of browning, once started, increased progressively, being most rapid in spite of the 20% of sulfite. If the model system (with or without sulfite) is heated for only 3 min and then cooled (*i.e.*, prior to the end of the induction period), no free radicals can be observed by ESR spectroscopy. However, on the addition of ascorbic acid to the sulfite-free system, an intense signal of the pyrazine radical cation was generated. With the 2% sulfite system, addition of ascorbic acid generated only 49% of the signal and, with the 20%

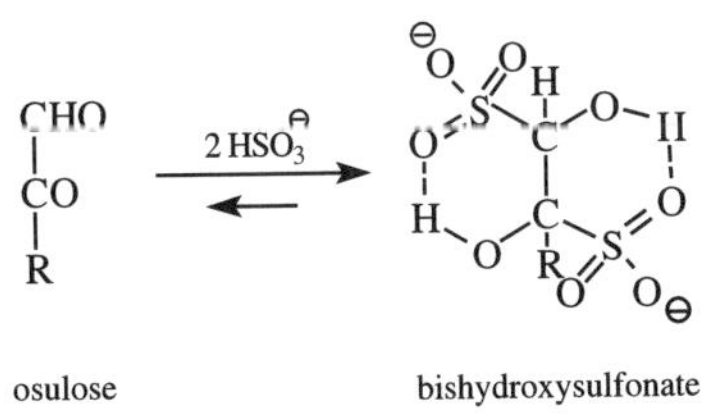

osulose → bishydroxysulfonate

Scheme 13.1

sulfite system, addition of ascorbic acid generated no signal at all. In a model system of glycolaldehyde and N^α-acetyllysine, 20% sulfite also suppressed the ESR signal. These results show that sulfite inhibits radical-mediated browning at a very early stage, most likely by blocking the radical precursors, glyoxal and glycolaldehyde (see Chapter 2).

Although sulfite occupies most of the stage in terms of modifying the Maillard reaction, other sulfur compounds have an effect too; for example, the use of N-acetylcysteine to reduce burnt off-flavours in canned liver sausage[272] (see Chapter 5).

5 Effect of a_w

As stated above, browning can be controlled to some extent by modifying the values of a_w, but the situation is not straightforward, because browning usually shows a maximum at intermediate values of a_w of 0.5–0.8. Some values for specific foods are given in Table 13.2.

At the Intermediate Stage, the Maillard reaction involves the loss of substantial amounts of water. Therefore, in an excess of water, i.e., in dilute solution, at high a_w values, browning is hindered by the dilution of the reagents. On the contrary, at low a_w values, the concentrations of reactants will have increased, but they will have begun to lose mobility. It is not surprising therefore that the Maillard reaction possesses an a_w for maximum reactivity. This was realised by Wolfrom and Rooney as early as 1953.[574]

The presence of other substances can also have an effect. Humectants, such as glycerol, move the maximum to lower values of a_w, for example, from 0.7–0.5, to an a_w below that of many intermediate moisture foods.[575] Recently, Sherwin and Labuza[576] have used a model formulation of glucose and sodium caseinate to compare the effects of the presence of about one-third of the dry weight of the reactants of glycerol (a liquid) and sorbitol (a solid). Whereas T_g curves showed increased plasticisation by both humectants, sorbitol produced no change in reaction rate at equal a_w compared with the control. Glycerol, however, gave high rates of browning over $a_w = 0.11$–0.78 that were higher than those of the control, the highest rate being at $a_w = 0.25$, which was 1.5 times higher than that of the control at its maximum at $a_w = 0.65$. These results were taken to imply that a solvent mechanism was operating.

Table 13.2 *Water activity for maximum rate of browning*

Food	a_w	Reference
Apricots (under nitrogen)	0.25–0.45	Eichner[623]
Whey powder	ca 0.44	Labuza[624]
Pork bites	0.52	Labuza[575]
Dried meat	0.57	Eichner[623]
Pork sausage	0.62	Labuza[575]
Dried milk	0.68	Eichner[623]
Pea soup mix	0.69	Labuza[575]
Potatoes	0.73	Eichner[623]
Corn chowder mix	0.79	Labuza[575]
Egg noodles	<0.75	Labuza[625]

Inhibition of Nonenzymic Browning in Foods

Leaving time and temperature aside for the moment, the concentration and drying of foods run into most trouble from browning as the food passes through these intermediate values of a_w. As regards shelf life, it is again food with such intermediate values of a_w that are most liable to deterioration from nonenzymic browning.

The interrelations of some of these factors are well illustrated (Figure 13.1) by the results obtained from air-drying of carrot cubes by Eichner and Wolf.[577] The process falls into four stages:

1. As the temperature of the cubes rises, a constant drying rate sets in, which lasts until the diffusion of water from the interior of the cubes begins to be limiting.
2. Water continues to be lost, but the rate of loss begins to decrease and the temperature rises quite steeply.
3. The Maillard reaction is initiated, as assessed by the formation of some Amadori compounds by means of ion-exchange amino acid analysis of aqueous extracts without prior hydrolysis.
4. The concentration of Amadori compounds reaches a maximum as browning itself takes off, indicating that the colourless Amadori compounds are being converted into coloured products.

If the drying process had been stopped at about 120 min, the cubes would have exhibited very little browning, but considerable amounts of Amadori

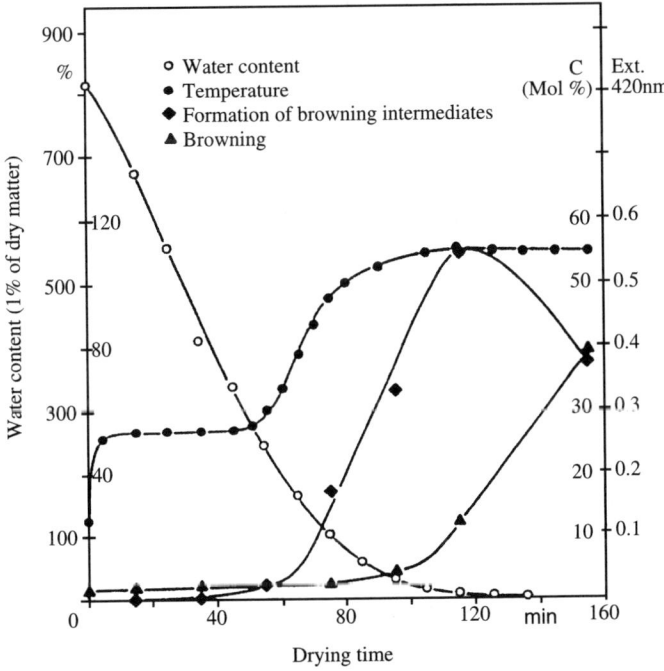

Figure 13.1 *Formation of browning intermediates and browning during air-drying of carrot cubes (air temperature = 110 °C). Reproduced from Eichner and Ciner-Doruk[577] with permission from Elsevier.*

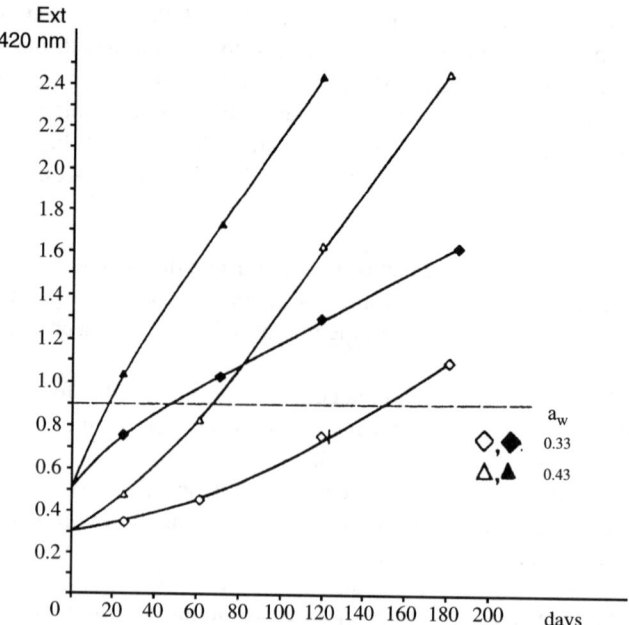

Figure 13.2 *The influence of heat treatment on the shelf life of freeze-dried tomato powder stored at 23 °C and different water activities ($\lambda_{ext,\ 420} = 0.9$ corresponds to the limit of acceptability). ♦, ▲, preheated at 40 °C and $a_w = 0.11$ (about 30% of glutamic acid has reacted to fructosylglutamic acid); ◇, △, unheated. Reproduced from Eichner and Ciner-Doruk[577] with permission from Elsevier.*

products would have been present. These have great significance, as can be seen from Figure 13.2.

This compares the browning of freeze-dried tomato powder, containing considerable amounts of Amadori products because of prior exposure to mild heating, with that of unexposed powder. The presence of Amadori compounds greatly curtails shelf life at 23 °C, browning starting immediately at the highest rate.

Recently,[578] the loss of available lysine has been determined in mixtures of lactose: casein (3:2; lactose: available lysine, *ca* 8:1) stored for 1 week at 37, 50, or 60 °C at $a_w = 0.33, 0.49, 0.52, 0.69, 0.85$, or 0.98. Lysine damage became considerable under each of the conditions used, reaching >60% at $a_w = 0.52$ after 300, 60, and 15 h at the three temperatures, respectively. The reaction rate was highest at $a_w = 0.52$ both at 37 and at 50 °C. The reaction rate had greater temperature dependence as the a_w decreased.

CHAPTER 14

Inhibition of the Maillard Reaction in Vivo

1 Introduction

In Chapter 13, six main ways of inhibiting the Maillard reaction in foods received attention: refrigeration, the use of sulfur dioxide, lowering of pH, adjusting a_w, and removing one or other of the reactants. Four of these are clearly not appropriate for *in vivo* systems, which normally cannot be refrigerated, have their pH or a_w adjusted, or be exposed to sulfur dioxide. The reactants (and intermediate products derived from them) therefore become the main targets.

According to Monnier *et al.*[579] potential strategies against the Maillard reaction *in vivo* could thus focus on seven targets:

- Glycaemia
    - Food restriction
    - Hypoglycaemic agents
- Amadori product
    - Fructosylamino acid oxidase
    - Fructose 3-phosphokinase
- Carboxymethyllysine and other AGEs
    - Decoy agents (sRAGE/lysozyme)
    - Receptor antagonists
    - Enzymically
- Dicarbonyl compounds (2-oxopropanal/deoxyglucosone)
    - Trapping agents
    - Detoxifying enzymes
- Oxidative stress
- Cell signalling
- Metabolic uncoupling

Monnier *et al.*[579] had further comments on each. In this chapter, attention is paid especially to trapping agents and intervention through enzymes, ending up with mention of a recently demonstrated hypoglycaemic agent.

2 Trapping Agents

The following are considered below: aminoguanidine, compounds related to aminoguanidine (including dimethyldiguanide), pyridoxamine, OPB-9195, phenacylthiazolium derivatives, tenilsetam, and flavonoids.

2.1 Aminoguanidine

Evidence that aminoguanidine (AG) is capable of reducing diabetes-induced arterial wall protein crosslinking to a great extent had already been put forward by Brownlee *et al.*[395] Incubation of BSA in the absence of glucose left its fluorescence (370/440 nm) unchanged, but, on incubation with glucose, the fluorescence associated with crosslinking increased within a few days and continued to do so progressively, but AG was able to give 90% inhibition of the increase. In contrast, the amount of Amadori compound formed, as measured by the incorporation of [^{14}C]glucose, remained essentially unchanged. In diabetic rats, *i.e.*, *in vivo*, the fluorescence of aortic tissue was 5.5 times that of normal rats and parenteral administration of 25 mg (kg b.w.)$^{-1}$ d^{-1} AG lowered the fluorescence to 1.3 times that of normal rats. This result was corroborated by measurements of the solubility of collagen fibrils from aortic tissue in acetic acid and after treatment with CNBr and with pepsin, which showed that crosslinking in diabetic rats is significantly alleviated by AG.

That reactive α-oxoaldehydes derived from sugar fragmentation and glycolysis are probably intermediates in the crosslinking processes is supported by the work of Oimomi *et al.*,[580] who took the next step by showing that, whereas incubation with 3-deoxyglucosone increased the fluorescent Maillard products from BSA about 10-fold, preincubation of the 3-DG with AG brought the fluorescence down by a factor of almost 5.

Administration of AG has also been shown to be able to normalise nerve blood supply and to improve conduction in streptozotocin-induced diabetic rats, thus showing potential for the treatment of diabetic neuropathy.[581]

Diabetic patients are 3–4 times more susceptible to atherosclerosis and vascular insufficiency. The situation is similar for those suffering from renal insufficiency, independent of diabetes. Makita *et al.*[582] found that there was a direct correlation ($P < 0.005$) between serum AGE-peptide levels and renal function, as assessed by creatinine clearance. This led Bucala *et al.*[145] to consider that the modification of the plasma proteins, such as LDL, may arise due to the reaction with reactive, circulating AGE-peptides rather than glucose. AGE-LDL did form readily *in vitro* when native LDL was incubated with either synthesised AGE-peptides or AGE-peptides isolated directly from the plasma of patients. Such AGE-LDL exhibited markedly impaired clearance kinetics when injected into transgenic mice, expressing human LDL receptor. Their data indicate that AGE modification could contribute to elevated LDL levels in patients with diabetes or renal insufficiency. This hypothesis was further supported by the administration of AG to diabetic patients, when circulating levels of LDL decreased by almost 30%. Levels of Hb-AGE decreased too.

The chemistry has been clarified by Hirsch *et al.*,[583] who have shown that triazines are formed from 1- and 3-dehydroglucosone, as depicted in Scheme 14.1. The conversion is

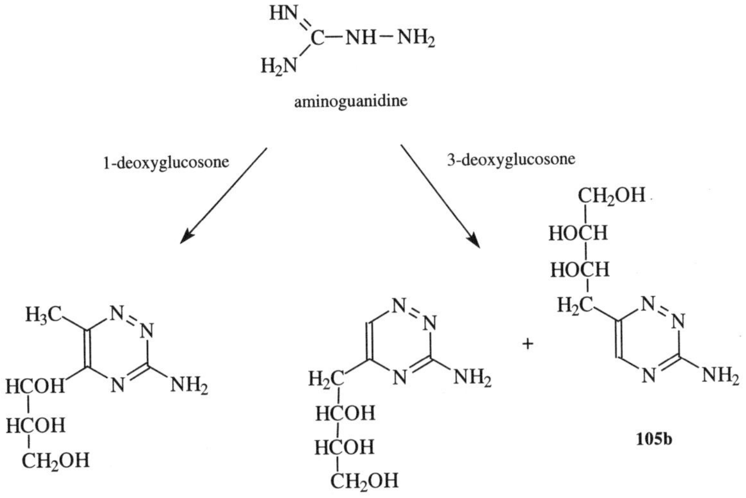

Inhibition of the Maillard Reaction in vivo 163

Scheme 14.1 *Formation of 3-amino-1,2,4-triazines from aminoguanidine and deoxyglucosones*[583]

Table 14.1 *pK_a values for some Amadori compounds derived from glucose*[583]

	Fructosyl-glycine	Fructosyl-lysine	Difructosyl-glycine
pK_{a1}	2.20	3.08	1.76
pK_{a2}	8.18	9.02	5.18

pK_{a1}, acid dissociation constant of carboxy group;
pK_{a2}, acid dissociation constant of amino group.

rapid, the triazines are stable, and glucose reacts with AG only slowly and reversibly. The Amadori compounds, fructosylglycine, fructosyllysine, and difructosylglycine, were incubated with AG in phosphate buffer at their pK_{a2} (see Table 14.1) at 37 °C for 30 d, the yields of triazine being about 17.5, 10, and 45%, respectively. The first two gave mainly **104**, but only somewhat less of (**105a** + **105b**), whereas the last gave largely (**105a** + **105b**) with only one fifth as much of **104**. Fructose and glucose incubated similarly at pH 7.0 were considerably more stable, leading to a yield of triazines less than 1%, with fructose giving more (**105a** + **105b**), the amount of **104** from either being negligible. One is tempted to attribute the higher yield from difructosylglycine to the lower pH, but, from additional experiments, the instability of the Amadori compounds does not seem to be a function of the pH of the medium or of their pK_{a2}.

Thornalley *et al.*[584] studied the kinetics of the reaction between aminoguanidine and glyoxal, 2-oxopropanal (MGO), and 3-deoxyglucosone in 50 mM phosphate buffer at pH 7.4 and 37 °C and found that the last reaction was the slowest. The reaction with MGO was complex. MGO exists free, as a hydrate, and as a dihydrate; the ratio between the forms is about 1:71:28, the free form being more reactive than the hydrate. Only at 20 mM MGO and AG were the two isomers formed in about equal amounts; at about 1 mM, the 5- was formed in 9-fold excess to the 6-isomer, whereas at high concentrations the ratio was 1:2. The ratio between the rates of reaction of MGO with water, AG, and cysteinyl thiol groups was about 1:0.21:488, implying that AG will not compete effectively with water for MGO as it is formed. Under pharmacologically relevant conditions (MGO, 100 nM; AG, 10–50 μM), the main product was calculated to be the 6-isomer (96%).

Agalou et al.[585] extended this kinetic study to the interaction of AG not only with glyoxal, MGO, and 3-deoxyglucosone, but also with hydroxypyruvaldehyde, glucosone, ribosone, erythrosone, and the corresponding 3-deoxyosones. The most rapid reactions took place with MGO, followed by hydroxypyruvaldehyde, and glyoxal in decreasing order.

Taguchi et al.[586] showed that the administration of AG causes the formation of the Schiff base of pyridoxal phosphate in the liver and kidney of mice, with concomitant lowering of pyridoxal phosphate itself. This led them[587] to prepare the Schiff base from pyridoxal and AG (PL-AG). The corresponding phosphate was insufficiently soluble in water to be useful. PL-AG was then found to inhibit AGE formation to a similar or higher degree than AG itself. The amounts of pyridoxal and its phosphate in the tissues of treated mice did not decrease at all. Taguchi et al.[588] therefore proceeded to examine the effect of PL-AG on the severity of nephropathy in STZ-induced diabetic mice, by administering 27 μmol d^{-1} for 9 weeks. Neither PL-AG nor AG itself altered glycaemic control. AG lessened the increase in glomerular volume, fractional mesangial volume, and glomerular basement thickness, but did not alter urinary albumin excretion (UAE). PL-AG, however, improved UAE by 78% over the diabetic control and was to some extent more effective in preventing the progression of renal pathology. Its in vitro antioxidant activity was also superior to that of AG, as assessed by the photooxidation of methyl orange in the presence of ZnO, H_2O_2-induced hydroxylation of benzoate, peroxidation of human LDL in the presence of Cu^{2+}, and peroxidation of erythrocyte membrane lipids induced by H_2O_2.

Although AG inhibited pentosidine formation at concentrations higher than 18 mM, Urios et al.[589] found that, at 1.6–18 mM, pentosidine formation was enhanced in Type I insoluble collagen (9 mg mL^{-1}, 200 mM phosphate buffer, pH 7.4, 250 mM glucose, 28 d, 37 °C). The formation of pentosidine was increased maximally at 8 mM (roughly doubled), but almost completely inhibited at 89 mM.

Liggins et al.[590] have investigated even lower concentrations of AG, because data from patients with end-stage renal disease, for example, suggest that oral doses of 1.2 g give plasma levels of the order of 0.13 mM.[145] When 0.6 mM BSA was incubated in 0.1 M glucose for 7 d, 1 mM AG completely blocked AGE crosslinking between preglycated and native protein. The formation of AGE fluorophores was blocked by 60% and that of protein-bound carbonyls by 10%. The fluorescence was very sensitive to AG concentration, becoming fully inhibited in only 4 mM AG. CML formation was similarly affected. On the other hand, protein-bound carbonyls reached 70% inhibition at 25 mM AG; further increases in AG concentration did not affect % inhibition.

2.2 Compounds Related to Aminoguanidine

Dimethyldiguanide (metformin) is a drug used in the therapy of Type II diabetes. It decreases blood glucose levels by increasing hepatic and peripheral tissue sensitivity

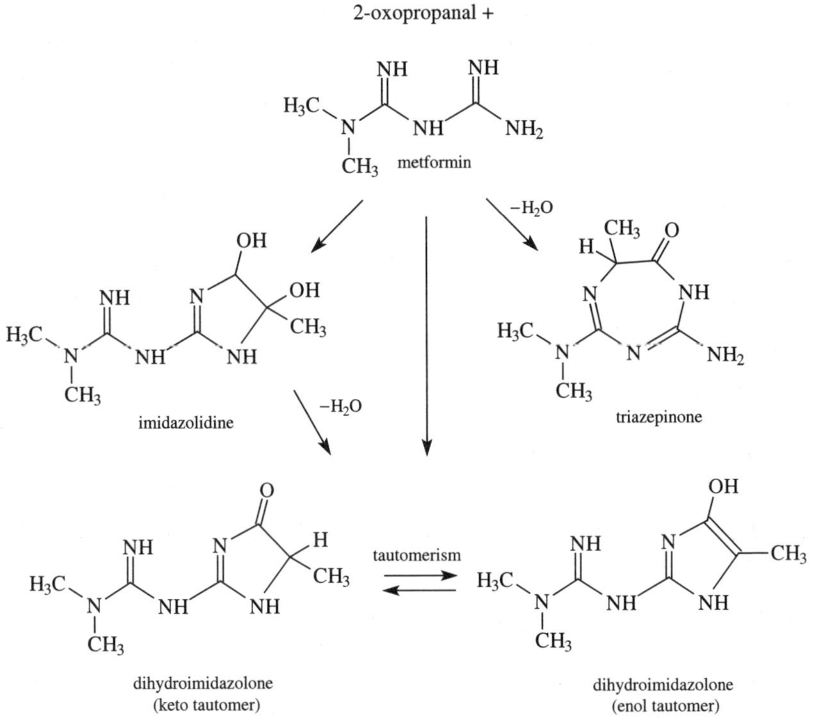

Scheme 14.2 *Reaction products of 2-oxopropanal with metformin*[591]

to insulin and decreases the blood plasma concentration of 2-oxopropanal. Battah et al.[591] therefore examined the rate of reaction of 2-oxopropanal with metformin at pH 7.4 and 37 °C, which gave dihydroimidazolone, triazepinone, and other products (see Scheme 14.2). Metformin proved not to be an efficient scavenger and so mechanisms other than the scavenging of 2-oxopropanal are likely to be responsible for the lowering of 2-oxopropanal levels in diabetics.

A series of heterocyclic *N*-amidino compounds has been synthesised by Niigata et al., comprising triazoles,[592] pyrazoles,[593–595] and indazoles.[596] All had inhibitory effects on the Maillard reaction, the IC_{50} values for the triazoles ranging from 6.6 to 30.0 μM, with low toxicity. The compounds were considered to have potential applications in the treatment of diabetic complications and aging-related diseases, as well as in skin medication, cosmetics, food products, and beverages.

2.3 Pyridoxamine

Pyridoxamine (PM) inhibits *in vitro* AGE formation from Amadori adducts on proteins, as well as advanced lipoxidation endproducts (ALEs). Baynes[597] has now reported that PM inhibits the increase in AGE/ALEs in the skin collagen of streptozotocin-induced diabetic Sprague–Dawley rats and Zucker obese, nondiabetic rats. Inhibition of AGE/ALE formation was accompanied by lower increases in triglycerides and cholesterol. PM also provided potent protection against early renal disease as evidenced by lowered plasma creatinine and urinary albumin. The combination of efficacy and low toxicity suggests that PM could be useful in retarding the progression of renal and other pathologies associated with diabetes and dyslipidaemia.

In a related study, streptozotocin-induced diabetic rats were treated with pyridoxamine, vitamin E, and enalapril (N-(1-[ethoxycarbonyl]-3-phenylpropyl)-Ala-Pro), an AGE/ALE inhibitor, an antioxidant, and an ACE-inhibitor, respectively.[598] Diabetic hyperglycaemia was accompanied by severe dyslipidaemia. Treatment with pyridoxamine was the most effective in reducing lipid abnormalities and in retarding nephropathy, retinopathy, and protein modification. Vitamin E was the next most effective treatment in retarding nephropathy, but did not affect retinopathy or AGE/ALE formation. Enalapril normalised blood pressure and retarded nephropathy and the accumulation of CML in the kidney, but did not affect dyslipidaemia and retinopathy. Thus pyridoxamine is the most effective therapy overall.

2.4 2-Isopropylidenehydrazono-4-oxothiazolidin-5-ylacetanilide (OPB-9195)

OPB-9195 belongs to a group of thiazoline derivatives, known as hypoglycaemic drugs, in spite of not lowering blood glucose levels. *In vitro*, OPB exerted inhibitory effects on AGE formation and AGE-derived crosslinks, as shown by ELISA and SDS PAGE, respectively.[599] The effects were observed at 10 mM as compared with 200 mM AG. In Otsuga-Long-Evans-Tokushima-Fatty rats, a model of noninsulin-dependent diabetes mellitus (Type II), administration of the drug prevented progression of glomerular sclerosis and AGE deposition in glomeruli. Elevation of circulating AGE levels and urinary excretion were significantly prevented, even in rats 56 weeks old and with persistent hyperglycaemia.

In streptozotocin-induced diabetic rats, OPB-9195 improved delayed motor-nerve conduction velocity by 60%, reduced stress-related DNA damage in the periphery of sciatic nerves, and reduced serum-AGE levels, but did not affect body weight, blood glucose levels, and glycated haemoglobin.[600]

Inhibition of the formation of CML and pentosidine from various AGE precursors and BSA was more efficient with OPB-9195 than with AG.[601] OPB-9195 also inhibited the formation of two ALEs (malondialdehyde-lysine and 4-hydroxynonenal-protein adduct) with an efficiency similar to that of AG. In glucose-based peritoneal dialysis fluid, OPB-9195 inhibited AGE formation, probably by trapping reactive carbonyls, such as GO, MGO, and 3-deoxyglucosone.

2.5 Phenacylthiazolium Derivatives

Phenacylthiazolium derivatives, relatively readily prepared from phenacyl bromide and an appropriate thiazole, react with dicarbonyls and make the C–C bond between them labile.[602] This is thought to take place as depicted in Scheme 14.3. Phenacylthiazolium compounds have shown promise in preclinical studies of vascular stiffness and have reached Phase II clinical trials for systolic hypertension.[603]

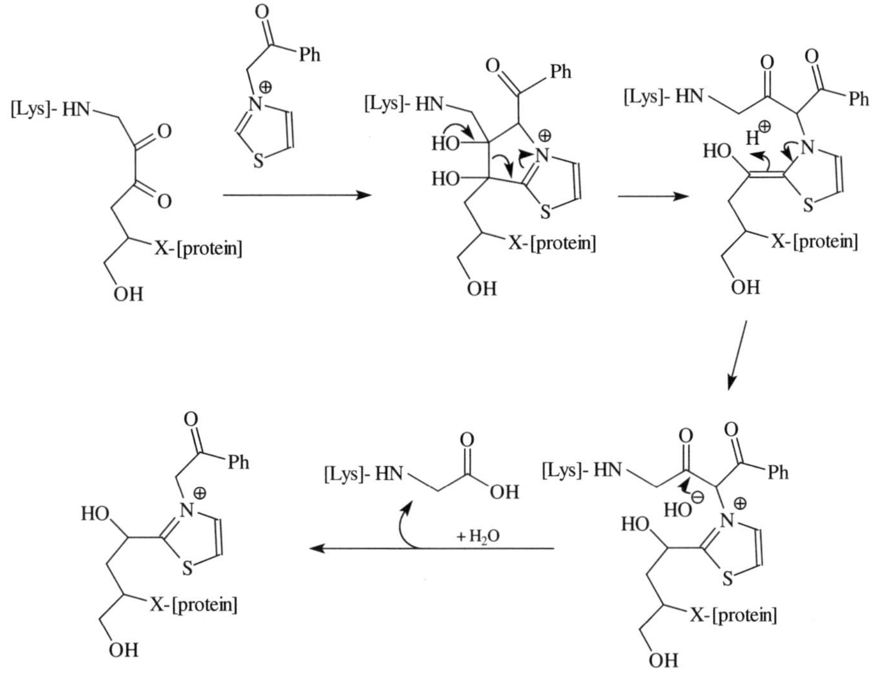

Scheme 14.3 *Proposed mechanism for cleavage of a protein crosslink by an N-phenacyl-thiazolium salt*[602]

Earlier, 3-phenacyl-4,5-dimethylthiazolium chloride (ALT-711) had been shown to reverse diabetes-induced increases in large artery stiffness in streptozotocin-treated rats,[604] age-related left ventricular stiffness in aged dogs,[605] and age-related deterioration in arterial and ventricular function in aged primates.[606]

2.6 3-(2-Thienyl)-2-piperazinone (tenilsetam)

Tenilsetam **106** was introduced as an anti-dementia drug and, since AGEs have been shown to be involved in Alzheimer lesions, it is worth exploring its potential as a Maillard reaction inhibitor. Accordingly, Shoda *et al.*[607] demonstrated that it inhibited glucose- and fructose-induced polymerisation of lysozyme in a concentration-dependent manner *in vitro*. Reduced enzymic digestion of collagen incubated with 100 mM glucose for 4 weeks was restored to the control level by coincubation with 100 mM tenilsetam. Streptozotocin-diabetic rats were treated with tenilsetam at 50 mg kg^{-1} d^{-1}, when elevated levels of AGE-derived fluorescence and pyrraline in the renal cortex and aorta were suppressed after 16 weeks. Thus tenilsetam is also active *in vivo*, and could possible play a therapeutic role in controlling diabetes.

2.7 Flavonoids

Kim and Kim[608] tested 16 flavonoids as glycation inhibitors in a model system of 250 mM fructose and 10 mg mL^{-1} BSA (pH 7.4, 37 °C), luteolin proving to be the

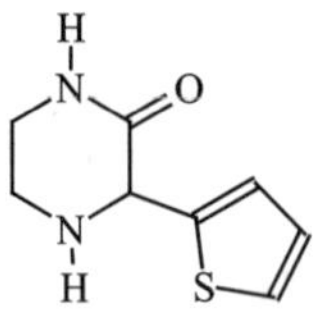

106
tenilsetam
3-(2-thienyl)-2-piperazinone

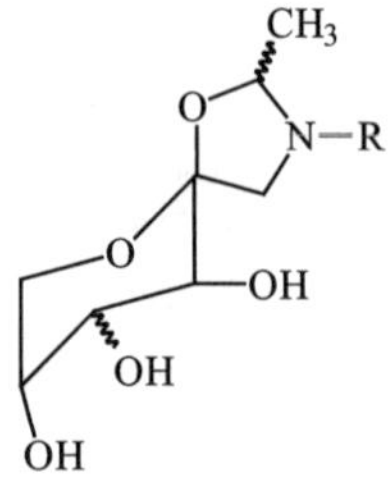

107
acetaldehyde derivative
of Amadori compound

most powerful ($IC_{50} = 7.0$ μM). Epigallocatechin gallate, kaempferol, quercitrin (quercetin 3-rhamnoside), and peltatoside (quercetin 3-arabinoglucoside) were also strong inhibitors, more effective than AG.

Rutin, quercetin 3-rutinoside, *i.e.*, 3-rhamnoglucoside, has been reported to suppress glycation,[609] but it is only sparingly soluble in water. Nagasawa *et al.*[610] have therefore examined the water-soluble 4^G-α-D-glucopyranosyl derivative (G-rutin), obtained from a sugar factory. Streptozotocin-treated rats were fed for 4 weeks on a 20% casein diet, when they showed an increase in fructosyllysine obtained on hydrolysis of their protein; however, when their diet had been supplemented with 0.2% G-rutin, the increase was reduced by 20% for kidney protein. Supplementation reduced AGE accumulation, as measured with an anti-AGE monoclonal antibody, in the serum and kidney proteins to the level of the control rats. Supplementation inhibited aldose reductase activity in the kidney, but not in the liver.

Another flavonoid inhibitor of the Maillard reaction is plantagoside (5,7,4′,5′-tetrahydroxyflavanone 3′-*O*-glucoside), from the seeds of *Plantago asiatica* L. Matsuura *et al.*[611] claim that it is 90 times more powerful than AG.

Diosmin, the 7-rutinoside of diosmetin, 3′,5,7-trihydroxy-4′-methoxyflavone, has been shown to decrease the formation of glycated collagen and haemoglobin in diabetic rats, as well as decreasing the formation of malondialdehyde.[612]

3 Intervention through Enzymes

3.1 Glyoxalase

The glyoxalase system is able to convert α-oxoaldehydes into the corresponding α-hydroxyacids, a reaction that lowers oxidative stress. Glyoxalase I (EC 4.4.1.5) mimetic activity has been associated with imidazole derivatives, such as histidine and carnosine, and their activity in liberating lactic acid from *S*-lactoylglutathione has been confirmed.[613] However, mammalian tissue displays about 4000 times this activity.

3.2 Amadoriases

The first report of an enzyme capable of catabolysing Amadori compounds is by Horiuchi *et al.*,[614] who were able to purify it about 40-fold to a single protein band with a yield of 35% from a crude extract of a *Corynebacterium*. It crystallised in rhombic plates, gave ~88 kDa on gel filtration and 44 kDa on SDS PAGE, and had noncovalently bound FAD as the prosthetic group. Its isoelectric point (IP) was 4.6 and the optimum pH in phosphate buffer was ~8.3. Fructosyl-α-L-amino acids were the preferred substrates, and N-fructosyl derivatives of other materials, such as β-amino acids, L-imino acids, D-amino acids, and alkylamines, were left virtually unchanged. The apparent K_m values for fructosylglycine and fructosylphenylalanine were 0.74 and 0.71 mM, respectively, with V_{max} = 8.78 and 14.9 s^{-1}, respectively. Hg^{2+} and Pb^{2+} were inhibitory.

Subsequently,[615] a similar enzyme was purified about 75-fold to a single protein band with a yield of 18% from a crude extract of an *Aspergillus*. It gave ~83 kDa on gel filtration and 43 kDa on SDS PAGE and had noncovalently bound FAD as the prosthetic group. Its IP was 6.8 and the optimum pH 7.7. Fructosyl-α-L-amino acids were the preferred substrates and, contrary to the Corynebacterium enzyme, N-fructosyl derivatives of ε- and α-D-amino acids were also oxidised, but at lower rates. The apparent K_m values for fructosylated glycine, β-alanine, and methylamine were 2.2, 5.9, and 220 mM, respectively. Metal ions, such as Hg^{2+}, as well as azide and p-chloromercuribenzoate, were inhibitory. The products from fructosylglycine were shown to be glucosone, glycine and H_2O_2 and the enzyme was designated fructosylamine:oxygen oxidoreductase (EC 1.5.3).

A similar enzyme from *Aspergillus terreus* GP1 was purified to homogeneity by Yoshida et al.[616] It was not able to utilise glycated proteins directly, but was able to grow on glycated HSA after treatment with a protease. cDNAs coding for the enzyme were cloned from cDNA libraries of *A. terreus* and *Penicillium janthinellum* AKU 3413, the coding region for both fungal enzymes corresponding to 437 amino acid residues. The sequence, GXGXXG, for binding the FAD cofactor in the N-terminal region and the C-terminal tripeptides, SKL and AKL from the two fungi, respectively, represent typical peroxisomal targeting signals.

In parallel work, Takahashi et al.[617] purified four fractions to homogeneity from an *Aspergillus* sp. soil strain selected on fructosyladamantanamine as the sole C source. They were designated Ia, Ib, Ic, and II, respectively. The first three corresponded to MM 51 kDa, whereas II gave 49 kDa. The K_m value for fructosyladamantanamine was 14.4–14.7 and 3.4 mM, optimum pH was 8.0 and 8.5, and the IP was 5.5–5.7 and 6.7, respectively. Thus, there are two distinct amadoriase isoenzymes. FAD was present in all and all had the FAD cofactor-binding sequence in the N-terminal region.

Takahashi et al.[618] further identified the primary structure by preparing a cDNA library from *A. fumigatus* induced with fructosylpropylamine and isolated a clone using a polyclonal Amadoriase II antibody. The structure comprised 438 amino acid residues, corresponding to 48.798 kDa. The identity of the Amadoriase II cDNA was further confirmed by expression in *Escherichia coli* cells with an inducible expression system. Northern-blotting analysis showed that Amadoriase II was induced by fructosylpropylamine in a dose-dependent manner. The sequence determined showed the enzyme to represent a new family of mammalian enzymes. The sequence exhibited 82 and 36% identity and 92 and 65% similarity, respectively, with the two sequences determined by Yoshida et al.[616] Amadori products have been implicated in the formation of H_2O_2, but the *in vivo* mechanism needs to be elucidated further.

Although metal-catalysed oxidation can lead to glucosone and H_2O_2, the existence of amadoriases suggests that Amadori products might well be involved.

The enzyme isolated by Gerhardinger et al.[619] from a soil *Pseudomonas* differed in that it produced free fructosamine, *i.e.*, it acts on the alkylamino bond rather than the sugar-amino bond of ε-fructosylaminocaproic acid. The apparent K_m values for fructosylaminocaproic acid, ε-fructosyllysine, fructosylglycine, and ribated lysine were 0.21, 2.73, 3.52, and 1.57 mM, respectively. The optimum pH was 6.5 in phosphate buffer. It was completely inhibited by Hg^{2+}, azide, and chloride, suggesting a role not only for SH groups, but also for Cu.

Stopped-flow kinetic studies of Amadoriase I using fructosylpropylamine and oxygen as substrates in 10 mM Tris hydrochloride buffer (pH 7.9) at 4 °C pointed to the pyranose form as being the active configuration. The redox potentials were found to be $+48$ and -52 mV for the oxidised enzyme/anionic quinone and anionic semiquinone/reduced enzyme reactions, respectively, at pH 7.0 and 25 °C.[620]

The reactivity of Amadori compounds can also be lowered nonenzymically, potentially *in vivo*. Thus, compounds, such as acetaldehyde, can stabilise Amadori compounds (see Structure **107**).[603] Indeed, diabetic rats fed an ethanol diet for 4 weeks showed a 52% decrease in Hb_{AGE} compared with diabetic rats not exposed to ethanol,[621] HbA_{1c} being left unaffected. This provides a possible mechanism to explain the French paradox.

3.3 Fructosamine-3-kinase (FN3K)

Szwergold et al.[397,398] proposed that intracellular nonenzymic glycation is controlled partly by means of FN3K, which phosphorylates FL, giving FL3P. This then decomposes, liberating lysine and producing 3-deoxyglucosone and P_i. FN3K has recently been purified, sequenced, and cloned. It is an entirely novel 35-kDa monomeric enzyme (309 amino acid residues), with no significant homology with any mammalian protein. In man, it has been found to be expressed in all tissues examined. Evidence for its effectiveness comes from demonstrations such as loss of its activity on lysis of erythrocytes. L-Glucose with intact cells gives a rate of haemoglobin glycation five times that of the D-isomer, because its product remains unaffected by the enzyme. There is evidence for an additional, but FN3K-independent deglycation system.

4 Hypoglycaemic Agents

4.1 Erythritol

Yokozawa et al.[622] have shown that when erythritol, which occurs naturally in algae, wine, sake, beer, pears, grapes, watermelon, and mushrooms, is administered orally to streptozotocin-induced diabetic rats at 400 mg $(kg\ b.w.)^{-1} d^{-1}$ for 10 d, the levels of HMF and TBA-reactive substances in the serum were significantly reduced; creatinine was also reduced. More than 90% of the erythritol was eliminated in the urine, *i.e.*, not metabolised.

5 Summary

There is no doubt that our understanding of the ramifications of the Maillard reaction *in vivo* has been and is being progressively improved, but, as yet, it has not led to recognised clinical approaches to the problems of diabetes, cataract formation, aging, development of Alzheimer's disease, and so on. However, the underlying science seems poised on the brink of providing a breakthrough.

Bibliography

1 Books of the International Maillard Symposia

A C. Eriksson (ed), *Maillard Reactions in Food: Chemical, Physiological and Technological Aspects*, Vol. 5, Pergamon Press, Oxford, 1981.
B G. R. Waller and M. S. Feather (eds), *The Maillard Reaction in Foods and Nutrition*, Vol. 215, American Chemical Society, Washington, DC, 1983.
C M. Fujimaki, M. Namiki, and H. Kato (eds), *Amino-Carbonyl Reactions in Food and Biological Systems*, Vol. 13, Elsevier, Amsterdam, 1986.
D P. A. Finot, H. U. Aeschbacher, R. F. Hurrell, and R. Liardon (eds), *The Maillard Reaction in Food Processing, Human Nutrition and Physiology*, Birkhauser Verlag, Basel, 1990.
E T. P. Labuza, G. A. Reineccius, V. M. Monnier, J. O'Brien, and J. W. Baynes (eds), *Maillard Reactions in Chemistry, Food, and Health*, Vol. 151, The Royal Society of Chemistry, Cambridge, 1994.
F J. O'Brien, H. E. Nursten, M. J. C. Crabbe, and J. M. Ames (eds), *The Maillard Reaction in Foods and Medicine*, Vol. 223, The Royal Society of Chemistry, Cambridge, 1998.
G S. Horiuchi, N. Taniguchi, F. Hayase, T. Kurata, and T. Osawa (eds), *The Maillard Reaction in Food Chemistry and Medical Science: Update for the Postgenomic Era*, Vol. 1245, Elsevier, Amsterdam, 2002.

2 References

Where reference is made below to the above books, they are designated by the associated capital letter.

1 L.-C. Maillard, Action des acides amines sur les sucres: formation des melanoidines par voie methodique, *C.R. Hebd. Seances Acad. Sci.*, 1912, **154**, 66–68.
2 S. Kawamura, Seventy years of the Maillard reaction, in *B*, 1983, 3–18.
3 L.-C. Maillard, Condensation des acides amines en presence de la glycerine; Cycloglycylglycine et polypeptides, *C.R. Hebd. Seances Acad. Sci.*, 1911, **153**, 1078–1080.
4 A. R. Ling, Malting, *J. Inst. Brewing*, 1908, **14**, 494–521.
5 R. O'Reilly, *The nature of the chemical groupings responsible for the colour of products of the Maillard reaction*, The University of Reading, 1982.
6 J. E. Hodge, Chemistry of browning reactions in model systems, *J. Agric. Food Chem.*, 1953, **1**, 928–943.
7 J. Mauron, The Maillard reaction in food; a critical review from the nutritional viewpoint, in *A*, 1981, 5–35.
8 M. Karel, Symptoms of the Maillard reaction, *personal communication*, 1961.

Bibliography

9 H. E. Nursten, Maillard browning reaction in dried foods, in *Concentration and Drying of Foods*, D. MacCarthy (ed), Elsevier Applied Science, London, 1986, 53–68.
10 J. M. Ames (ed), *Melanoidins in Food and Health*, Vol. 1, European Communities, Luxembourg, 2000.
11 J. M. Ames (ed), *Melanoidins in Food and Health*, Vol. 2, European Communities, Luxembourg, 2001.
12 V. Fogliano and T. Henle (eds), *Melanoidins in Food and Health*, Vol. 3, European Communities, Luxembourg, 2002.
13 G. Vegarud and F. J. Morales (eds), *Melanoidins in Food and Health*, Vol. 4, European Communities: Luxembourg, 2003.
14 J. W. Baynes and V. M. Monnier (eds), *The Maillard Reaction in Aging, Diabetes, and Nutrition*, Alan Liss, New York, 1989.
15 R. Ikan (ed), *The Maillard Reaction: Consequences for the Chemical and Life Sciences*, Wiley, Chichester, 1996.
16 S. E. Fayle and J. A. Gerrard, *The Maillard Reaction*, Royal Society of Chemistry, Cambridge, 2002.
17 T. M. Reynolds, Chemistry of nonenzymic browning. I. The reaction between aldoses and amines, *Adv. Food Res.*, 1965, **14**, 1–52.
18 T. M. Reynolds, Chemistry of nonenzymic browning. II, *Adv. Food Res.*, 1965, **14**, 167–283.
19 M. Namiki, Chemistry of Maillard reactions: recent studies on the browning reaction mechanism and the development of antioxidants and mutagens, *Adv. Food Res.*, 1988, **38**, 115–183.
20 F. Ledl and E. Schleicher (translator: H. E. Nursten), New aspects of the Maillard reaction in foods and in the human body, *Angew. Chem. Int. Edn English*, 1990, **29**, 565–594.
21 I. Blank, T. Davidek, S. Devaud, and N. Clety, Analysis of Amadori compounds by high performance anion exchange chromatography-pulse amperometric detection, in *G*, 2002, 263–267.
22 K. Eichner, M. Reutter, and R. Wittmann, Detection of Amadori compounds in heated foods, in *Thermally Generated Flavors: Maillard, Microwave, and Extrusion Process*, T. H. Parliment, M. J. Morello, R. J. McGorrin (eds), American Chemical Society, Washington, DC, 1994, 42–54.
23 K. Eichner, M. Reutter, and R. Wittmann, Detection of Maillard reaction intermediates by high pressure liquid chromatography (HPLC) and gas chromatography, in *D*, 1990, 63–77.
24 N. Ide, K. Ryu, K. Ogasawara, T. Sasaoka, H. Matsuura, S.-I. Sumi, H. Sumiyoshi, and B. H. S. Lau, Antioxidants in processed garlic. I. Fructosyl arginine identified in aged garlic extract, in *G*, 2002, 447–448.
25 N. Ide, M. Ichikawa, K. Ryu, K. Ogasawara, J. Yoshida, S. Yoshida, T. Sasaoka, S.-I. Sumi, and H. Sumiyoshi, Antioxidants in processed garlic. II. Tetrahydro-b-carboline derivatives identified in aged garlic extract, in *G*, 2002, 449–450.
26 V. A. Yaylayan and A. Huyghues-Despointes, Chemistry of Amadori rearrangement products: Analysis, synthesis, kinetics, reactions, and spectroscopic properties, *CRC Crit. Rev. Food Sci. Nutr.*, 1994, **34**, 321–369.
27 M. L. Sanz, M. D. del Castillo, N. Corzo, and A. Olano, Formation of Amadori compounds in dehydrated fruits, *J. Agric. Food Chem.*, 2001, **49**, 5228–5231.
28 D. J. McWeeny, The role of carbohydrate in non-enzymic browning, in *Molecular Structure and Function of Food Carbohydrate*, G. G. Birch, and L. F. Green (eds), Applied Science, London, 1973, 21–32.
29 E. J. Birch, J. Lelievre, and E. L. Richards, Thermal analysis of 1-deoxy-1-glycino-D-fructose and 1-β-alanino-1-deoxy-D-fructose, *Carbohydr. Res.*, 1980, **83**, 263–272.

30 R. Tressl and D. Rewicki, Heat generated flavors and precursors, in *Flavor Chemistry: Thirty Years of Progress*, R. Teranishi, E. L. Wick, and I. Hornstein (eds), Kluwer/Plenum, New York, 1999, 305–325.
31 P. A. Finot, Toxicology of nonenzymatic browning, in *Encyclopedia of Food Science and Nutrition*, 2nd edn., Vol. 2, B. Caballero, L. C. Trugo, and P. M. Finglas (eds), Academic Press, London, 2003, 673–678.
32 E. Ferrer, A. Alegria, R. Farre, P. Abellan, F. Romero, and G. Clemente, Evolution of available lysine and furosine contents in milk-based infant formulas throughout shelf-life storage period, *J. Sci. Food Agric.*, 2003, **83**, 465–472.
33 T. Henle, H. Walter, and H. Klostermeyer, Evaluation of the extent of the early Maillard-reaction in milk products by direct measurement of the Amadori-product lactuloselysine, *Z. Lebensm. Unters. Forsch.*, 1991, **193**, 119–122.
34 W. L. Claeys, A. M. van Loey, and M. E. Hendricks, Kinetics of hydroxymethylfurfural, lactulose and furosine formation in milk with different fat content, *J. Dairy Res.*, 2003, **70**, 85–90.
35 M. Rada-Mendoza, A. Olano, and M. Villamiel, Furosine as indicator of Maillard reaction in jams and fruit-based infant foods, *J. Agric. Food Chem.*, 2002, **50**, 4141–4145.
36 M. L. Sanz, M. D. del Castillo, N. Corzo, and A. Olano, Presence of 2-furoylmethyl derivatives in hydrolysates of processd tomato products, *J. Agric. Food Chem.*, 2000, **48**, 468–471.
37 A. Hidalgo, M. Rossi, and C. Pompei, Furosine as a freshness parameter of shell eggs, *J. Agric. Food Chem.*, 1995, **43**, 1673–1677.
38 P. Resmini and L. Pellegrino, Evaluation of the advanced Maillard reaction in dried pasta, in *E*, 1994, 418.
39 M. D. del Castillo, M. L. Sanz, M. J. Vicente-Arana, and N. Corzo, Study of 2-furoylmethyl amino acids in processed foods by HPLC-mass spectrometry, *Food Chem.*, 2002, **79**, 261–266.
40 J. A. Rufian-Henares, Guerra-Hernández, and B. García-Villanova, Maillard reaction in enteral formula processing: furosine, loss of o-phthalaldehyde reactivity, and fluorescence, *Food Res. Int.*, 2002, **35**, 527–533.
41 J. A. Rufián-Henares, García-Villanova, and E. Guerra-Hernández, Furosine content, loss of o-phthalaldehyde reactivity, fluorescence and colour in stored enteral formula, *Int. J. Dairy Techn.*, 2002, **55**, 121–126.
42 E. Marconi, M. F. Caboni, M. C. Messia, and G. Panfili, Furosine: a suitable marker for assessing the freshness of royal jelly, *J. Agric. Food Chem.*, 2002, **50**, 2825–2829.
43 J. Leclère and I. Birlouez-Aragon, The fluorescence of advanced Maillard products is a good indicator of lysine damage during the Maillard reaction, *J. Agric. Food Chem.*, 2001, **49**, 4682–4687.
44 X. Li and S. C. Ricke, Influence of soluble lysine maillard reaction products on Escherichia coli amino acid lysine auxotroph growth-based assay, *J. Food Sci.*, 2002, **67**, 2126–2128.
45 E. Guerra-Hernández, A. Ramirez-Jiménez, and B. García-Villanova, Glucosylisomaltol, a new indicator of browning reaction in baby cereals and bread, *J. Agric. Food Chem.*, 2002, **50**, 7282–7287.
46 M. Akagawa, T. Miura, and K. Suyama, Factors influencing the early stage of the Maillard reaction, in *G*, 2002, 395–396.
47 S. J. French, W. J. Harper, N. M. Kleinholz, R. B. Jones, and K. B. Green-Church, Maillard reaction induced lactose attachment to bovine β-lactoglobulin: electrospray ionization and matrix-assisted laser desorption/ionization examination, *J. Agric. Food Chem.*, 2002, **50**, 820–823.

48 F. Guyomarc'h, F. Warin, D. D. Muir, and J. Leaver, Lactosylation of milk proteins during manufacture and storage of skim milk powders, *Int. Dairy J.*, 2000, **10**, 863–872.
49 F. J. Moreno, R. López-Fandiño, A. Olano, Characterization and functional properties of lactosyl caseinomacropeptide conjugates, *J. Agric. Food Chem.*, 2002, **50**, 5179–5184.
50 N. Ahmed, O. K. Argirov, H. S. Minhas, C. A. A. Cordeiro, and P. J. Thornalley, Assay of advanced glycation endproducts (AGEs): surveying AGEs by chromatographic assay with derivatization by 6-quinolyl-N-hydroxysuccinimidyl-carbamate and application to N^ε-carboxymethyl-lysine and N^ε-(1-carboxyethyl)-lysine modified albumin, *Biochem. J.*, 2002, **364**, 1–14.
51 M. U. Ahmed, S. R. Thorpe, and J. W. Baynes, Identification of N-carboxymethyllysine as a degradation product of fructoselysine in glycated protein, *J. Biol. Chem.*, 1986, **261**, 4889–4894.
52 T. Davidek, N. Clety, S. Aubin, and I. Blank, Degradation of the Amadori compound N-(1-deoxy-D-fructos-1-yl)glycine in aqueous model systems, *J. Agric. Food Chem.*, 2002, **50**, 5472–5479.
53 I. Tosun and N. S. Ustun, Nonenzymic browning during storage of white hard grape pekmez (Zile pekmesi), *Food Chem.*, 2003, **80**, 441–443.
54 M. S. Feather, Amine-assisted sugar dehydration reactions, in *A*, 1981, 37–45.
55 K. Eichner, R. Schnee, and M. Heinzler, Indicator compounds and precursors for cocoa aroma formation, in *Thermally Generated Flavors: Maillard, Microwave, and Extrusion Processes*, T. H. Parliment, M. J. Morello, R. J. McGorrin (eds), Vol. 543, American Chemical Society: Washington, DC, 1994, ACS Symposium Series, 218–227.
56 M. Pischetsrieder, C. Schoetter, and T. Severin, Formation of an aminoreductone during the Maillard reaction of lactose with N-acetyllysine or proteins, *J. Agric. Food Chem.*, 1998, **46**, 928–931.
57 H. Weenen and W. Apeldoorn, Carbohydrate cleavage in the Maillard reaction. In *Flavour Science: Recent Developments*, Vol. 197, A. J. Taylor and D. S. Mottram (eds), Royal Society of Chemistry, Cambridge, 1996, 211–216.
58 S. J. Meade and J. A. Gerrard, The structure-activity relationships of dicarbonyl compounds and their role in the Maillard reaction, in *G*, 2002, 455–456.
59 V. A. Yaylayan and A. Wnorowski, The role of beta-hydroxyamino acids in the Maillard reaction — transamination route to Amadori products, in *G*, 2002, 195–200.
60 M. A. Glomb and G. Lang, Isolation and characterization of glyoxal-arginine modifications, *J. Agric. Food Chem.*, 2001, **49**, 1493–1501.
61 P. F. G. de Sa, J. M. Treubig Jr, P. R. Brown, and J. A. Dain, The use of capillary electrophoresis to monitor Maillard reaction products (MRP) by glyceraldehyde and epsilon amino group of lysine, *Food Chem.*, 2001, **72**, 379–384.
62 F. H. Stadtman, C. O. Chichester, and G. Mackinney, Carbon dioxide production in the browning reaction, *J. Am. Chem. Soc.*, 1952, **74**, 3194–3196.
63 A. Strecker, A note concerning a peculiar oxidation by alloxan, *Annalen*, 1862, **123**, 363–365.
64 A. Schonberg and R. Moubacher, The Strecker degradation of α-amino acids, *Chem. Rev.*, 1952, **50**, 261–277.
65 G. P. Rizzi, The Strecker degradation and its contribution to food flavor, in *Flavor Chemistry: Thirty Years of Progress*, R. Teranishi, E. L. Wick, and I. Hornstein (eds), Kluwer/Plenum, New York, 1999, 335–343.
66 I. D. Morton, P. Akroyd, and C. G. May, Flavoring substances, *U.S. Patent* 1960, 2934437, via *Chem. Abstr.*, 1960, **54**, 17746a.

67 I. Blank, S. Devaud, and L. B. Fay, New aspects of the formation of 3(2H)-furanones through the Maillard reaction, in *Flavour Science: Recent Developments*, Vol. 197, A. J. Taylor and D. S. Mottram (eds), Royal Society of Chemistry, Cambridge, 1996, 188–193.

68 K. Suyama, M. Akagawa, and T. Sasaki, Oxidative deamination of lysine residue in plasma protein from diabetic rat: α-dicarbonyl-mediated mechanism, in *G*, 2002, 243–248.

69 L. Benzing-Purdie, J. A. Ripmeester, and C. I. Ratcliffe, Effects of temperature on Maillard reaction products, *J. Agric. Food Chem.*, 1985, **33**, 31–33.

70 L. Benzing-Purdie and C. I. Ratcliffe, A study of the Maillard reaction by ^{13}C and ^{15}N CP-MAS NMR: Influence of time, temperature, and reactants on major products, in *C*, 1986, 193–205.

71 K. Olsson, P. A. Pernemalm, and O. Theander, Reaction products and mechanism in some simple model systems, in *A*, 1981, 47–55.

72 V. Fogliano, R. C. Borrelli, and S. M. Monti, Characterization of melanoidins from different carbohydrate amino acids model system, in *Melanoidins in Food and Health*, Vol. 2, J. M. Ames (ed), European Communities, Luxembourg, 2001, 65–72.

73 T. Hofmann, Studies on the influence of the solvent on the contribution of single Maillard reaction products to the total color of browned pentose/alanine-solutions — a quantitative correlation by using the color activity concept, *J. Agric. Food Chem.*, 1998, **46**, 3912–3917.

74 R. C. Borrelli, V. Fogliano, S. M. Monti, and J. M. Ames, Characterization of melanoidins from a glucose-glycine model system, *Eur. Food Res. Technol.*, 2002, **215**, 210–215.

75 H. Kato and F. Hayase, An approach to estimate the chemical structure of melanoidins, in *G*, 2002, 3–7.

76 R. Tressl, G. T. Wondrak, R. P. Kruger, and D. Rewicki, New melanoidin-like Maillard polymers from 2-deoxypentoses, *J. Agric. Food Chem.*, 1998, **46**, 104–110.

77 R. Tressl, G. T. Wondrak, L.-A. Garbe, R. P. Kruger, and D. Rewicki, Pentoses and hexoses as sources of new melanoidin-like Maillard polymers, *J. Agric. Food Chem.*, 1998, **46**, 1765–1776.

78 G. T. Wondrak, R. Tressl, and D. Rewicki, Maillard reaction of free and nucleic acid-bound 2-deoxy-D-ribose and D-ribose with ω-amino acids, *J. Agric. Food Chem.*, 1997, **45**, 321–327.

79 R. C. Borrelli, A. Visconti, C. Menella, M. Anese, and V. Fogliano, Chemical characterization and antioxidant properties of coffee melanoidins, *J. Agric. Food Chem.*, 2002, **50**, 6527–6533.

80 T. Hofmann, On the preparation of glucose/glycine standard melanoidins and their separation by using dialysis, ultrafiltration and gel permeation chromatography, in *Melanoidins in Food and Health*, Vol. 2, J. M. Ames (ed), European Communities, Luxembourg, 2001, 11–21.

81 B. Cämmerer, I. J. Fuchs, and L. W. Kroh, Antioxidative activity of melanoidins — radical and oxygen scavenging properties, in *Melanoidins in Food and Health*, Vol. 2, J. M. Ames (ed), European Communities, Luxembourg, 2001, 159–164.

82 B. Cämmerer, V. Jalyschkov, and L. W. Kroh, Carbohydrate structures as part of the melanoidin skeleton, in *G*, 2002, 269–273.

83 H. Ottinger and T. Hofmann, Influence of roasting on the melanoidin spectrum in coffee beans and instant coffee, in *Melanoidins in Food and Health*, Vol. 2, J. M. Ames (ed), European Communities, Luxembourg, 2001, 119–125.

84 B. L. Wedzicha and M. T. Kaputo, Melanoidins from glucose and glycine: Composition, characteristics and reactivity towards sulphite ion, *Food Chem.*, 1992, **63**, 359–367.

Bibliography

85 J. M. Ames, B. Caemmerer, J. Velisek, K. Cejpek, C. Obretenov, and M. Cioroi, The nature of melanoidins and their investigation, in *Melanoidins in Food and Health*, Vol. 1, J. M. Ames (ed), European Communities, Luxembourg, 2000, 13–29.

86 M. Anese, L. Manzocco, and E. Maltini, Determination of the glass transition temperatures of "solution A" and HMW melanoidins and estimation of viscosities by the WLF equation: a preliminary study, in *Melanoidins in Food and Health*, Vol. 2, J. M. Ames (ed), European Communities, Luxembourg, 2001, 137–141.

87 K. A. Tehrani, M. Kersiene, A. Adams, R. Venskutonis, and N. de Kimpe, Thermal degradation studies of glucose/glycine melanoidins, *J. Agric. Food Chem.*, 2002, **50**, 4062–4068.

88 H. Mitsuda, K. Yasumoto, and K. Yokoyama, Studies on the free radical in amino-carbonyl reaction, *Agric. Biol. Chem.*, 1965, **29**, 751–756.

89 M. Namiki, T. Hayashi, and S. Kawakishi, Free radicals developed in the amino-carbonyl reaction of sugars with amino acids, *Agric. Biol. Chem.*, 1973, **37**, 2935–2936.

90 M. Namiki and T. Hayashi, Development of novel free radicals during amino-carbonyl reaction of sugars with amino acids, *J. Agric. Food Chem.*, 1975, **23**, 487–491.

91 M. Namiki and T. Hayashi, Formation of novel free radical products in an early stage of Maillard reaction, in *A*, 1981, 81–91.

92 T. Hayashi and M. Namiki, Formation of two-carbon sugar fragment at an early stage of the browning reaction of sugar with amine, *Agric. Biol. Chem.*, 1980, **44**, 2575–2580.

93 T. Hayashi, S. Mase, and M. Namiki, Formation of three-carbon sugar fragment at an early stage of the browning reaction of sugar with amines or amino acids, *Agric. Biol. Chem.*, 1986, **50**, 1959–1964.

94 T. Hofmann, W. Bors, and K. Stettmaier, Studies on radical intermediates in the early stage of the non-enzymatic browning of carbohydrates and primary amino acids, *J. Agric. Food Chem.*, 1999, **47**, 379–390.

95 M. Namiki and T. Hayashi, A new mechanism of the Maillard reaction involving sugar fragmentation and free radical formation, in *B*, 1983, 21–46.

96 J. P. O'Meara, E. K. Truby, and T. M. Shaw, Free radicals in roasted coffee, *Food Res.*, 1957, **22**, 96–100.

97 E. C. Pascual, B. A. Goodman, and C. Yeretzian, Characterization of free radicals in soluble coffee by electron paramagnetic resonance spectroscopy, *J. Agric. Food Chem.*, 2002, **50**, 6114–6122.

98 J. M. Ames and A. Apriyantono, Effects of pH on the volatile compounds formed in a xylose-lysine model system, in *Thermally Generated Flavors: Maillard, Microwave, and Extrusion Processes*, T. H. Parliment, M. J. Morello, and R. J. McGorrin (eds), American Chemical Society, Washington, DC, 1994, 228–239.

99 A. Arnoldi and G. Boschin, Low molecular weight coloured compounds from Maillard reaction model systems, in *Melanoidins in Food and Health*, Vol. 2, J. M. Ames (ed), European Communities, Luxembourg, 2001, 23–29.

100 T. Davidek, I. Blank, N. Clety, and S. Aubin, The fate of N-(1-deoxy-D-fructose-1-yl)glycine in aqueous model systems, in *G*, 2002, 375–376.

101 R. A. Lawrie, *Meat Science*, 6th edn, Woodhead Publishing, Cambridge, 1998.

102 F. Dransfield, G. R. Nute, D. S. Mottram, T. G. Rowan, and T. L. J. Lawrence, Pork quality from pigs fed on low glucosinolate rapeseed meal: influence of level in the diet, sex, and ultimate pH, *J. Sci. Food Agric.*, 1985, **36**, 546–556.

103 D. S. Mottram and F. B. Whitfield, Aroma volatiles from meatlike Maillard systems, in *Thermally Generated Flavors: Maillard, Microwave, and Extrusion Processes*, T. H. Parliment, M. J. Morello, R. J. McGorrin (eds), American Chemical Society, Washington, DC, 1994, 180–191.

104 D. S. Mottram and A. Leseigneur, The effect of pH on the formation of aroma volatiles in meat-like Maillard systems, in *Flavour Science and Technology*, Y. Bessière and A. F. Thomas (eds), Wiley, Chichester, 1990, 121–124.
105 V. M. Hill, D. A. Ledward, and J. M. Ames, Influence of high hydrostatic pressure and pH on the rate of Maillard browning in a glucose-lysine system, *J. Agric. Food Chem.*, 1996, **44**, 594–598.
106 N. S. Isaacs and M. Coulson, Effect of pressure on processes modeling the Maillard reaction, *J. Phys. Org. Chem.*, 1996, **9**, 639–644 (*via Chem. Abstr.*, 1996, **125**, 245925h).
107 M. Bristow and N. S. Isaacs, The effect of high pressure on the formation of volatile products in a model Maillard reaction, *J. Chem. Soc. Perkin Trans. 2*, 1999, 2213–2218.
108 T. Tamaoka, N. Itoh, and R. Hayashi, High pressure effect on Maillard reaction, *Agric. Biol. Chem.*, 1991, **55**, 2071–2074.
109 O. Frank, I. Heberle, P. Schieberle, and T. Hofmann, Influence of high hydrostatic pressure on the formation of intense chromophores formed from pentoses and primary amino acids, in *G*, 2002, 387–388.
110 F. J. Moreno, E. Molina, A. Olano, and R. Lopez-Fandiño, High pressure effects on Maillard reaction between glucose and lysine, *J. Agric. Food Chem.*, 2003, **51**, 394–400.
111 F. J. Moreno, M. Villamiel, and A. Olano, Effect of high pressure on isomerization and degradation of lactose in alkaline media, *J. Agric. Food Chem.*, 2003, **51**, 1894–1896.
112 U. Schwarzenbolz, H. Klostermeyer, and T. Henle, Maillard reaction under high hydrostatic pressure: studies on the formation of protein-bound amino acid derivatives, in *G*, 2002, 223–227.
113 H. E. Nursten and R. O'Reilly, Coloured compounds formed by the interaction of glycine and xylose, *Food Chem.*, 1986, **20**, 45–60.
114 M. A. Glomb and R. Tschirnich, Detection of α-dicarbonyl compounds in Maillard reaction systems and in vivo, *J. Agric. Food Chem.*, 2001, **49**, 5543–5550.
115 K. M. Biemel, O. Reihl, J. Conrad, and M. O. Lederer, Formation pathways for lysine-arginine cross-links derived from hexoses and pentoses by Maillard processes, in *G*, 2002, 255–261.
116 H. E. Nursten, Key mechanistic problems posed by the Maillard reaction, in *D*, 1990, 145–153.
117 S.-J. Ge and T.-C. Lee, Kinetic significance of the Schiff base reversion in the early-stage Maillard reaction of a phenylalanine-glucose aqueous model system, *J. Agric. Food Chem.*, 1997, **45**, 1619–1623.
118 B. L. Wedzicha and L. P. Leong, Modelling of the Maillard reaction: rate constants for individual steps in the reaction, in *F*, 1998, 141–146.
119 S. Mundt, B. L. Wedzicha, M. A. J. S. van Boekel, A kinetic model for the maltose-glycine reaction, in *G*, 2002, 465–467.
120 S. I. F. S. Martins and M. A. J. S. van Boekel, Key intermediates in early stage Maillard reaction: kinetic analysis, in *G*, 2002, 469–470.
121 M. A. J. S. van Boekel and S. I. F. S. Martins, Fate of glycine in the glucose-glycine reaction: a kinetic analysis, in *G*, 2002, 289–293.
122 L. Bates, J. M. Ames, D. B. MacDougall, and P. C. Taylor, Laboratory reaction cell to model Maillard color development in a starch-glucose-lysine system, *J. Food Sci.*, 1998, **68**, 991–996.
123 C. M. Brands and M. A. J. S. van Boekel, Kinetic modelling of Maillard reaction browning: effect of heating temperature, in *Melanoidins in Food and Health*, Vol. 2, J. M. Ames (ed), European Communities, Luxembourg, 2001, 143–144.

124 C. M. Brands and M. A. J. S. van Boekel, Reactions of monosaccharides during heating of sugar-casein systems: building a reaction network model, *J. Agric. Food Chem.,* 2001, **49**, 4667–4675.
125 G. A. Reineccius, The influence of Maillard reactions on the sensory properties of foods, in *D,* 1990, 157–170.
126 H. D. Stahl and T. H. Parliment, Formation of Maillard products in the proline-glucose model system: high-temperature short-time kinetics, in *Thermally Generated Flavors: Maillard, Microwave, and Extrusion Processes,* T. H. Parliment, M. J. Morello, and R. J. McGorrin (eds), American Chemical Society, Washington, DC, 1994, 251–262.
127 R. Tressl, B. Helak, and D. Rewicki, Maltoxazine, a tricyclic compound from malt, *Helv. Chim. Acta.,* 1982, **65**, 483–489.
128 G. A. Reineccius, Kinetics of flavor formation during Maillard browning, in *Flavor Chemistry: Thirty Years of Progress,* R. Teranishi, E. L. Wick, and I. Hornstein (eds), Kluwer Academic/Plenum, New York, 1999, 345–352.
129 M. Peleg, R. Engel, C. Gonzales-Martinez, and M. G. Corradini, Non-Arrhenius and non-WLF kinetics in food systems, *J. Sci. Food Agric.,* 2002, **82**, 1346–1355.
130 F. Jousse, T. Jongen, W. Agterof, S. Russell, and P. Braat, Simplified kinetic scheme of flavor formation by the Maillard reaction, *J. Food Sci.,* 2002, **67**, 2534–2542.
131 L. N. Bell, D. E. Touma, K. L. White, and Y.-H. Chen, Glycine loss and Maillard browning as related to the glass transition in a model food system, *J. Food Sci.,* 1998, **63**, 625–628.
132 I. D. Craig, R. Parker, N. M. Rigby, P. Cairns, and S. G. Ring, Maillard reaction kinetics in model preservation systems in the vicinity of the glass transition: experiment and theory, *J. Agric. Food Chem.,* 2001, **49**, 4706–4712.
133 S. M. Lievonen and Y. H. Roos, Nonenzymatic browning in amorphous food models: effects of glass transition and water, *J. Food Sci.,* 2002, **67**, 2100–2106.
134 S. M. Lievonen, T. J. Laaksonen, and Y. H. Roos, Nonenzymatic browning in food models in the vicinity of the glass transition: effects of fructose, glucose, and xylose as reducing sugar, *J. Agric. Food Chem.,* 2002, **50**, 7034–7041.
135 N. van Chuyen, T. Kurata, and M. Fujimaki, Studies on the reaction of dipeptides with glyoxal, *Agric. Biol. Chem.,* 1973, **37**, 327–334.
136 N. van Chuyen, T. Kurata, and M. Fujimaki, Formation of *N*-[2(3-alkylpyrazin-2-on-1-yl)acyl]amino acids or -peptides on heating tri- or tetrapeptides with glyoxal, *Agric. Biol. Chem.,* 1973, **37**, 1613–1618.
137 P. M. T. de Kok and E. A. E. Rosing, Reactivity of peptides in Maillard reaction, in *Thermally Generated Flavors: Maillard, Microwave, and Extrusion Processes,* T. H. Parliment, M. J. Morello, and R. J. McGorrin (eds), American Chemical Society, Washington, DC, 1994, 158–179.
138 Y. Chen and C.-T. Ho, Effects of carnosine on volatile generation from Maillard reaction of ribose and cysteine, *J. Agric. Food Chem.,* 2002, **50**, 2372–2378.
139 R. Bucala, Z. Makita, T. Koschinsky, A. Cerami, and H. Vlassara, Lipid advanced glycosylation: Pathway for lipid oxidation *in vivo, Proc. Natl. Acad. Sci. USA,* 1993, **91**, 6434–6438.
140 W. C. Fountain, J. R. Requena, A. J. Jenkins, T. J. Lyons, B. Smyth, J. W. Baynes, and S. R. Thorpe, Quantification of *N*-(glucitol)ethanolamine and *N*-(carboxymethyl)serine: two products of nonenzymatic modification of aminophospholipids formed *in vivo, Anal. Biochem.,* 1999, **272**, 48–55.
141 A. Ravandi, A. Kuksis, L. Marai, J. J. Myher, G. Steiner, G. Lewisa, and H. Kamido, Isolation and identification of glycated aminophospholipids from red cells and plasma of diabetic blood, *FEBS Lett.,* 1996, **381**, 77–81.

142 J. R. Requena, M. U. Ahmed, C. W. Fountain, T. P. Degenhardt, S. Reddy, C. Perez, T. J. Lyons, A. J. Jenkins, J. W. Baynes, and S. R. Thorpe, Carboxymethylethanolamine, a biomarker of phospholipid modification during Maillard reaction in vivo, *J Biol. Chem.*, 1997, **272**, 17473–17479.

143 T. Miyazawa, J.-H. Oak, M. Yamada, and K. Nakagawa, Synthesis and UV-analysis of glycated Amadori-phospholipids, in *G*, 2002, 285–288.

144 A. Ravandi, A. Kuksis, and N. A. Shaikh, Glucosylated glycerophosphoethanolamines are the major LDL glycation products and increase LDL susceptibility to oxidation: Evidence of their presence in atherosclerotic lesions, *Arterioscler. Thromb. Vasc. Biol.*, 2000, **20**, 467–477.

145 R. Bucala, Z. Makita, G. Vega, S. Grundy, T. Koschinsky, A. Cerami, and H. Vlassara, Modification of low density lipoprotein by advanced glycation end products contributes to the dyslipidemia of diabetes and renal insufficiency, *Proc. Natl. Acad. Sci. USA*, 1994, **91**, 9441–9445.

146 S. M. Poling, R. D. Plattner, and D. Weisleder, N-(1-Deoxy-D-fructos-1-yl)fumonisin B1, the initial reaction product of Fumonisin B1 and D-glucose, *J. Agric. Food Chem.*, 2002, **50**, 1318–1324.

147 Y. Lu, L. Clifford, C. C. Hauck, S. Hendrich, G. Osweiler, and P. A. Murphy, Characterization of Fumonisin B1-glucose reaction kinetics and products, *J. Agric. Food Chem.*, 2002, **50**, 4726–4733.

148 M. M. Costelo, L. S. Jackson, M. A. Hanna, B. H. Reynolds, and L. B. Bullerman, Loss of fumonisin B1 in extruded and baked corn-based foods with sugars, *J. Food Sci.*, 2001, **66**, 416–421.

149 F. B. Whitfield, Volatiles from interactions of Maillard reactions and lipids, *Crit. Rev. Food Nutr.*, 1992, **31**, 1–58.

150 J. Pokorný, Browning from lipid-protein interactions, in *A*, 1981, 421–428.

151 D. S. Mottram and R. A. Edwards, The role of triglycerides and phospholipids in the aroma of cooked beef, *J. Sci. Food Agric.*, 1983, **34**, 517–522.

152 D. S. Mottram, Flavor compounds formed during the Maillard reaction, in *Thermally Generated Flavors: Maillard, Microwave, and Extrusion Processes*, T. H. Parliment, M. J. Morello, and R. J. McGorrin (eds), American Chemical Society, Washington, DC, 1994, 104–126.

153 F. B. Whitfield, D. S. Mottram, S. Brock, D. J. Puckey, and L. J. Salter, Effect of phospholipid on the formation of volatile heterocyclic compounds in heated aqueous solutions of amino acids and ribose, *J. Sci. Food Agric.*, 1988, **42**, 261–272.

154 L. J. Farmer and D. S. Mottram, Interaction of lipid in the Maillard reaction between cysteine and ribose: the effect of triglyceride and three phospholipids on the volatile products, *J. Sci. Food Agric.*, 1990, **53**, 505–525.

155 T. P. Labuza, S. R. Tannenbaum, and M. Karel, Water content and stability of low-moisture and intermediate-moisture foods, *Food Technol.*, 1970, **24**, 543–544, 546–548, 550.

156 T. Uematsu, L. Párkanyiová, T. Endo, C. Matsuyama, T. Yano, M. Miyahara, H. Sakurai, and J. Pokorný, Effect of the unsaturation degree on browning reactions of peanut oil and other edible oils with proteins under storage and frying conditions, in *G*, 2002, 445–446.

157 R. Tressl, C. T. Piecchotta, D. Rewicki, and E. Krause, Modification of peptide lysine during Maillard reaction of D-glucose and D-lactose, in *G*, 2002, 203–209.

158 K. Hasenkopf, B. Ronner, H. Hiller, and M. Pischetsrieder, Analysis of glycated and ascorbylated proteins by gas chromatography-mass spectrometry, *J. Agric. Food Chem.*, 2002, **50**, 5697–5703.

159 T. Kislinger, A. Humeny, C. C. Peich, X. Zhang, T. Niwa, M. Pischetsrieder, and C.-M. Becker, Relative quantification of N^{ε}-(carboxymethyl)lysine, imidazolone A, and the

Amadori product in glycated lysozyme by MALDI-TOF mass spectrometry, *J. Agric. Food Chem.*, 2003, **51**, 51–57.

160 J. W. C. Brock, D. J. S. Hinton, W. E. Cotham, T. O. Metz, S. R. Thorpe, J. W. Baynes, and J. M. Ames, Proteomic analysis of the site specificity of glycation and carboxymethylation of ribonuclease, *J. Proteome Res.*, 2003, **2**, 506–513.

161 N. Ahmed, and P. J. Thornalley, Chromatographic assay of glycation adducts in human serum albumin glycated in vitro by derivatization with 6-aminoquinolyl-*N*-hydroxysuccinimidyl-carbamate and intrinsic fluorescence, *Biochem. J.*, 2002, **364**, 15–24.

162 D. J. S. Hinton and J. M. Ames, Analysis of glycated protein by capillary electrophoresis, in *G*, 2002, 471–474.

163 F. Chevalier, J.-M. Chobert, C. Genot, and T. Haertle, Scavenging free radicals, antimicrobial, and cytotoxic activities of the Maillard reaction products of β-lactoglobulin glycated with several sugars, *J. Agric. Food Chem.*, 2001, **49**, 5031–5038.

164 A. Scaloni, V. Perillo, P. Franco, E. Fedele, R. Froio, L. Ferrara, and P. Bergamo, Characterization of heat-induced lactosylation products in caseins by immunoenzymatic and mass spectrometric methodologies, *Biochem. Biophys. Acta*, 2002, **1598**, 30–39.

165 A. Hollnagel and L. W. Kroh, 3-Deoxypentosulose: an α-dicarbonyl compound predominating in nonenzymatic browning of oligosaccharides in aqueous solution, *J. Agric. Food Chem.*, 2002, **50**, 1659–1664.

166 L. Pellegrino, P. Resmini, I. de Noni, and S. Cattaneo, Occurrence of glucosyl-β-pyranone and other AGEs from 1-deoxyosone pathway in cereal-based foods, in *G*, 2002, 461–462.

167 T. Severin and V. Krönig, Studien zur Maillard-Reaktion. IV. Struktur eines farbigen Produktes aus Pentosen, *Chem. Mikrobiol. Technol. Lebensm.*, 1972, **1**, 156–157.

168 F. Ledl and T. Severin, Braunungsreaktionen von Pentosen mit Aminen. Untersuchungen zur Maillard-Reaktion. XIII, *Z. Lebensm. Unters. Forsch.*, 1978, **167**, 410–413.

169 F. Ledl and T. Severin, Formation of coloured compounds from hexoses, *Z. Lebensm. Unters. Forsch.*, 1982, **175**, 262–265.

170 H. Lerche, M. Pischetsrieder, and T. Severin, Maillard reaction of D-glucose: identification of a colored product with conjugated pyrrole and furanone rings, *J. Agric. Food Chem.*, 2002, **50**, 2984–2986.

171 F. Ledl, U. Krönig, T. Severin, and H. Lotter, Studies on the Maillard reaction. XVIII. Isolation of N-containing coloured compounds, *Z. Lebensm. Unters. Forsch.*, 1983, **177**, 267–270; *Food Sci. Technol. Abstr.*, 1984, **16**, 7A487.

172 F. Ledl, J. Hiebl, and T. Severin, Studies on the Maillard reaction. XIX. Formation of coloured β-pyrones from pentoses and hexoses, *Z. Lebensm. Unters. Forsch.*, 1983, **177**, 353–355; *Food Sci. Technol. Abstr.*, 1985, **17**, 2A55.

173 S. B. Banks, J. M. Ames, and H. E. Nursten, Isolation and characterisation of 4-hydroxy-2-hydroxymethyl-3-(2'-pyrrolyl)-2-cyclopenten-1-one from a xylose/lysine reaction mixture, *Chem. Ind.*, 1988, 433–434.

174 A. J. Tomlinson, J. A. Mlotkiewicz, and I. A. S. Lewis, An investigation of the compounds produced by spray-drying glucose and glycine, *Food Chem.*, 1993, **48**, 373–379.

175 J. G. Farmar, P. C. Ulrich, and A. Cerami, Novel pyrroles from sulfite-inhibited Maillard reactions: insight into the mechanism of inhibition, *J. Org. Chem.*, 1988, **53**, 2346–2349.

176 M. J. Lane and H. E. Nursten, The variety of odors produced in Maillard model systems and how they are influences by reaction conditions, in *B*, 1983, 141–158.

177 T. Hofmann, Characterization of chemical structure of novel colored Maillard reaction products from furan-2-carboxaldehyde and amino acids, *J. Agric. Food Chem.*, 1998, **46**, 932–940.

178 T. Hofmann, O. Frank, M. Kemeny, E. Bernardy, M. Habermeyer, U. Weyand, S. Meiers, and D. Marko, Studies on the inhibition of tumor cell growth and microtubule assembly by 3-hydroxy-4-[(E)-(2-furyl)methylidene]methyl-3-cyclopentene-1,2-dione, an intensely colored Maillard product formed from carbohydrates and L-proline, in *G*, 2002, 401–402.

179 T. Hofmann, Acetylformoin — a chemical switch in the formation of colored Maillard reaction products from hexoses and primary and secondary amino acids, *J. Agric. Food Chem.*, 1998, **46**, 3918–3928.

180 T. Hofmann, 4-Alkylidene-2-imino-5-[4-alkylidene-5-oxo-1,3-imidazol-2-inyl]azamethylidine-1,3-imidazolidine — A novel colored substructure in melanoidins formed by Maillard reactions of bound arginine with glyoxal and furan-2-carboxaldehyde, *J. Agric. Food Chem.*, 1998, **46**, 3896–3901.

181 W. A. W. Mustapha, S. E. Hill, J. M. V. Blanshard, and W. Derbyshire, Maillard reactions: Do the properties of liquid matrices matter? *Food Chem.*, 1998, **62**, 441–449.

182 G. P. Rizzi, Chemical structure of colored Maillard reaction products, *Food Rev. Intern.*, 1997, **13**, 1–28.

183 H. Kato and H. Tsuchida, Estimation of melanodin structure by pyrolysis and oxidation, in *A*, 1981, 147–156.

184 M. S. Feather and D. Nelson, Maillard polymers derived from D-glucose, D-fructose, 5-(hydroxymethyl)-2-furaldehyde, and glycine and methionine, *J. Agric. Food Chem.*, 1984, **32**, 1428–1432.

185 U. Lessig and W. Baltes, Model experiments on the Maillard reaction. VI. Structural studies on selected melanoidins, *Z. Lebensm. Unters. Forsch.*, 1981, **173**, 435–444; via *Food Sci. Technol. Abstr.*, 1982, **14**, 6A528.

186 F. Hayase, Y. Takahashi, S. Tominaga, M. Miura, T. Gomyo, and H. Kato, Identification of blue pigment formed in a D-xylose-glycine reaction system, *Biosci. Biotech. Biochem.*, 1999, **63**, 1512–1514.

187 R. S. Hannan and C. H. Lea, The reaction between proteins and reducing sugars in the "dry" state. VI. The reactivity of the terminal amino groups of lysine in model systems, *Biochem. Biophys. Acta*, 1952, **9**, 293–305; via *Chem. Abstr.*, 1953, **47**, 641.

188 A. V. Clark and S. R. Tannenbaum, Isolation and characterization of pigments from protein-carbonyl browning systems: Isolation, purification, and properties, *J. Agric. Food Chem.*, 1970, **18**, 891–894.

189 A. V. Clark and S. R. Tannenbaum, Studies on limit-peptide pigments from glucose-casein browning systems using radioactive glucose, *J. Agric. Food Chem.*, 1973, **21**, 40–43.

190 A. V. Clark and S. R. Tannenbaum, Isolation and characterization of pigments from protein-carbonyl systems. Models for two insulin-glucose pigments, *J. Agric. Food Chem.*, 1974, **22**, 1089–1093.

191 T. Hofmann, Studies on melanoidin-type colourants generated from the Maillard reaction of casein and furan-2-carboxaldehyde — chemical characterisation of a red coloured domaine, *Z. Lebensm. Unters. Forsch.*, 1998, **206**, 251–258.

192 C. M. J. Brands, B. L. Wedzicha, and M. A. J. S. van Boekel, The use of radiolabelled sugar to estimate the extinction coefficient of melanoidins formed in heated sugar-casein systems, in *G*, 2002, 249–253.

193 V. Fogliano, S. M. Monti, T. Musella, G. Randazzo, and A. Ritieni, Formation of coloured Maillard reactionproducts in a gluten-glucose model system, *Food Chem.*, 1999, **66**, 19–25.

194 T. Kurata, M. Fujimaki, and Y. Sakurai, Red pigment produced by the reaction of dehydro-L-ascorbic acid with alpha-amino acid, *Agric. Biol. Chem.*, 1973, **37**, 1471–1477.

195 E. P. a. C. Directive, Colours for use in foodstuffs, *Official J.*, 1994, **36/94**, 13–29.

196 W. Kamuf, A. Nixon, O. Parker, and G. C. Barnum, Overview of caramel colors, *Cereal Foods World,* 2003, **48**, 64–69.
197 R. Hardt and W. Baltes, The analysis of caramel colours. Part 1. Differentiation of the classes of caramel colours by Curie-point pyrolysis-capillary gas chromatography-mass spectrometry, *Z. Lebensm. Unters. Forsch.,* 1987, **185**, 275–280.
198 A. Dross and W. Baltes, Uber die Fraktionierung von Zuckercouleur-Inhaltsstoffen nach ihrer Molmasse, *Z. Lebensm. Unters. Forsch.,* 1989, **188**, 540–544.
199 L. Royle and C. M. Radcliffe, Analysis of caramels by capillary electrophoresis and ultrafltration, *J. Sci. Food Agric.,* 1999, **79**, 1709–1714.
200 L. Royle, J. M. Ames, L. Castle, H. E. Nursten, and C. M. Radcliffe, A new method for the identification and quantification of Class IV caramels using capillary electrophoresis and its application to soft drinks, *J. Sci. Food Agric.,* 1998, **76**, 579–587.
201 J. S. Coffey, H. E. Nursten, J. M. Ames, and L. Castle, A liquid chromatographic method for the estimation of Class III caramel added to foods, *Food Chem.,* 1997, **58**, 259–267.
202 R. Wang and S. A. Schroeder, The effect of caramel coloring on the multiple degradation pathways of aspartame, *J. Food Sci.,* 2000, **65**, 1100–1106.
203 M. Manley-Harris and G. N. Richards, A novel fructoglucan from the thermal polymerization of sucrose, *Carbohydrate Res.,* 1993, **240**, 183–196.
204 M. Rychlik, P. Schieberle, and W. Grosch, *Compilation of Odor Thresholds, Odor Qualities and Retention Indices of Key Food Odorants,* Deutsche Forschungsanstalt für Lebensmittelchemie and Institut für Lebensmittelchemie der Technischen Universität München, Garching, 1998.
205 H. E. Nursten, Workshop on volatile products, in *A,* 1981, 491–496.
206 R. Teranishi, E. L. Wick, and I. Hornstein (eds), *Flavor Chemistry: Thirty Years of Progress,* Kluwer/Plenum, New York, 1999.
207 G. Reineccius, Instrumental methods of analysis, in *Food Flavour Technology,* A. J. Taylor (ed), Sheffield Academic Press, Sheffield, 2002, 210–251.
208 W. M. Coleman III, SPME-GC-MS detection analysis of Maillard reaction products, in *Applications of Solid Phase Microextraction,* J. Pawliszyn (ed), Royal Society of Chemistry, Cambridge, 1999, 585–608.
209 T. H. Parliment, A concerted procedure for the generation, concentration, fractionation, and sensory evaluation of Maillard reaction products, in *Flavor Chemistry: Thirty Years of Progress,* R. Teranishi, E. L. Wick, and I. Hornstein (eds), Kluwer/Plenum, New York, 1999, 43–54.
210 L. B. Fay, A. Newton, H. Simian, F. Robert, D. Douce, P. Hancock, M. Green, and I. Blank, Potential of gas chromatography-orthogonal acceleration time-of-flight mass spectrometry (GC-oaTOFMS) in flavor research, *J. Agric. Food Chem.,* 2003, **51**, 2708–2713.
211 W. Engel, T. Hofmann, and P. Schieberle, Characterization of 3,4-dihydroxy-3-hexen-2,5-dione as the first open-chain caramel-like smelling flavor compound, *Eur. Food Res. Technol.,* 2001, **213**, 104–106.
212 A. Wnorowski and V. A. Yaylayan, Influence of pyrolytic and aqueous-phase reactions on the mechanism of formation of Maillard products, *J. Agric. Food Chem.,* 2000, **48**, 3549–3554.
213 T. Hofmann and P. Schieberle, Acetylformoin – an important progenitor of 4-hydroxy-2,5-dimethyl-3(2H)-furanone and 2-acetyltetrahydropyridine during thermal food processing, in *Flavour 2000: Perception, Release, Evaluation, Formation, Acceptance, Nutrition/Health,* M. Rothe (ed), Eigenverlag: Bergholz-Rehbrücke, 2001, 311–322.
214 A. Kobayashi, Sotolon: Identification, formation, and effects on flavor, in *Flavor Chemistry: Trends and Developments,* R. Teranishi, R. G. Buttery, and F. Shahidi (eds),

American Chemical Society, Washington, DC, 1989, 49–59, via *Chem. Abstr.*, 1989, **111**, 76632p.
215 J. A. Maga, Pyrroles in foods, *J. Agric. Food Chem.*, 1981, **29**, 691–694.
216 G. Vernin and C. Párkányi, Mechanisms of formation of heterocyclic compounds in Maillard and pyrolysis reactions, in *Chemistry of Heterocyclic Compounds in Flavours and Aromas*, G. Vernin (ed), Ellis Horwood, Chichester, 1982, 151–207.
217 H. Kato and M. Fujimaki, Formation of *N*-substituted pyrrole-2-aldehydes in the browning reaction between D-xylose and amino compounds, *J. Food Sci.*, 1968, **33**, 445.
218 H. Shigematsu, S. Shibata, T. Kurata, H. Sato, and M. Fujimaki, 5-Acetyl-2,3-hydro-1*H*-pyrrolizines and 5,6,7,8- tetrahydroindolizin-8-ones, odor constituents formed on heating proline with D-glucose, *J. Agric. Food Chem.*, 1975, **23**, 233–237.
219 R. Tressl, B. Helak, H. Koppler, and D. Rewicki, Formation of 2-(1-pyrrolidinyl)-2-cyclopentenones and cyclopent(b)azepin-8(1*H*)-ones as proline specific Maillard products, *J. Agric. Food Chem.*, 1985, **33**, 1132–1137.
220 R. Tressl, K. G. Grunewald, E. Kersten, and D. Rewicki, Formation of pyrroles and tetrahydroindolizin-6-ones as hydroxyproline-specific Maillard products from glucose and rhamnose, *J. Agric. Food Chem.*, 1985, **33**, 1137–1142.
221 P. Schieberle, The role of free amino acids present in yeast as precursors of the odorants 2-acetyl-1-pyrroline and 2-acetyltetrahydropyridine in wheat bread crust, *Z. Lebensm. Unters. Forsch.*, 1990, **191**, 206–209.
222 J. Kerler, J. G. M. van der Ven, and H. Weenen, α-Acetyl-*N*-heterocycles in the Maillard reaction, *Food Rev. Int.*, 1997, **13**, 553–575.
223 R. Tressl, B. Helak, E. Kersten, and D. Rewicki, Formation of proline- and hydroxyproline-specific Maillard products from [1-^{13}C]glucose, *J. Agric. Food Chem.*, 1993, **41**, 547–553.
224 R. G. Buttery, L. C. Ling, and B. O. Juliano, 2-Acetyl-1-pyrroline: an important aroma component of cooked rice, *Chem. Ind.*, 1982, 958–959.
225 R. G. Buttery, L. C. Ling, B. O. Juliano, and J. G. Turnbaugh, Cooked rice aroma and 2-acetyl-1-pyrroline, *J. Agric. Food Chem.*, 1983, **31**, 823–826.
226 G. Jianming, Identification of 2-acetylpyridine in Xiangjing-8618 rice and in Yahonkaoluo leaves, *Food Chem.*, 2002, **78**, 163–166.
227 P. Schieberle, Quantitation of important roast-smelling odorants in popcorn by stable isotope dilution analysis and model studies on flavor formation during popping, *J. Agric. Food Chem.*, 1995, **43**, 2442–2448.
228 I. R. Hunter, M. K. Walden, J. R. Scherer, and R. E. Lundin, Preparation and properties of 1,4,5,6-tetrahydro-2-acetopyridine, a cracker-odor constituent of bread aroma, *Cereal Chem.*, 1969, **46**, 189–195.
229 J. E. Hodge, F. D. Mills, and B. E. Fisher, Compounds of browned flavor derived from sugar-amine reactions, *Cereal Sci. Today*, 1972, **17**, 34.
230 N. G. De Kimpe, W. S. Dhooge, Y. Shi, M. A. Keppens, and M. M. Boelens, On the Hodge mechanism of the formation of the bread flavor component 6-acetyl-1,2,3,4-tetrahydropyridine from proline and sugars, *J. Agric. Food Chem.*, 1994, **42**, 1739–1742.
231 J. A. Maga and C. E. Sizer, Pyrazines in foods, *CRC Crit. Rev. Food Technol.*, 1973, **4**, 39–115.
232 J. A. Maga, Pyrazines in flavour, in *Food Flavours: Part A. Introduction*; I. D. Morton and A. J. MacLeod (eds), Elsevier, Amsterdam, 1982, 283–323.
233 J. A. Maga, Pyrazine update, *Food Rev. Int.*, 1992, **8**, 479–558.
234 T. Shibamoto and R. A. Bernhard, Investigation of pyrazine formation pathways in glucose-ammonia model systems, *Agric. Biol. Chem.*, 1977, **41**, 143–153.
235 W. Baltes and G. Bochmann, Model reactions on roast aroma formation. IV. Mass spectrometric identification of pyrazines from the reaction of serine and threonine with

sucrose under the conditions of coffee roasting, *Z. Lebensm. Unters. Forsch.*, 1987, **184**, 485–493.
236 P. E. Koehler and G. V. Odell, Factors affecting the formation of pyrazine compounds in sugar-amine reactions, *J. Agric. Food Chem.*, 1970, **18**, 895–898.
237 G. P. Rizzi, A mechanistic study of alkylpyrazine formation in model systems, *J. Agric. Food Chem.*, 1972, **20**, 1081–1085.
238 H. Weenen, S. B. Tjan, P. J. de Valois, N. Bouter, A. Pos, and H. Vonk, Mechanism of pyrazine formation, in *Thermally Generated Flavors: Maillard, Microwave, and Extrusion Processes*, T. H. Parliment, M. J. Morello, R. J. McGorrin (eds), American Chemical Society, Washington, DC, 1994, 142–157.
239 C.-T. Ho and J. Chen, Generation of volatile compounds from Maillard reaction of serine, threonine, and glutamine with monosaccharides, in *Flavor Chemistry: Thirty Years of Progress*, R. Teranishi, E. L. Wick, and I. Hornstein (eds), Kluwer/Plenum, New York, 1999, 327–333.
240 R. Scarpellino and R. J. Soukup, Key flavors from heat reactions of food ingredients, in *Flavor Science: Sensible Principles and Techniques*, T. E. Acree and R. Teranishi (eds), American Chemical Society, Washington, DC, 1993, 309–335.
241 H. V. Izzo, T. G. Hartman, and C.-T. Ho, Ammonium bicarbonate and pyruvaldehyde as flavor precursors in extruded food systems, in *Thermally Generated Flavors: Maillard, Microwave, and Extrusion Processes*, T. H. Parliment, M. J. Morello, and R. J. McGorrin (eds), American Chemical Society, Washington, DC, 1994, 328–333.
242 V. A. Yaylayan and A. Keyhani, Elucidation of the mechanism of pyrrole formation during thermal degradation of ^{13}C-labeled L-serines, *Food Chem.*, 2001, **74**, 4–9.
243 V. A. Yaylayan and L. J. W. Haffenden, Mechanism of imidazole and oxazole formation in [^{13}C-2]-labelled glycine and alanine model systems, *Food Chem.*, 2003, **81**, 403–409.
244 U. S. Gi and W. Baltes, Pyridoimidazoles, histidine-specific reaction products, in *Thermally Generated Flavors: Maillard, Microwave, and Extrusion Processes*, T. H. Parliment, M. J. Morello, and R. J. McGorrin (eds), American Chemical Society, Washington, DC, 1994, 263–269.
245 G. MacLeod, The scientific and technological basis of meat flavours, in *Developments in Food Flavours*, G. G. Birch and M. G. Lindley (eds), Elsevier Applied Science, London, 1986, 191–223.
246 Y. Zheng and C.-T. Ho, Kinetics of the release of hydrogen sulfide from cysteinme and glutathione during thermal treatment, in *Sulfur Compounds in Foods*, C. J. Mussinan and M. E. Keelan (eds), American Chemical Society, Washington, DC, 1994, 138–146.
247 R. Tressl, E. Kersten, C. Nittka, and D. Rewicki, Formation of sulfur-containing flavor compounds from [^{13}C]-labeled sugars, cysteine, and methionine, in *Sulfur Compounds in Foods*, C. J. Mussinan and M. E. Keelan (eds), American Chemical Society, Washington, DC, 1994, 224–235.
248 I. Flament, Coffee, cocoa, and tea, in *Volatile Compounds in Foods and Beverages*, H. Maarse (ed), Dekker, New York, 1991, 617–669.
249 W. Baltes and C. Song, New aroma compounds in wheat bread, in *Thermally Generated Flavors: Maillard, Microwave, and Extrusion Processes*, T. H. Parliment, M. J. Morello, R. J. McGorrin (eds), American Chemical Society, Washington, DC, 1994, 192–205.
250 C. Cerny and T. Davidek, Formation of aroma compounds from ribose and cysteine during Maillard reaction, *J. Agric. Food Chem.*, 2003, **51**, 2714–2721.
251 H.-D. Belitz and W. Grosch, *Food Chemistry*, Springer, Berlin, 1987.
252 J. A. Maga, The role of sulfur compounds in food flavor. Part II: thiophens, *CRC Crit. Rev. Food Technol.*, 1975, **6**, 241–270.

253 G. Vernin and Vernin, Genevieve. Heterocyclic aroma compounds in foods: occurrence and organoleptic properties, in *Chemistry of Heterocyclic Compounds in Flavours and Aromas*, G. Vernin and C. Párkányi (eds), Ellis Horwood, Chichester, 1982, 92–97.
254 T. Hofmann and P. Schieberle, Evaluation of the key odorants in a thermally treated solution of ribose and cysteine by aroma extract dilution techniques, *J. Agric. Food Chem.*, 1995, **43**, 2187–2194.
255 D. S. Mottram, M. S. Madruga, and F. B. Whitfield, Some novel meatlike aroma compounds from the reactions of alkanediones with hydrogen sulfide and furanthiols, *J. Agric. Food Chem.*, 1995, **43**, 189–193.
256 T. A. Bolton, G. A. Reineccius, R. Liardon, and T. Huynh Ba, Role of cysteine in the formation of 2-methyl-3-furanthiol in a thiamine-cysteine model system, in *Thermally Generated Flavors: Maillard, Microwave, and Extrusion Processes*, T. H. Parliment, M. J. Morello, and R. J. McGorrin (eds), American Chemical Society, Washington, DC, 1994, 270–278.
257 G. MacLeod and M. Seyyedain-Ardebili, Natural and simulated meat flavors (with particular reference to beef), *CRC Crit. Rev. Food Sci. Nutr.*, 1981, **14**, 309–437.
258 M. Sakaguchi and T. Shibamoto, Formation of heterocyclic compounds from the reaction of cysteamine and D-glucose, acetaldehyde, or glyoxal, *J. Agric. Food Chem.*, 1978, **26**, 1179–1183.
259 J. A. Maga, The role of sulfur compounds in food flavor. Part I: thiazoles, *CRC Crit. Rev. Food Technol.*, 1975, **6**, 153–176.
260 G. Vernin, Recent progress in food flavors: the role of heterocyclic compounds, *Ind. Alim. Agric.*, 1980, **97**, 433–449.
261 T. Hayashi and T. Shibamoto, Analysis of methyl glyoxal in foods and beverages, *J. Agric. Food Chem.*, 1985, **33**, 1090–1093.
262 T. Hofmann and P. Schieberle, Studies on the formation and stability of the roast-flavor compound 2-acetyl-2-thiazoline, *J. Agric. Food Chem.*, 1995, **43**, 2946–2950.
263 T. Hofmann and P. Schieberle, Studies on intermediates generating the flavour compounds 2-methyl-3-furanthiol, 2-acetyl-2-thiazoline and sotolon by Maillard-type reactions, in *Flavour Science: Recent Developments*, A. J. Taylor and D. S. Mottram (eds), Royal Society of Chemistry, Cambridge, 1996, 182–187.
264 R. B. Rhlid, Y. Fleury, I. Blank, L. B. Fay, D. H. Welti, F. A. Vera, and M. A. Juillerat, Generation of roasted notes based on 2-acetyl-2-thiazoline and its precursor, 2-(1-hydroxyethyl)-4,5-dihydrothiazole, by combined bio and thermal approaches, *J. Agric. Food Chem.*, 2002, **50**, 2350–2355.
265 W. Engel and P. Schieberle, Identification and quantitation of key aroma compounds formed in Maillard-type reactions of fructose with cysteamine or isothiaproline (1,3-thiazolidine-2-carboxylic acid), *J. Agric. Food Chem.*, 2002, **50**, 5394–5399.
266 W. Engel and P. Schieberle, Structural determination and odor characterization of *N*-(2-mercaptoethyl)-1,3-thiazolidine, a new intense porcorn-like-smelling odorant, *J. Agric. Food Chem.*, 2002, **50**, 5391–5393.
267 E. J. Mulders, Volatile components from the non-enzymic browning reaction of the cysteine/cystine-ribose system, *Z. Lebensm. Unters. Forsch.*, 1973, **152**, 193–201.
268 T. Shibamoto and H. Yeo, Flavor in the cysteine-glucose model system prepared in microwave and conventional ovens, in *Thermally Generated Flavors: Maillard, Microwave, and Extrusion Processes*, T. H. Parliment, M. J. Morello, and R. J. McGorrin (eds), American Chemical Society, Washington, DC, 1994, 457–465.
269 M. Guntert, J. Bruning, R. Emberger, M. Kopsel, W. Kuhn, T. Thielmann, and P. Werkhoff, Identification and formation of some selected sulfur-containing flavor compounds in various meat model systems, *J. Agric. Food Chem.*, 1990, **38**, 2027–2041.

270 T. Hofmann, R. Hassner, and P. Schieberle, Determination of the chemical structure of the intense roasty, popcorn-like odorant 5-acetyl-2,3-dihydro-1,4-thiazine, *J. Agric. Food Chem.*, 1995, **43**, 2195–2198.

271 C. Hilmes and A. Fischer, Role of amino acids and glucose in development of burnt off-flavours in liver sausage during heat processing, *Meat Sci.*, 1997, **47**, 249–258, via *Food Sci. Technol. Abstr.*, 1998, 05S0817.

272 C. Hilmes and A. Fischer, Inhibitory effect of sulfur-containing amino acids on burnt off-flavours in canned liver sausages, *Meat Sci.*, 1997, **46**, 199–210, via *Food Sci. Technol. Abstr.*, 1997, 12S0192.

273 J. E. Hodge, Origin of flavor in foods: Nonenzymatic browning reactions, in *The Chemistry and Physiology of Flavors*, H. W. Schultz, E. A. Day, and L. M. Libbey (eds), The AVI Publishing Co, Westport, CO, 1967, 465–491.

274 H. E. Nursten, The mechanism of formation of 3-methylcyclopent-2-en-2-olone, in *F*, 1998, 65–68.

275 A. Arnoldi and G. Boschin, Flavors from the reaction of lysine and cysteine with glucose in the presence of lipids, in *Thermally Generated Flavors: Maillard, Microwave, and Extrusion Processes*, T. H. Parliment, M. J. Morello, and R. J. McGorrin (eds), American Chemical Society, Washington, DC, 1994, 240–250.

276 H. Maarse (ed), *Volatile Compounds in Foods and Beverages*, Dekker, New York, 1991.

277 D. D. Roberts and T. E. Acree, Gas chromatography-olfactometry of glucose-proline Maillard reaction products, in *Thermally Generated Flavors*, T. H. Parliment, M. J. Morello, and R. J. McGorrin (eds), American Chemical Society, Washington, DC, 1994, 71–79.

278 H. E. Nursten, Volatiles produced by the Maillard reaction, in *Frontiers of Flavour Science*, P. Schieberle and K.-H. Engel (eds), Deutsche Forschungsanstalt für Lebensmittelchemie, Garching, 2000, 475–480.

279 R. Harper, D. G. Land, N. M. Griffiths, and E. C. Bate-Smith, Odour qualities: a glossary of usage, *Br. J. Psychol.*, 1968, **59**, 231–252.

280 A. Dravnieks, Odor quality: semantically generated multidimensional profiles are stable, *Science*, 1982, **218**, 799–801.

281 V. A. Yaylayan, N. G. Forage, and S. Mandeville, Microwave and thermally induced Maillard reactions, in *Thermally Generated Flavors: Maillard, Microwave, and Extrusion Processes*, T. H. Parliment, M. J. Morello, R. J. McGorrin (eds), American Chemical Society, Washington, DC, 1994, 449–456.

282 J. Kerler, and C. Winkel, The basic chemistry and process conditions underpinning reaction flavour production, in *Food Flavour Technology*, A. J. Taylor (ed), Sheffield Academic Press, Sheffield, 2002, 27–59.

283 H. Ottinger and T. Hofmann, Quantitative model studies on the efficiency of precursors in the formation of cooling-active 1-pyrrolidinyl-2-cyclopenten-1-ones and bitter-tasting cyclopenta-[b]azepin-8(1H)-ones, *J. Agric. Food Chem.*, 2002, **50**, 5156–5161.

284 O. Frank, M. Jezussek, and T. Hofmann, Sensory activity, chemical structure, and synthesis of Maillard generated bitter-tasting 1-oxo-2,3-dihydro-1H-indolizinium-6-olates, *J. Agric. Food Chem.*, 2003, **51**, 2693–2699.

285 O. Frank, and T. Hofmann, Reinvestigation of the chemical structure of bitter-tasting quinizolate and homoquinizolate and studies on their Maillard-type formation pathways using suitable ^{13}C-labeling experiments, *J. Agric. Food Chem.*, 2002, **50**, 6027–6036.

286 H. Ottinger, T. Soldo, and T. Hofmann, Discovery and structure determination of a novel Maillard-derived sweetness enhancer by application of the comparative taste dilution analysis (cTDA), *J. Agric. Food Chem.*, 2003, **51**, 1035–1041.

287 H. Ottinger, A. Bareth, and T. Hofmann, Characterization of natural "cooling" compounds formed from glucose and L-proline in dark malt by application of taste dilution analysis, *J. Agric. Food Chem.*, 2001, **49**, 1336–1344.
288 H. Ottinger, T. Soldo, and T. Hofmann, Systematic studies on structure and physiological activity of cyclic α-keto enamines, a novel class of "cooling" compounds, *J. Agric. Food Chem.*, 2001, **49**, 53383–55390.
289 K.-G. Lee and T. Shibamoto, Toxicology and antioxidant activities of non-enzymic browning reaction products: Review, *Food Rev. Int.*, 2002, **18**, 151–175.
290 M. Friedman, Food browning and its prevention: an overview, *J. Agric. Food Chem.*, 1996, **44**, 631–653.
291 V. Faist, K. Krome, J. M. Ames, and H. F. Erbersdobler, Effects of non-enzymic browning products formed by roasting glucose/glycine and glucose/casein mixtures on nadph-cytochrome c-reductase and glutathione-S-transferase in Caco-2 cells, in *Melanoidins in Food and Health*, Vol. 2, J. M. Ames (ed), European Communities, Luxembourg, 2001, 95–106.
292 V. Faist, T. Hofmann, H. Zill, J. W. Baynes, S. R. Thorpe, K. Sebekova, R. Schinzel, A. Heidland, E. Wenzel, and H. F. Erbersdobler, Effects of dietary N^ε-carboxymethyllysine on expression of the biotransformation enzyme, glutathione-S-transferase, in the rat, in *G*, 2002, 313–320.
293 T. Hofmann, H. F. Erbersdobler, I. Kruse, and V. Faist, Molecular weight distribution of non-enzymatic browning products in Japanese soy sauce and studies on their effects on NADPH-cytochrome c-reductase and glutathione-S-transferase in intestinal cells, in *G*, 2002, 485–486.
294 C.-C. Chen, T.-H. Tseng, J.-D. Hsu, and C.-J. Wang, Tumor-promoting effect of GGN-MRP extract from the Maillard reaction products of glucose and glycine in the presence of sodium nitrite in C3H10T1/2 cells, *J. Agric. Food Chem.*, 2001, **49**, 6063–6067.
295 E. M. P. Widmark, Presence of cancer-producing substances in roasted food, *Nature*, 1939, **143**, 984.
296 D. S. Mottram, B. L. Wedzicha, and A. T. Dodson, Acrylamide is formed in the Maillard reaction, *Nature*, 2002, **419**, 448–449.
297 E. Tareke, P. Rydberg, P. Karlsson, S. Eriksson, and M. Törnqvist, Analysis of acrylamide, a carcinogen formed in heated foodstuffs, *J. Agric. Food Chem.*, 2002, **50**, 4998–5006.
298 R. H. Stadler, I. Blank, N. Varga, F. Robert, J. Hay, P. A. Guy, M.-C. Robert, and S. Riediker, Acrylamide from Maillard reaction products, *Nature*, 2002, **419**, 449–450.
299 F. L. Martin and J. M. Ames, Formation of Strecker aldehydes and pyrazines in a fried potato model system, *J. Agric. Food Chem.*, 2001, **49**, 3885–3892.
300 M. Nagao, M. Honda, Y. Seino, T. Yahagi, and T. Sugimura, Mutagenicities of smoke condensates and the charred surface of fish and meat, *Cancer Lett.*, 1977, **2**, 221–226.
301 T. Sugimura, T. Kawachi, M. Nagao, T. Yahagi, Y. Seino, T. Okamoto, K. Shudo, T. Kosuge, K. Tsuji, K. Wakabayashi, Y. Iitaka, and A. Itai, Mutagenic principle(s) in tryptophan and phenylalanine pyrolysis products, *Proc. Japan Acad.*, 1977, **53**, 58–61.
302 M. Jägerstad, K. Skog, and A. Solyakov, Effects of possible binding of potential human carcinogens in cooked foods to melanoidins, in *Melanoidins in Food and Health*, Vol. 1, J. M. Ames (ed), European Communities, Luxembourg, 2000, 89–92.
303 J. S. Felton, M. A. Malfatti, M. G. Knize, C. P. Salmon, E. C. Hopmans, and B. W. Wu, Health risks of heterocyclic amines, *Mut. Res.*, 1997, **376**, 37–41.
304 J. S. Felton and M. G. Knize, Carcinogens in cooked foods: How do they get there and do they have an impact on human health? in *F*, 1998, 11–18.

305 A. Solyakov, K. Skog, and M. Jägerstad, Possible binding of carcinogenic/mutagenic heterocyclic amines to melanoidins, in *Melanoidins in Foods and Health*, Vol. 2, J. M. Ames (ed), European Communities, Luxembourg, 2001, 117–118.
306 M. J. Barnes and J. H. Weisburger, In vitro binding of the food mutagen 2-amino-3-methylimidazo(4,5-f)quinoline to dietary fiber, *J. Natl. Cancer Inst.*, 1983, **70**, 757–760.
307 M. Jägerstad, A. L. Reutersward, R. Öste, A. Dahlqvist, S. Grivas, K. Olsson, and T. Nyhammar, Creatinine and Maillard reaction products as precursors of mutagenic compounds formed in fried beef, in *B*, 1983, 507–519.
308 C. Negishi, K. Wakabayashi, M. Tsuda, S. Sato, T. Sugimura, H. Saito, M. Maeda, and M. Jägerstad, Formation of 2-amino-3,7,8-trimethylimidazo[4,5-f]quinoxaline, a new mutagen, by heating a mixture of creatinine, glucose and glycine, *Mut. Res. Lett.*, 1984, **140**, 55–59.
309 T. Nyhammar, S. Grivas, K. Olsson, and M. Jägerstad, Isolation and identification of beef mutagens (IQ compounds) from heated model systems of creatinine, fructose and glycine or alanine, in *C*, 1986, 323–327.
310 S. Zöchling, and M. Murkovic, Formation of the heterocyclic aromatic amine PhIP: Idenfication of precursors and intermediates, *Food Chem.*, 2002, **79**, 125–134.
311 K. Skog, A. Solyakov, P. Arvidsson, and M. Jägerstad, Screening for toxic Maillard reaction products in meat flavours and bouillons, in *F*, 1998, 444.
312 N. Kinae, K. Kujirai, C. Kajimoto, M. Furugori, S. Masuda, and K. Shimoi, Formation of mutagenic and carcinogenic heterocyclic amines in model systems without heating, in *G*, 2002, 341–345.
313 L. M. Tikkanen, T. M. Sauri, and K. J. Latva-Kala, Screening of heat-processed Finnish foods for the mutagens 2-amino-3,4,8-dimethylimidazo[4,5-f]quinoxaline, 2-amino-3,8-dimethylimidazo[4,5-f]quinoxaline, and 2-amino-1-methyl-6-phenylimidazo[4,5-b]pyridine, *Food Chem. Toxic.*, 1993, **31**, 717–721.
314 B. Zimmerli, P. Rhyn, O. Zoller, and J. Schlatter, Occurrence of heterocyclic aromatic amines in the Swiss diet: analytical method, exposure estimation and risk assessment, *Food Add. Contam.*, 2001, **18**, 533–551.
315 R. D. Klassen, D. Lewis, B. P.-Y. Lau, and N. P. Sen, Heterocyclic aromatic amines in cooked hamburgers and chicken obtained from local fast food outlets in the Ottawa region, *Food Res. Int.*, 2002, **35**, 837–847.
316 K. Skog, A. Eneroth, and M. Svanberg, Effect of different cooking methods on the formation of food mutagens in meat, *Int. J. Food Sci. Technol.*, 2003, **38**, 313–323.
317 M. G. Knize, C. P. Salmon, and J. S. Felton, Meat surface effects: Marinating before grilling can inhibit or promote the formation of heterocyclic amines, in *F*, 1998, 417.
318 J. H. Weisburger, Specific Maillard reactions yield powerful mutagens and carcinogens, in *E*, 1994, 335–340.
319 E. B. Brittebo, K. Skog, and M. Jägerstad, Binding of the food mutagen PhIP in pigmented tissues of mice, *Carcinogenesis*, 1992, **13**, 2263–2269, via *Chem. Abstr.*, 1993, **118**, 54067a.
320 C. M. J. Brands, G. M. Alink, M. A. J. S. van Boekel, and W. M. F. Jongen, Mutagenicity of heated sugar-casein systems: Effect of the Maillard reaction, in *Melanoidins in Food and Health*, Vol. 2, J. M. Ames (ed), European Communities, Luxembourg, 2001, 175–180.
321 G. C. Yen and C.-M. Liao, Effects of Maillard reaction products on DNA damage in human cells and their possible mechanisms, in *G*, 2002, 321–325.
322 B. L. Pool, H. Roeper, S. Roeper, and K. Romruen, Mutagenicity studies on *N*-nitrosated products of the Maillard browning reaction: *N*-nitroso-fructose-amino acids, *Food Chem. Toxic.*, 1984, **22**, 797–801, via *Chem. Abstr.*, 1985, **102**, 60899j.

323 H. Mi, K. Hiramoto, K. Kujirai, K. Ando, Y. Ikarashi, and K. Kikugawa, Effect of food reductones, 2,5-dimethyl-4-hydroxy-3(2H)-furanone (DMHF) and hydroxyhydroquinone (HHQ), on lipid peroxidation and Type IV and I allergy responses of mouse, *J. Agric. Food Chem.,* 2001, **49**, 4950–4955.

324 J. Gasic-Milenkovic, S. Dukic-Stefanovic, K. Nowick, W. Conrad, and G. Münch, Oxidative stress and re-entry of neurones into the cell cycle: can advanced glycation endproducts derived from food cause double trouble in the brain? in *Melanoidins in Food and Health*, J. M. Ames (ed), European Communities, Luxembourg, 2001, 107–115.

325 T. Koschinsky, C. J. He, T. Mitsuhashi, R. Bucala, C. Liu, C. Buenting, K. Heitmann, and H. Vlassara, Orally absorbed reactive glycation products (glycotoxins): an environmental risk factor in diabetic nephropathy, *Proc. Natl. Acad. Sci. USA,* 1997, **94**, 6474–6479.

326 H. Satoh, M. Togo, M. Hara, T. Miyata, K. Han, H. Maekawa, N. Ohno, Y. Hashimoto, K. Kurokawa, and T. Watanabe, Advanced glycation endproducts stimulate mitogen-activated protein kinase and proliferation in rabbit vascular smooth muscle cells, *Biochem. Biophys. Res. Comm.,* 1997, **239**, 111–115.

327 D. Ruggiero-Lopez, N. Rellier, M. Lecomte, M. Lagarde, and N. Wiernsperger, Growth modulation of retinal microvascular cells by early and advanced glycation products, *Diabetes Res. Clin. Pract.,* 1997, **34**, 135–142.

328 G.-C. Yen and P.-P. Hsieh, Possible mechanisms of antimutagenic effect of Maillard reaction products prepared from xylose and lysine, *J. Agric. Food Chem.,* 1994, **42**, 133–137.

329 H. Kato, I. E. Lee, N. van Chuyen, S. B. Kim, and F. Hayase, Inhibition of nitrosamine formation by nondialyzable melanoidins, *Agric. Biol. Chem.,* 1987, **51**, 1333–1338.

330 H. Einarsson, S. G. Snygg, and C. Eriksson, Inhibition of bacterial growth by Maillard reaction products, *J. Agric. Food Chem.,* 1983, **31**, 1043–1047.

331 H. Einarsson, The effect of pH and temperature on the antibacterial effect of Maillard reaction products, *Lebensm. Wiss. Technol.,* 1987, **20**, 56–58.

332 H. Einarsson, T. Eklund, and I. F. Nes, Inhibitory mechanisms of Maillard reaction products, *Microbios,* 1988, **53**, 27–36.

333 R. E. Oste, D. L. Brandon, A. H. Bates, and M. Friedman, Effect of the Maillard reaction of the Kunitz soybean trypsin inhibitor on its interaction with monoclonal antibodies, *J. Agric. Food Chem.,* 1990, **38**, 258–261.

334 J. O'Brien and P. A. Morrissey, Nutritional and toxicological aspects of the Maillard browning reaction in foods, *Crit. Rev. Food Sci. Nutr.,* 1989, **28**, 211–248.

335 R. J. van Barneveld, E. S. Batterham, and B. W. Norton, The effect of heat on amino acids for growing pigs. 1. A comparison of ileal and faecal digestibilities of amino acids in raw and heat-treated field peas (*Pisum sativum* cultivar Dundale), *Br. J. Nutr.,* 1994, **72**, 221–241.

336 R. J. van Barneveld, E. S. Batterham, and B. W. Norton, The effect of heat on amino acids for growing pigs. 2. Utilization of ileal-digestible lysine from heat-treated field peas (*Pisum sativum* cultivar Dundale), *Br. J. Nutr.,* 1994, **72**, 243–256.

337 R. J. van Barneveld, E. S. Batterham, and B. W. Norton, The effect of heat on amino acids for growing pigs. 3. The availability of lysine from heat-treated field peas (*Pisum sativum* cultivar Dundale), *Br. J. Nutr.,* 1994, **72**, 257–275.

338 R. J. van Barneveld, E. S. Batterham, D. C. Skingle, and B. W. Norton, The effect of heat on amino acids for growing pigs. 4. Nitrogen balance and urine, serum and plasma composition of growing pigs fed on raw or heat-treated field peas (*Pisum sativum*), *Br. J. Nutr.,* 1995, **73**, 259–273.

339 N. Terasawa, M. Murata, and S. Homma, Separation of model melanoidin into components with copper chelating Sepharose 6B column chromatography and comparison of chelating activity, *Agric. Biol. Chem.,* 1991, **55**, 1507–1514.

Bibliography

340 A. N. Wijewickreme, D. D. Kitts, and T. D. Durance, Reaction conditions influence the elementary composition and metal chelating affinity of nondialyzable model Maillard products, *J. Agric. Food Chem.*, 1997, **45**, 4577–4583.

341 C. Delgado-Andrade, I. Seiquer, and M. P. Navarro, Copper metabolism in rats fed diets containing Maillard reaction products, *J. Food Sci.*, 2002, **67**, 855–860.

342 H. F. Erbersdobler, A. Brandt, E. Scharrer, and B. von Wangenheim, Transport and metabolism studies with fructose amino acids, in *A*, 1981, 257–263.

343 H. F. Erbersdobler and V. Faist, Metabolism of Amadori products in rats and humans, in *Melanoidins in Food and Health*, Vol. 2, J. M. Ames (ed), European Communities, Luxembourg, 2001, 165–174.

344 K. Lee and H. F. Erbersdobler, Balance experiments on human volunteers with epsilon-fructoselysince (FL) and Lysinoalanine (LAL), in *E*, 1994, 358–363.

345 A. Niederwieser, P. Giliberti, and A. Matasovic, N^ε-1-Deoxyfructosyl-lysine in urine after ingestion of a lactose free, glucose containing milk formula, *Pediatr. Res.*, 1975, **9**, 867–872.

346 A. G. Wynne, C. Sauter, J. M. Ames, and G. R. Gibson, Evaluation of the microbial degradation of melanoidins and the implications for human gut health, Vol. 2, in *Melanoidins in Food and Health*, J. M. Ames (ed), European Communities, Luxembourg, 2001, 181–186.

347 G. R. Gibson and M. B. Roberfroid (ed), *Colonic Microbiota, Nutrition and Health*, Kluwer Academic, Dordrecht, 1999.

348 S. Rahbar, O. Blumenfeld, and H. M. Ranney, Unusual hemoglobin in patients with diabetes mellitus, *Biochem. Biophys. Res. Comm.*, 1969, **36**, 838–843, via *Chem. Abstr.*, 1969, **71**, 99722x.

349 M. C. De Rosa, M. T. Sanna, I. Messana, M. Castgnola, A. Galtieri, E. Tellone, R. Scatena, B. Botta, M. Botta, B. Giardina, Glycated human hemoglobin (HbA_{1c}): functional characteristics and molecular modeling studies, *Biophys. Chem.*, 1998, **72**, 323–335, via *Chem. Abstr.*, 1998, **129**, 91907x.

350 H. E. Nursten, The Maillard reaction in food and nutrition, in *Flavour 2000: Perception, Release, Evaluation, Formation, Acceptance, Nutrition/Health*, M. Rothe (ed), Eigenverlag, Bergholz-Rehbrücke, 2001, 297–309.

351 T. P. Degenhardt, E. Brinkmann-Frye, S. R. Thorpe, and J. W. Baynes, Role of carbonyl stress in aging and age-related diseases, in *F*, 1998, 3–10.

352 S. R. Thorpe and J. W. Baynes, CML: a brief history, in *G*, 2002, 91–99.

353 M. A. Glomb and C. Pfahler, Amides are novel protein modifications formed by physiological sugars, *J. Biol. Chem.*, 2001, **276**, 41638–41647.

354 T. Henle, U. Schwarzenbolz, A. W. Walter, and H. Klostermeyer, Protein-bound Maillard compounds in foods: Analytical and technological aspects, in *F*, 1998, 178–183.

355 R. Bucala, Advanced glycosylation: Role in atherosclerosis. In *The Glycation Hypothesis of Atherosclerosis*, C. A. L. S. Colaco (ed), Landes Biosciences/Springer, Georgetown, Texas/Heidelberg, 1997, 89–107.

356 M. Lindenmeier, V. Faist, and T. Hofmann, Structural and functional characterization of pronyl-lysine, a novel protein modification in bread crust melanoidins showing in vitro antioxidative and Phase I/II enzyme modulating activity, *J. Agric. Food Chem.*, 2002, **50**, 6997–7006.

357 F. Tessier, V. M. Monnier, and J. A. Kornfield, Characterization of novel chromophores, fluorophores and cross-links from glyceraldehyde, lysine and arginine, in *G*, 2002, 303–311.

358 H. Odani, K. Iijima, M. Nakata, S. Miyata, Y. Yasuda, S. Irie, K. Maeda, and D. Fujimoto, Identification of N^ω-carboxymethylarginine, as a new advanced glycation endproduct in serum proteins of diabetic patients, in *G*, 2002, 295–301.

359 F. Hayase, N. Nagashima, T. Koyama, S. Sagara, and Y. Takahashi, Reaction mechanisms operating in 3-deoxyglucosone-protein systems, in *F*, 1998, 262–267.
360 F. Hayase, Y. Takahashi, S. Sasaki, S. Shizuuchi, and H. Watanabe, 3-Deoxyosone-related advanced glycation end products in foods and biological systems, in *G*, 2002, 217–221.
361 E. A. Abordo and P. J. Thornalley, Pro-inflammatory cytokine synthesis by human monocytes induced by proteins minimally modified by methylglyoxal, in *F*, 1998, 357–362.
362 U. Schwarzenbolz, T. Henle, and H. Klostermeyer, Studies on the reaction of glyoxal with protein-bound arginine, in *F*, 1998, 443.
363 M. A. Glomb and R. H. Nagaraj, Protein modification by glyoxal and methylglyoxal during the Maillard reaction of higher sugars, in *F*, 1998, 250–255.
364 T. Oya-Ito, S. Kawakishi, N. Shibata, M. Kobayashi, T. Osawa, and K. Uchida, A novel monoclonal antibody against methylglyoxal-arginine adduct, in *G*, 2002, 397–399.
365 P. S. Padayatti, A. S. Ng, K. Uchida, M. A. Glomb, and R. H. Nagaraj, Argpyrimidine, a blue fluorophore in human lens proteins: High levels in brunescent cataractous lenses, *Invest. Ophthal. Visual Sci.*, 2001, **42**, 1299–1304.
366 J. M. Onorato, S. R. Thorpe, and J. W. Baynes, Immunohistochemical and ELISA assays for biomarkers of oxidative stress in aging and disease, *Ann. N.Y. Acad. Sci.*, 1998, **854**, 277–290.
367 T. Hofmann, W. Bors, and K. Stettmaier, Radical-assisted melanoidin formation during thermal processing of foods as well as under physiological conditions, *J. Agric. Food Chem.*, 1999, **47**, 391–396.
368 F. Tessier, M. Obrenovich, and V. M. Monnier, Structure and mechanism of formation of human lens fluorophore LM-1, *J. Biol. Chem.*, 1999, **274**, 20796–20804.
369 L. Graham, R. H. Nagaraj, R. Peters, L. M. Sayre, and V. M. Monnier, Structure and biological significance of pentodilysine, a novel fluorescent advanced Maillard reaction protein crosslink, in *F*, 1998, 410.
370 M. E. Westwood and P. J. Thornalley, Glycation and advanced glycation endproducts, in *The Glycation Hypothesis of Atherosclerosis*, C. A. L. S. Colaco (ed), Landes Bioscience/Springer, Austin, TX, 1997, 57–87.
371 T. Knerr, H. Lerche, M. Pischetsrieder, and T. Severin, Formation of a novel colored product during the Maillard reaction of D-glucose, *J. Agric. Food Chem.*, 2001, **49**, 1966–1970.
372 C. A. L. S. Colaco (ed), *The Glycation Hypothesis of Atherosclerosis*, Landes Bioscience/Springer, Austin, TX, 1997.
373 M. A. Friedlander, R. A. Rodby, E. J. Lewis, and D. Hricik, Serum 'free' pentosidine levels and urinary excretion predict deteriorating renal function in diabetic nephropathy, in *F*, 1998, 408.
374 R. G. Paul, T. J. Sims, N. C. Avery, and A. J. Bailey, Identification and inhibition of glycation cross-links impairing the function of collagenous tissues, in *F*, 1998, 437.
375 J. teKoppele, J. de Groot, N. Verzijl, and R. A. Bank, Nonenzymic glycation as risk factor in osteoarthritis, in *F*, 1998, 447.
376 K. Miyazaki, R. Nagai, and S. Horiuchi, Formation of pentosidine-like AGE structure from creatine, in *G*, 2002, 457–458.
377 K. M. Biemel, O. Reihl, J. Conrad, and M. O. Lederer, Formation pathways for lysine-arginine cross-links derived from hexoses and pentoses by Maillard processes, *J. Biol. Chem.*, 2001, **276**, 23405–23412.
378 M. O. Lederer and R. G. Klaiber, Cross-linking of proteins by Maillard processes: Characterization and detection of lysine-arginine cross-links derived from glyoxal and methylglyoxal, *Bioorg. Med. Chem.*, 1999, **7**, 2499–2507.

379 M. O. Lederer and H. P. Bühler, Cross-linking of proteins by Maillard processes — characterization and detection of a lysine-arginine cross-link derived from glucose, *Bioorg. Med. Chem.*, 1999, **7**, 1081–1088.
380 Y. Al-Abed and R. Bucala, A novel AGE crosslink exhibiting immunological cross-reactivity with AGEs formed *in vivo*, in *F*, 1998, 239–244.
381 F. Gerum, M. O. Lederer, and T. Severin, Cross-linking of proteins by Maillard processes: Model reaction of an Amadori compound with N^α-acetyl-L-arginine, in *F*, 1998, 409.
382 R. Tressl, G. Wondrak, E. Kersten, and D. Rewicki, Structure and potential crosslinking reactivity of a new pentose-specific Maillard product, *J. Agric. Food Chem.*, 1994, **42**, 2692–2697.
383 T. Osawa, T. Oya, H. Kumon, Y. Morimitsu, H. Kobayashi, M. Akiba, and N. Kakimoto, A novel type of advanced glycation endproduct found in diabetic rats, in *F*, 1998, 434.
384 T. Henle, U. Schwarzenbolz, and H. Klostermeyer, Irreversible crosslinking of casein during storage of UHT-treated skim milk, in *Heat Treatments & Alternative Methods*, International Dairy Federation, Brussels, 1996, 290–298.
385 G. Boschin, A. D'Agostina, and A. Arnoldi, A convenient synthesis of some cross-linked amino acids and their diastereoisomeric characterization, *Food Chem.*, 2002, **78**, 325–331.
386 K. Iijima, H. Odani, K. Maeda, D. Fujimoto, and S. Irie, Immunological detection of N^ω-carboxymethylarginine, in *G*, 2002, 349–351.
387 T. Araki, Y. Chijiiwa, R. Nagai, N. Araki, and S. Horiuchi, Application of the amino acid analysis for the detection of AGE-proteins of the Maillard reaction, in *G*, 2002, 391–393.
388 R. Nagai, T. Araki, and S. Horiuchi, Preparation of specific antibody against CML, one of the major AGE structures, in *G*, 2002, 479–480.
389 T. Oya, N. Hattori, Y. Mizuno, S. Miyata, S. Maeda, T. Osawa, and K. Uchida, Methylglyoxal modification of protein: chemical and immunochemical characterization of methylglyoxal-arginine adducts, *J. Biol. Chem.*, 1999, **274**, 18492–18502.
390 H. Watanabe, S. Iwaki, K. Aida, and F. Hayase, Formation and determination of α-dicarbonyls and an AGE cross-link, pyrropyridine in glycated proteins and in vivo, in *G*, 2002, 153–156.
391 K. Sekine, M. Uchiyama, H. Kawakami, K. Yoshiharar, T. Kuragano, M. Miura, and M. Beppu, Chromatographic quantification of pentosidine and pyrraline in patients with diabetic complications, in *G*, 2002, 427–428.
392 P. Chellan and R. H. Nagaraj, Early glycation products produce pentosidine cross-links on native proteins: Novel mechanism of pentosidine formation and propagation of glycation, *J. Biol. Chem.*, 2001, **276**, 3895–3903.
393 L. Kennedy and J. W. Baynes, Nonenzymatic glycosylation and the chronic complications of diabetes: An overview, *Diabetologia*, 1984, **26**, 93–98.
394 H. Vlassara, R. Bucala, and L. Striker, Pathogenic effects of advanced glycosylation: Biochemical, biologic and clinical implications for diabetes and aging, *Lab. Invest.*, 1994, **70**, 138–151.
395 M. Brownlee, H. Vlassara, A. Kooney, P. Ulrich, and A. Cerami, Aminoguanidine prevents diabetes-induced arterial wall protein crosslinking, *Science*, 1986, **232**, 1629–1632.
396 Z. Makita, H. Vlassara, E. Rayfield, K. Cartwright, E. Friedman, R. Rodby, A. Cerami, and R. Bucala, Hemoglobin-AGE: A circulating marker of advanced glycosylation, *Science*, 1992, **258**, 651–653.
397 B. S. Szwergold, K. Taylor, S. Lal, B. Su, F. Kappler, and T. R. Brown, Identification of a novel protein kinase activity specific for Amadori adducts on glycated proteins, *Diabetes*, 1997, **46**, 108A.

398 B. S. Szwergold, S. K. Howell, and P. J. Beisswenger, Nonenzymatic glycation/enzymatic deglycation: A novel hypothesis on the etiology of diabetic complications, in *G*, 2002, 143–152.
399 J. W. Baynes, The role of oxidative stress in the development of complications in diabetes, *Diabetes,* 1991, **40**, 405–412.
400 S. P. Wolff, Z. Y. Jiang, and J. V. Hunt, Protein glycation and oxidative stress in diabetes mellitus and ageing, *Free Radical Biol. Med.,* 1991, **10**, 339–352.
401 K. J. Wells-Knecht, D. V. Zyzak, J. E. Litchfield, S. R. Thorpe, and J. W. Baynes, Mechanism of autoxidative glycosylation: Identification of glyoxal and arabinose as intermediates in autoxidative modification of proteins by glucose, *Biochemistry,* 1995, **34**, 3702–3709.
402 T. Hayashi and M. Namiki, Role of sugar fragmentation in the Maillard reaction, in *C*, 1986, 29–38.
403 K. J. Wells-Knecht, S. R. Thorpe, and J. W. Baynes, Pathways of formation of glycoxidation products during glycation of collagen, *Biochemistry,* 1995, **34**, 15132–15141.
404 M. A. Glomb and V. M. Monnier, Mechanism of protein modification by glyoxal and glycolaldehyde, reactive intermediates of the Maillard reaction, *J. Biol. Chem.,* 1995, **270**, 10017–10026.
405 M. X. Fu, J. R. Requena, A. J. Jenkins, T. J. Lyons, J. W. Baynes, and S. R. Thorpe, The advanced glycation end-product, N^ε-(carboxymethyl)lysine (CML), is a product of both lipid peroxidation and glycoxidation reactions, *J. Biol. Chem.,* 1996, **271**, 9982–9986.
406 M. M. Anderson, J. R. Requena, S. L. Hazen, M. X. Fu, S. R. Thorpe, and J. W. Heinecke, A pathway for the generation of advanced glycosylation end products by the myeloperoxidase system of activated macrophages, *Circulation,* 1997, **8**, I-37 (Abstr.).
407 C. M. Hayashi, R. Nagai, K. Miyazaki, F. Hayase, T. Araki, T. Ono, and S. Horiuchi, Conversion of Amadori product of the Maillard reaction to N^ε-(carboxymethyl)lysine by short-term heating process, in *G*, 2002, 409–410.
408 T. Niwa, T. Katsuzaki, S. Miyazaki, T. Miyazaki, Y. Ishizaki, F. Hayase, N. Tatemichi, and Y. Takei, Immunohistochemical detection of imidazolone, a novel advanced glycation end product, in kidneys and aortas of diabetic patients, *J. Clin. Invest.,* 1997, **99**, 1272–1280.
409 P. J. Thornalley, Advanced glycation and the development of diabetic complications. Unifying the involvement of glucose, methylglyoxal and oxidative stress, *Endocrinol. Metab.,* 1996, **3**, 149–166.
410 J. E. Litchfield, S. R. Thorpe, and J. W. Baynes, Oxygen is not required for the browning and crosslinking of protein by pentoses: Relevance to Maillard reactions in vivo, *Int. J. Biochem. Cell Biol.,* 1999, **31**, 1297–1305.
411 S. Lal, B. S. Szwergold, A. H. Taylor, W. C. Randall, F. Kappler, K. Wells-Knecht, J. W. Baynes, and T. R. Brown, Metabolism of fructose-3-phosphate in the diabetic rat lens, *Arch. Biochem. Biophys.,* 1995, **318**, 191–199.
412 R. Liardon, D. de Weck-Gaudard, G. Philippossian, and P.-A. Finot, Identification of N^ε-carboxymethyllysine: A new Maillard reaction product, in rat, *J. Agric. Food Chem.,* 1987, **35**, 427–431.
413 S. K. Wadman, P. K. de Bree, F. J. van Sprang, J. P. Kamerling, J. Haverkamp, and J. F. G. Vliegenthart, N^ε-(Carboxymethyl)lysine, a constituent of human urine, *Clin. Chim. Acta,* 1975, **59**, 313–320.
414 G. H. Chiang, High-performance liquid chromatographic determination of ε-pyrrolelysine in processed food, *J. Agric. Food Chem.,* 1988, **36**, 506–509.
415 Y. Al-Abed, T. Mitsuhashi, P. Ulrich, and R. Bucala, Novel modification of N^α-BOC-arginine and N^α-CBZ-lysine by methylglyoxal, *Bioorg. Med. Chem. Lett.,* 1996, **6**, 1577–1578.

Bibliography

416 J. A. Gerrard, P. K. Brown, and S. E. Fayle, Maillard crosslinking of food proteins I: The reaction of glutaraldehyde, formaldehyde and glyceraldehyde with ribonuclease, *Food Chem.*, 2002, **79**, 343–349.

417 J. W. Baynes and S. R. Thorpe, Role of oxidative stress in diabetic complications: A new perspective on an old paradigm, *Diabetes,* 1999, **48**, 1–9.

418 A. A. Booth, R. G. Khalifa, P. Todd, and B. G. Hudson, In vitro kinetic studies of formation of antigenic advanced glycation end products (AGEs). Novel inhibition of post-Amadori glycation pathways, *J. Biol. Chem.,* 1997, **272**, 5430–5437.

419 J. W. Baynes, From life to death — the struggle between chemistry and biology during aging; the Maillard reaction as an amplifier of genomic damage, *Biogerontology,* 2000, **1**, 235–246.

420 E. B. Frye, T. P. Degenhardt, S. R. Thorpe, and J. W. Baynes, Role of the Maillard reaction in aging tissue proteins: Advanced glycation end product-dependent increase in imidazolium cross-links in human lens, *J. Biol. Chem.,* 1998, **273**, 18714–18719.

421 M. U. Ahmed, E. B. Frye, T. P. Degenhardt, S. R. Thorpe, and J. W. Baynes, N^ε-(Carboxyethyl)lysine, a product of the chemical modification of proteins by methylglyoxal, increases with age in human lens proteins, *Biochem. J.,* 1997, **324**, 565–570.

422 D. G. Dyer, J. A. Blackledge, S. R. Thorpe, and J. W. Baynes, Formation of pentosidine during nonenzymatic browning of proteins by glucose. Identification of glucose and carbohydrates as possible precursors of pentosidine *in vivo*, *J. Biol. Chem.,* 1991, **266**, 11654–11660.

423 M. C. Wells-Knecht, T. G. Huggins, D. G. Dyer, S. R. Thorpe, and J. W. Baynes, Oxidized amino acids in lens protein with age. Measurement of *o*-tyrosine and dityrosine in the aging human lens, *J. Biol. Chem.,* 1993, **268**, 12348–12352.

424 B. Buckingham and K. M. Reiser, Relationship between content of lysyl oxidase-dependent cross-links in skin collagen, nonenzymatic glycosylation, and long-term complications in type I diabetes mellitus, *J. Clin. Invest.,* 1990, **86**, 1046–1054.

425 N. Verzijl, J. DeGroot, E. Oldehinkel, R. A. Bank, S. R. Thorpe, J. W. Baynes, M. T. Bayliss, J. W. J. Bijlsma, F. P. J. G. Lafeber, and J. M. TeKoppele, Age-related accumulation of Maillard reaction products in human articular cartilage collagen, *Biochem. J.,* 2000, **350**, 381–387.

426 N. Verzijl, J. DeGroot, S. R. Thorpe, R. A. Bank, J. N. Shaw, T. J. Lyons, J. W. J. Bijlsma, F. P. J. G. Lafeber, J. W. Baynes, and J. M. TeKoppele, Effect of collagen turnover on accumulation of advanced glycations end products, *J. Biol. Chem.,* 2000, **275**, 39027–39031.

427 A. Maroudas, G. Palla, and E. Gilav, Racemization of aspartic acid in human articular cartilage, *Connect. Tissue Res.,* 1992, **28**, 161–169.

428 S. Ohtani and K. Yamamoto, Age estimation using racemization of amino acid in human dentin, *J. Forensic Sci.,* 1991, **36**, 792–800, *Chem. Abstr.*, 1992, **117**, 126029u.

429 X. Ling, R. Nagai, N. Sakashita, M. Takeya, K. Takahashi, and S. Horiuchi, Immunohistochemical distribution and quantitative biochemical detection of advanced glycation end products in rats from fetal to adult life, in *G*, 2002, 137–142.

430 T. Jono, R. Nagai, K. Miyazaki, N. Ahmed, P. J. Thornalley, T. Kitamura, and S. Horiuchi, Detection of 3-deoxyglucosone-derived AGE structures in vitro, in *G*, 2002, 239–242.

431 H. Siren, P. Laitinen, U. Turpeinen, and P. Karppinen, Direct monitoring of glycohemoglobin A_{1c} in the blood samples of diabetic patients by capillary electrophoresis: Comparison with an immunoassay method, *J. Chromatogr., A* 2002, **979**, 201–207.

432 R. H. Nagaraj, T. S. Kern, D. R. Sell, J. Fogarty, and R. L. Engerman, Evidence of a glycemic threshold for the formation of pentosidine in diabetic dog lens but not in collagen, *Diabetes,* 1996, **45**, 587–594, via *Chem. Abstr.*, 1996, **125**, 55251n.

433 Y. Morimitsu, K. Kubota, T. Tashiro, E. Hashizume, T. Kamiya, and T. Osawa, Inhibitory effect of anthocyanins and colored rice on diabetic cataract formation in the rat lens, in *G*, 2002, 503–508.
434 E. Spoerl and T. Seiler, Techniques for stiffening the cornea, *J. Refract. Surg.*, 1999, **15**, 711–713.
435 Y. Izuhara, T. Miyata, Y. Ueda, and K. Kurokawa, Accumulation of carbonyls accelerates the formation of two advanced glycation endproducts: Carbonyl stress in uremia, in *G*, 2002, 381–382.
436 A. Moh, N. Sakata, A. Noma, N. Uesugi, S. Takebayashi, R. Nagai, and S. Seikoh, Glycoxidation and lipoperoxidation in the collagen of the myocardium in hemodialysis patients, in *G*, 2002, 429.
437 A. Noma, N. Sakata, Y. Yamamoto, K. Okamoto, A. Moh, S. Takebayashi, R. Nagai, and S. Horiuchi, An increase in elastin-associated pentosidine of aorta in hemodialysis patients, in *G*, 2002, 431–432.
438 K. Yoshimura, M. Nishimura, T. Hasegawa, H. Terawaki, T. Nakazato, K. Sakamoto, S. Arita, K. Nakajima, H. Kashiwabara, K. Hamaguti, R. Nagai, K. Horiuchi, and K. Yamada, Effect of successful renal transplantation on coronary AGE accumulation of uremic heart, in *G*, 2002, 183–191.
439 N. Uesugi, N. Sakata, S. Horiuchi, J. Meng, and S. Takebayashi, Glycoxidation induces vascular smooth muscle cell injury in diabetes through mediation of membrane attack complement, in *G*, 2002, 439–440.
440 S. Agalou, N. Karachalias, P. J. Thornalley, B. Tucker, and A. B. Dawnay, Estimation of α-oxoaldehydes formed from the degradation of glycolytic intermediates and glucose fragmentation in blood plasma of human subjects with uraemia, in *G*, 2002, 181–182.
441 R. Inagi, T. Miyata, Y. Ueda, A. Yoshino, M. Nangaku, C. van Ypersele de Strihou, and K. Kutokawa, Efficient lowering of carbonyl stress by the glyoxalase in peritoneal dialysis, in *G*, 2002, 359–360.
442 D. J. Millar, P. J. Thornalley, C. Holmes, and A. Dawnay, In vitro kinetics of AGE formation with PD fluid resembles that of glucose degradation products rather than glucose, in *G*, 2002, 475–477.
443 K. Yoshihara, Y. Nagayama, H. Horiguchi, S.-i. Yoshida, S. Tohyoh, S. Takahashi, H. Maruyama, N. Saito, and M. Beppu, Acceleration of pentosidine formation by medication, in *G*, 2002, 425–426.
444 H. J. Prochaska and P. Talahay, Regulatory mechanisms of monofunctional and bifunctional anticarcinogenic enzyme inducers in murine liver, *Cancer Res.*, 1988, **48**, 4776–4782, via *Chem. Abstr.*, 1988, **109**, 163043z.
445 S. M. Monti, R. G. Bailey, and J. M. Ames, The influence of pH on the non-volatile reaction products of aqueous Maillard model systems by HPLC with diode array detection, *Food Chem.*, 1998, **62**, 369–375.
446 M. Anese, and M. C. Nicoli, Comparison among different methodologies currently used for assessing the antioxidant activity of foods, in *Melanoidins in Food and Health*, Vol. 2, J. M. Ames (ed), European Communities, Luxembourg, 2001, 53–63.
447 C. Rice-Evans, Methods to quantify antioxidant activity of tea/tea extracts *in vitro*, *Crit. Rev. Food Nutr.*, 2001, **41**, 405–407.
448 I. F. F. Benzie and J. J. Strain, The ferric reducing ability of plasma (FRAP) as a measure of "antioxidant power": the FRAP assay, *Anal. Biochem.*, 1996, **239**, 70–76.
449 M. Cioroi, Antioxidative effect of Maillard reaction products in coffee brew, in *Melanoidins in Food and Health*, Vol. 3, V. Fogliano and T. Henle (eds), European Communities, Luxembourg, 2002, 159–162.

450 J. T. Tanner and S. A. Barnett, Methods of analysis for infant formula: Food and Drug Administration and Infant Formula Council collaborative study, *J. Assoc. Off. Anal. Chem.*, 1985, **68**, 514–522.

451 T. Shimamura, A. Takamori, H. Ukeda, S. Nagata, and M. Sawamura, Relationship between reduction of tetrazolium salt XTT and DNA strand breakage with aminosugars, *J. Agric. Food Chem.*, 2000, **48**, 1204–1209.

452 J. W. Hamilton and A. L. Tappel, Evaluation of antioxidants by a rapid polarographic method, *J. Am. Oil Chem. Soc.*, 1963, **40**, 52–54.

453 C. H. Lea, Methods for determining peroxide in lipids, *J. Sci. Food Agric.*, 1952, **3**, 586–594.

454 B. Brand and K. Eichner, Antioxidative properties of melanoidins of different origin, in *Melanoidins in Food and Health*, Vol. 3, V. Fogliano and T. Henle (eds), European Communities, Luxembourg, 2002, 143–158.

455 D. Bright, G. G. Stewart, and H. Patino, A novel assay for antioxidant potential of specialty malts, *J. Am. Soc. Brew. Chem.*, 1999, **57**, 133–137.

456 J. I. Gray, Measurement of lipid oxidation: A review, *J. Am. Oil Chem. Soc.*, 1978, **55**, 539–546.

457 F. Tubaro, E. Micossi, and F. Ursini, The antioxidant capacity of complex mixtures by kinetic analysis of crocin bleaching inhibition, *J. Am. Oil Chem. Soc.*, 1996, **73**, 173–179.

458 K. Yanagimoto, K.-G. Lee, H. Ochi, and T. Shibamoto, Antioxidative activity of heterocyclic compounds formed in Maillard reaction products, in *G*, 2002, 335–340.

459 V. Fogliano, V. Verdee, G. Randazzo, and A. Ritieni, Method for measuring antioxidant activity and its application to monitoring the antioxidant capacity of wines, *J. Agric. Food Chem.*, 1999, **47**, 1035–1040.

460 D.-O. Kim, K. W. Lee, J. C. Lee, and C. Y. Lee, Vitamic C equivalent oxidant capacity (VCEAC) of phenolic phytochemicals, *J. Agric. Food Chem.*, 2002, **50**, 3713–3717.

461 D. L. Berner, J. A. Conte, and G. A. Jacobson, Rapid method for determining antioxidant activity and fat stability, *J. Am. Oil Chem. Soc.*, 1974, **51**, 292–296.

462 R. J. DeLange and A. N. Glazer, Phycoerythrin fluorescence-based assay for peroxy radicals: A screen for biologically relevant protective agents, *Anal. Biochem.*, 1989, **177**, 300–306.

463 G. P. Rizzi, Electrochemical study of the Maillard reaction, *J. Agric. Food Chem.*, 2003, **51**, 1728–1731.

464 D. Huang, B. Ou, M. Hampsch-Woodill, J. A. Flanagan, and R. L. Prior, High-throughput assay of oxygen radical absorbance capacity (ORAC) using a multichannel liquid handling system coupled with a microplate fluorescence reader in 96-well format, *J. Agric. Food Chem.*, 2002, **50**, 4437–4444.

465 K.-I. Yeum, G. Aldini, H.-Y. Chung, N. I. Krinsky, and R. M. Russell, The activities of antioxidant nutrients in human plasma depend on the localization of attacking radical species, *J. Nutr.*, 2003, **133**, 2688–2691.

466 B. Cämmerer, M. Anese, B. Brand, M. Cioroi, C. Liégeois, and G. E. Vegarud, Antioxidative activity of melanoidins, in *Melanoidins in Food and Health*, Vol. 1, J. M. Ames (ed), European Communities, Luxembourg, 2000, 49–60.

467 H. Iwainsky and C. Franzke, Zur antioxydativen Wirkung der Melanoide. III, *Deutsch. Lebensm. Rundschau*, 1956, **52**, 129–133, via *Food Sci. Abstr.*, 1956, **28**, 1894.

468 F. Bressa, N. Tesson, M. D. Rosa, A. Sensidoni, and F. Tubaro, Antioxidant effect of Maillard reaction products: Application to a butter cookie of a competition kinetics analysis, *J. Agric. Food Chem.*, 1996, **44**, 692–695.

469 H. Lingnert and C. E. Eriksson, Antioxidative effect of Maillard reaction products, in *A*, 1981, 453–466.

470 A. N. Wijewickreme and D. D. Kitts, Influence of reaction conditions on the oxidative behavior of model Maillard reaction products, *J. Agric. Food Chem.*, 1997, **45**, 4571–4576.
471 G.-C. Yen, L. C. Tsai, and J.-D. Lii, Antimutagenic effect of Maillard browning products obtained from amino acids and sugars, *Food Chem. Toxic.*, 1992, **30**, 127–132.
472 A. N. Wijewickreme and D. D. Kitts, Oxidative reactions of model Maillard reaction products and α-tocopherol in a flour-lipid mixture, *J. Food Sci.*, 1998, **63**, 466–471.
473 N. Yamaguchi, Y. Koyama, and M. Fujimaki, Fractionation and antioxidative activity of browning reaction products between D-xylose and glycine, in *A*, 1981, 429–439.
474 N. van Chuyen, N. Utsunomiya, A. Hidaka, and H. Kato, Antioxidative effect of Maillard reaction products in vivo, in *D*, 1990, 285–290.
475 M. Ninomiya, T. Matsuzaki, and H. Shigematsu, Formation of reducing substances in Maillard reaction between D-glucose and γ-aminobutyric acid, *Biosci. Biotech. Biochem.*, 1992, **56**, 806–807.
476 A. J. Bedinghaus and H. W. Ockerman, Antioxidative Maillard reaction products from reducing sugars and free amino acids in cooked ground pork patties, *J. Food Sci.*, 1995, **60**, 992–995.
477 K. Kawashima, H. Itoh, and I. Chibata, Antioxidant activity of browning oroducts prepared from low molecular carbonyl compounds and amino acids, *J. Agric. Food Chem.*, 1977, **25**, 202–204.
478 G. R. Waller, R. W. Beckel, and B. O. Adeleye, Conditions for the synthesis of antioxidative arginine-xylose Maillard reaction products, in *B*, 1983, 125–140.
479 T. Obretenov, S. Ivanov, and D. Peeva, Antioxidative activity of Maillard reaction products obtained from hydrolysates, in *C*, 1986, 281–290.
480 G. Moon, M. Lee, Y. Lee, and G. Trakoontivakorn, Main component of soy sauce representing antioxidative activity, in *G*, 2002, 509–510.
481 H. Lingnert and G. Hall, Formation of antioxidative Maillard reaction products during food processing, in *C*, 1986, 273–279.
482 M. Murakami, A. Shigeeda, K. Danjo, T. Yamaguchi, H. Takamura, and T. Matoba, Radical-scavenging activity and brightly colored pigments in the early stage of the Maillard reaction, *J. Food Sci.*, 2002, **67**, 93–96.
483 L. Manzocco, S. Calligaris, and M. C. Nicoli, Assessment of pro-oxidant activity of foods by kinetic analysis of crocin bleaching, *J. Agric. Food Chem.*, 2002, **50**, 2767–2771.
484 C. Puscasu and I. Birlouez-Aragon, Intermediary and/or advanced Maillard products exhibit prooxidant activity on Trp: in vitro study on α-lactalbumin, *Food Chem.*, 2002, **78**, 399–406.
485 H. Lingnert, C. E. Eriksson, and G. R. Waller, Characterization of antioxidative Maillard reaction products from histidine and glucose, in *B*, 1983, 335–345.
486 K. Eichner, Antioxidative effect of Maillard reaction intermediates, in *A*, 1981, 441–451.
487 B. Brand and K. Eichner, Reducing, radical scavenging and antioxidative properties of model melanoidins, in *Melanoidins in Food and Health*, J. M. Ames (ed), Vol. 2, European Communities, Luxembourg, 2001, 151–158.
488 K.-H. Wagner, S. Derkits, M. Herr, W. Schuh, and I. Elmadfa, Antioxidative potential of melanoidins isolated from a roasted glucose-glycine model, *Food Chem.*, 2002, **78**, 375–382.
489 Y. Yoshimura, T. Iijima, T. Watanabe, and H. Nakazawa, Antioxidative effect of Maillard reaction products using glucose-glycine model system, *J. Agric. Food Chem.*, 1997, **45**, 4106–4109.
490 F. Hayase, Scavenging of active oxygen by melanoidins, in *The Maillard Reaction: Consequences for the Chemical and Life Sciences*, R. Ikan (ed), Wiley, Chichester, 1996, 89–104.

491 G.-C. Yen and P.-P. Hsieh, Antioxidative activity and scavenging effects on active oxygen of xylose-lysine Maillard reaction products, *J. Sci. Food Agric.*, 1995, **67**, 415–420.
492 M. C. Nicoli, M. Anese, M. T. Parpinel, S. Franceschi, and C. R. Lerici, Study on loss and/or formation of antioxidants during food processing and storage, *Cancer Lett.*, 1997, **114**, 1–4, via *Chem. Abstr.*, 1997, **126**, 316597m.
493 M. C. Nicoli, M. Anese, L. Manzocco, and C. R. Lerici, Antioxidant properties of coffee brews in relation to the roasting degree, *Lebensm. Wiss. Technol.*, 1997, **30**, 292–297.
494 F. J. Morales and M.-B. Babbel, Melanoidins exert a weak antiradical activity in watery fluids, *J. Agric. Food Chem.*, 2002, **50**, 4657–4561.
495 M. Anese and M. C. Nicoli, Antioxidant properties of ready-to-drink coffee brews, *J. Agric. Food Chem.*, 2003, **51**, 942–946.
496 M. Richelle, I. Tavazzi, and E. Offord, Comparison of the antioxidant activity of commonly consumed polyphenolic beverages (coffee, cocoa, and tea) prepared per cup serving, *J. Agric. Food Chem.*, 2001, **49**, 3438–3442.
497 P. Bersuder, M. Hole, and G. Smith, Antioxidants from a heated histidine-glucose model system. Investigation of the copper(II) binding ability, *J. Am. Oil Chem. Soc.*, 2001, **78**, 1079–1082.
498 S. M. Antony, I. Y. Han, J. R. Rieck, and P. L. Dawson, Antioxidative effect of Maillard reaction products added to turkey meat during heating by addition of honey, *J. Food Sci.*, 2002, **67**, 1719–1724.
499 K. D. Ross, Reduction of the azo food dyes FD&C Red 2 (amaranth) and FD&C Red 40 by thermally degraded D-fructose and D-glucose, *J. Agric. Food Chem.*, 1975, **23**, 475–478.
500 K. Yanagimoto, K. G. Lee, H. Ochi, and T. Shibamoto, Antioxidant activity of heterocyclic compounds found in coffee volatiles produced by Maillard reaction, *J. Agric. Food Chem.*, 2002, **50**, 5480–5484.
501 E. Dworschák and L. Szabó, Formation of antioxidative materials in the preparation of meals, in *C*, 1986, 311–319.
502 S. Katayama, J. Shima, and H. Saeki, Solubility improvement of shellfish muscle protein by reaction with glucose and its soluble state in low-ionic-strength medium, *J. Agric. Food Chem.*, 2002, **50**, 4327–4332.
503 N. Matsudomi, K. Nakano, A. Soma, and Ochi, A. Improvement of gel properties of dried egg white by modification with galactomannan through Maillard reaction, *J. Agric. Food Chem.*, 2002, **50**, 4113–4118.
504 J. A. Gerrard and P. K. Brown, Protein cross-linking in food: mechanisms, consequences, applications, in *G*, 2002, 211–215.
505 M. Petracco, *Espresso* coffee foam: A Maillard-mediated phenomenon? Description and analytical characterisation, in *Melanoidins in Food and Health*, J. M. Ames (ed), Vol. 2, European Communities, Luxembourg, 2001, 31–42.
506 M. Wahyuni, M. J. C. Crabbe, and J. M. Ames, Ribonuclease A/glucose-6-phosphate interactions. Monitoring by capillary electrophoresis and effect on emulsion activity, in *G*, 2002, 463–464.
507 J. Al-Hakkak and S. Kavale, Improvement of emulsification properties of sodium caseinate by conjugating to pectin through the Maillard reaction, in *G*, 2002, 491–499.
508 M. Buglione and J. Lozano, Nonenzymatic browning and chemical changes during grape juice storage, *J. Food Sci.*, 2002, **67**, 1538–1543.
509 T. Hofmann and P. Schieberle, Chemical interactions between odor-active thiols and melanoidins involved in the aroma staling of coffee beverage, *J. Agric. Food Chem.*, 2002, **50**, 319–326.

510 T. Hofmann, M. Czerny, S. Calligaris, and P. Schieberle, Model studies on the influence of coffee melanoidins on flavor volatiles of coffee beverages, *J. Agric. Food Chem.*, 2001, **49**, 2382–2386.

511 A. G. Miller and J. A. Gerrard, What residues are required for protein crosslinking and how does this process affect enzyme function? in *G*, 2002, 451–452.

512 T. Griffith and J. A. Johnson, Relation of the browning reaction to storage stability of sugar cookies, *Cereal Chem.*, 1957, **34**, 159–169.

513 N. Yamaguchi, Y. Yokoo, and Y. Koyama, Studies on the browning reaction products yielded by reducing sugar and amino acids. I. Effect of browning reaction products on the stability of fats contained in biscuits and cookies, *Nippon Shok. Kogyo Gakk.*, 1964, **11**, 184–189, via *Chem. Abstr.*, 1966, **64**, 14862h.

514 N. Yamaguchi, S. Naito, Y. Yokoo, and M. Fujimaki, Application of protein hydrolysates to biscuits as antioxidant, *Nippon Shok. Kogyo Gakk.*, 1980, **27**, 56–59, via *Chem. Abstr.*, 1980, **93**, 6338h.

515 H. Lingnert, Antioxidative Maillard reaction products. III. Application in cookies, *J. Food Proc. Preserv.*, 1980, **4**, 219–233.

516 R. H. Anderson, D. H. Moran, T. E. Huntley, and J. L. Holahan, Responses of cereals to antioxidants, *Food Technol.*, 1963, **17**, 1587–1592.

517 Y. Tomita, Antioxidant activity of amino-carbonyl reaction products. 5. Application tests of reaction products of tryptophan with glucose, *Kagoshima Daigaku Nogakubu Gakujutsu Hokoku*, 1972, **22**, 115–121, via *Chem. Abstr.*, 1973, **78**, 96215h.

518 J. Hauri, F. Escher, A. Denzler, and H. Neukom, The influence of processing conditions on the storage stability of drum-dried cereal flakes, *Lebensm. Wiss. Technol.*, 1982, **15**, 235–241.

519 H. S. Cheigh, J. S. Lee, and C. Y. Lee, Antioxidant characteristics of melanoidin-related products fractionated from fermented soybean sauce, *J. Korean Soc. Food Nutr.*, 1993, **22**, 570–575, via *Chem. Abstr.*, 1994, **120**, 268556z.

520 M. Anese, L. Manzocco, M. C. Nicoli, and C. R. Lerici, Antioxidant properties of tomato juice as affected by heating, *J. Sci. Food Agric.*, 1999, **79**, 750–754.

521 J. D. Findlay, C. Higginbottom, J. A. B. Smith, and C. H. Lea, The effect of the pre-heating temperature on the bacterial count and storage life of whole milk powder spray-dried by the Krause process, *J. Dairy Res.*, 1946, **14**, 378–399.

522 L. Vandewalle and A. Huyghebaert, The antioxidant activity of the non-enzymatic browning reaction in sugar-protein systems, *Med. Fac. Landbouww. Rijksuniv. Gent.*, 1980, **45**, 1277–1286, via *Food Sci. Technol. Abstr.*, 1983, **15**, 2A132.

523 E. Binder, F. Becker, J. Grubhofer, and W. Scholz, Oxidation stability and sensoric condition of high heat-, medium heat- and low heat spray-dried whole milk, *Oesterr. Milchwirtsch.*, 1981 (17, Beilage 2), **36**, 9–16, via *Food Sci. Technol. Abstr.*, 1982, **14**, 12P1854.

524 A. P. Hansen and F. L. Hemphill, Utilization of heat to increase shelf life of blended acid whey and buttermilk powder for frozen desserts, *J. Dairy Sci.*, 1984, **67**, 54–55.

525 C. Franzke and H. Iwainsky, Zur antioxydativen Wirkung der Melanoide. I, *Deutsch. Lebensm. Rundschau*, 1954, **50**, 251–254, via *Food Sci. Abstr.*, 1955, **27**, 717.

526 D. V. Josephson and C. D. Dahle, Heating makes butterfat keep, *Food Ind.*, 1945, **17**, 630–633, via *Chem. Abstr.*, 1945, **39**, 4987.

527 C. D. Evans, H. A. Moser, P. M. Cooney, and J. E. Hodge, Amino-hexose-reductones as antioxidants, *J. Am. Oil Chem. Soc.*, 1958, **35**, 84–88.

528 M. Maleki, Effect of nonenzymic browning in the presence of glucose and glycine on the development of rancidity in corn oil, *Fette, Seifen, Anstrichm.*, 1973, **75**, 103–104; via *Chem. Abstr.*, 1973, **79**, 17141a.

529 K. Taguchi, K. Iwami, M. Kawabata, and F. Ibuki, Antioxidant effects of wheat gliadin and hen's egg white in powder model systems: protection against oxidative deterioration of safflower oil and sardine oil, *Agric. Biol. Chem.*, 1988, **52**, 539–545.
530 M. W. Zipser and B. M. Watts, Lipid oxidation in heat-sterilized beef, *Food Technol.*, 1961, **15**, 445–447.
531 K. Sato, G. R. Hegarty, and H. K. Herring, The inhibition of warmed-over flavor in cooked meats, *J. Food Sci.*, 1973, **38**, 398–403.
532 M. Dagerskog, B. Karlström, and N. Bengtsson, Influence of degree of precooking on quality of frozen sliced beef and patties, *Proc. Europ. Mtg Meat Res. Workers, Malmo*, 1976, **22**, Jl:1-Jl:7, via *Food Sci. Technol. Abstr.*, 1977, **9**, 6S1022.
533 M. A. Einerson and G. A. Reineccius, Inhibition of warmed-over flavor in retorted turkey by antioxidants formed during processing, *J. Food Proc. Preserv.*, 1977, **1**, 279–291.
534 M. Tanaka, C. W. Kuei, W. Yuji, and T. Taguchi, Application of antioxidative Maillard reaction products from histidine and glucose to sardine products, *Nippon Suisan Gakk.*, 1988, **54**, 1409–1414, via *Chem. Abstr.*, 1988, **109**, 169145f.
535 M. J. Perkins, Spin trapping, *Adv. Phys. Org. Chem.*, 1980, **17**, 1–64.
536 L.-C. Maillard, Formation d'humus et de combustibles mineraux sans intervention de l'oxygene atmospherique, des microorganismes, des hautes temperatures, ou des fortes pressions, *C. R. Hebd. Seances Acad. Sci.*, 1912, **155**, 1554–1556.
537 L.-C. Maillard, Formation des matieres humiques par action de polypeptides sur les sucres, *C. R. Hebd. Seances Acad. Sci.*, 1913, **156**, 1159–1160.
538 E. Tipping, *Cation Binding by Humic Substances*, Cambridge University Press, Cambridge, 2002.
539 R. Ikan, Y. Rubinsztain, A. Nissenbaum, and I. R. Kaplan, Geochemical aspects of the Maillard reaction, in *The Maillard Reaction: Consequences for the Chemical and Life Sciences*, R. Ikan (ed), Wiley, Chichester, 1996, 1–25.
540 A. Jokic, A. I. Frenkel, and P. M. Huang, Effect of light on birnessite catalysis of the Maillard reaction and its implication in humification, *Can. J. Soil Sci.*, 2001, **81**, 277–283.
541 P. Arfaioli, O. L. Pantani, M. Bosetto, and G. G. Ristori, Influence of clay minerals and exchangeable cations on the formation of humic-like substances (melanoidins) from D-glucose and L-tyrosine, *Clay Minerals*, 1999, **34**, 487–497.
542 J. Burdon, Are the traditional concepts of structures of humic substances realistic? *Soil Sci.*, 2001, **166**, 752–769.
543 R. P. Evershed, H. A. Bland, P. F. van Bergen, J. F. Carter, M. C. Horton, and P. A. Rowley-Conwy, Volatile compounds in archaeological plant remains and the Maillard reaction during decay of organic matter, *Science*, 1997, **278**, 432–433.
544 Anon. Yellowing of textiles, *J. Soc. Dyers Col.*, 1986, **102**, 139.
545 L. Trezl, P. Bako, V. Horvath, I. Rusznak, and L. Toke, Adaption of Maillard reaction on keratin type proteins with a special focus on the reaction of glucose based crown ethers, in *E*, 1994, 411.
546 J. A. Johnson and R. M. Fusaro, Alteration of skin surface protein with dihydroxyacetone: A useful application of the Maillard browning reaction, in *E*, 1994, 114–119.
547 T. J. Painter, Concerning the wound-healing properties of *Sphagnum* holocellulose: the Maillard reaction in pharmacology, *J. Ethno-Pharmacol.*, 2003, **88**, 145–148.
548 R. C. George, R. J. Barbuch, E. W. Huber, and B. T. Regg, Investigation into the yellowing on aging of Sabril(R) tablet cores, *Drug Dev. Ind. Pharm.*, 1994, **20**, 3023–3032, via *Chem. Abstr.*, 1994, **121**, 263649t.
549 R. O. Macedo, T. G. do Nascimento, and J. W. E. Veras, Comparison of generic hydrochlorothiazide formulations by means of TG and DSC coupled to a photovisual

system, *J. Therm. Anal. Calorim.*, 2001, **64**, 757–763, via *Chem. Abstr.*, 2001, **135**, 335072p.

550 M. Otsuka, T. Kurata, and N. Arakawa, Isolation and characterization of an intermediate product in the degradation of 2,3-diketo-L-gulonic acid, *Agric. Biol. Chem.*, 1986, **50**, 531–533.

551 M. S. Feather, Dicarbonyl sugar derivatives and their role in the Maillard reaction, in *Thermally Generated Flavors: Maillard, Microwave, and Extrusion Processes*; T. H. Parliment, M. J. Morello, and R. J. McGorrin (eds), American Chemical Society, Washington, DC, 1994, 127–141.

552 K. M. Clegg and A. D. Morton, Carbonyl compounds and the non-enzymic browning of lemon juice, *J. Sci. Food Agric.*, 1965, **16**, 191–198.

553 B. L. Wedzicha, *Chemistry of Sulphur Dioxide in Foods*, Elsevier Applied Science, London, 1984.

554 K. M. Clegg, Non-enzymic browning of lemon juice, *J. Sci. Food Agric.*, 1964, **15**, 878–885.

555 K. M. Clegg, Citric acid and the browning of solutions containing ascorbic acid, *J. Sci. Food Agric.*, 1966, **17**, 546–549.

556 M. C. Manso, F. A. R. Oliveira, J. C. Oliveira, and J. M. Friar, Modelling ascorbic acid thermal degradation and browning in orange juice under aerobic conditions, *Int. J. Food Sci. Technol.*, 2001, **36**, 303–312.

557 M. Murata, Y. Shinoda, and S. Homma, Browning of model orange juice solution and changes in the components, in *G*, 2002, 459–460.

558 E. Arena, B. Fallico, and E. Maccarone, Thermal damage in blood orange juice: Kinetics of 5-hydroxymethyl-2-furancarboxaldehyde formation, *Int. J. Food Sci. Techn.*, 2001, **36**, 145–151.

559 C. Obretenov, J. Demyttenaere, K. A. Tehrani, A. Adams, M. Kersiene, and N. De Kimpe, Flavor release in the presence of melanodins prepared from L-(+)-ascorbic acid, *J. Agric. Food Chem.*, 2002, **50**, 4244–4250.

560 H. Sakurai, H. Koga, G. Ishikawa, T. Endo, C. Matsuyama, T. Yano, N. Ohta, H. Kumagai, H. T. T. Nguyen, and J. Pokorny, Formation of bitter substances in solutions containing vitamin C and aspartame, in *G*, 2002, 383–385.

561 Y. Nishikawa, B. Dmochowska, J. Madaj, M. Satake, P. L. Rinaldi, and V. M. Monnier, Impairment of vitamin C metabolism in STZ diabetic rats revealed with 6-deoxy-6-fluoroascorbic acid, in *G*, 2002, 417–418.

562 Y. Otsuka, E. Ueta, T. Yamamoto, Y. Tadokoro, E. Suzuki, E. Nanba, and T. Kurata, Effect of streptozotocin-induced diabetes on rat liver mRNA level of antioxidant enzymes, in *G*, 2002, 421–423.

563 M. S. Feather and J. F. Harris, Dehydration reactions of carbohydrates, *Adv. Carbohydrate Chem. Biochem.*, 1973, **28**, 161–224.

564 M. Karel and T. P. Labuza, Nonenzymatic browning in model systems containing sucrose, *J. Agric. Food Chem.*, 1968, **16**, 717–719.

565 E.-H. Ajandouz, L. S. Tchiakpe, F. Dalle Ore, A. Benajiba, and A. Puigserver, Effects of pH on caramelization and Maillard reaction kinetics in fructose-lysine model systems, *J. Food Sci.*, 2001, **66**, 926–931.

566 E.-H. Ajandouz and A. Puigserver, Nonenzymatic browning reaction of essential amino acids: effect of pH on caramelization and Maillard reaction kinetics, *J. Agric. Food Chem.*, 1999, **47**, 1786–1793.

567 K. Heyns, R. Stute, and H. Paulsen, Bräunungsreaktionen und Fragmentierungen von Kohlenhydraten. I. Die flüchtigen Abbauprodukte der Pyrolyse von D-Glucose, *Carbohydr. Res.* 1966, **2**, 132–149.

568 R. H. Walter and I. S. Fagerson, Volatile compounds from heated glucose, *J. Food Sci.,* 1968, **33**, 294–297.
569 E. P. a. C. Directive, Food additives other than colours and sweeteners, *Off. J.* 1995, **2/95**, 1–40.
570 Z. Jiang and B. Ooraikul, Reduction in nonenzymatic browning in potato chips and French fries with glucose oxidase, *J. Food Proc. Preserv.,* 1989, **13**, 175–186.
571 H. S. Burton, D. J. McWeeny, and D. O. Biltcliffe, Sulphites and aldose-amino reactions, *Chem. Industr.,* 1962, 219–221.
572 H. S. Burton, D. J. McWeeny, and D. O. Biltcliffe, Non-enzymic browning: The role of unsaturated carbonyl compounds as intermediates and of SO_2 as an inhibitor of browning, *J. Sci. Food Agric.,* 1963, **14**, 911–920.
573 B. L. Wedzicha and D. J. McWeeny, Concentrations of some sulphonates derived from sulphite in certain foods and preliminary studies on the nature of other sulphite derived products, *J. Sci. Food Agric.,* 1975, **26**, 327–335.
574 M. L. Wolfrom and C. S. Rooney, Chemical interactions of amino compounds and sugars. VIII. Influence of water, *J. Am. Chem. Soc.,* 1953, **75**, 5435–5436.
575 T. P. Labuza and M. Saltmarch, The nonenzymatic browning reaction as affected by water in foods, in *Water Activity: Influences on Food Quality*, L. B. Rockland and G. F. Stewart (eds), Academic Press, New York, 1981, 605–650.
576 C. P. Sherwin and T. P. Labuza, Role of moisture in Maillard browning reaction rate in intermediate moisture foods: Comparing solvent phase and matrix properties, *J. Food Sci.,* 2003, **68**, 588–594.
577 K. Eichner and M. Ciner-Doruk, Formation and decomposition of browning intermediates and visible sugar-amine browning reactions, in *Water Activity: Influences on Food Quality*, L. B. Rockland and G. F. Stewart (eds), Academic Press, New York, 1981, 567–603.
578 L. S. Malec, A. S. Pereyra Gonzales, G. B. Naranjo, and M. S. Vigo, Influence of water activity and storage temperature on lysine availability of a milk like system, *Food Res. Int.,* 2002, **35**, 349–353.
579 V. M. Monnier, D. R. Sell, X. Wu, and K. Rutter, The prospects of health and longevity from inhibition of the Maillard reaction in vivo, in *G*, 2002, 9–19.
580 M. Oimomi, N. Igaki, M. Sakai, T. Ohara, S. Babu, and H. Kato, The effects of aminoguanidine on 3-deoxyglucosone in the Maillard reaction, *Agric. Biol. Chem.,* 1989, **53**, 1727–1728.
581 M. Kihara, J. D. Schmelzer, J. F. Poduslo, G. L. Curran, K. K. Nickander, and P. A. Low, Aminoguanidine effects on nerve blood flow, vascular permeability, electrophysiology, and oxygen free radicals, *Proc. Natl Acad. Sci., USA* 1991, **88**, 6107–6111.
582 Z. Makita, S. Radoff, E. J. Rayfield, Z. Yang, E. Skolnik, V. Delaney, E. A. Friedman, A. Cerami, and H. Vlassara, Advanced glycosylation end products in patients with diabetic nephropathy, *New England J. Med.,* 1991, **325**, 836–842, via *Biol. Abstr.*, 1991, **92**, 142693.
583 J. Hirsch, V. V. Mossine, and M. S. Feather, Detection of some dicarbonyl intermediates arising from the degradation of Amadori compounds (the Maillard reaction), *Carbohydr. Res.,* 1995, **273**, 171–177
584 P. J. Thornalley, A. Yurek-George, and O. K. Argirov, Kinetics and mechanism of the reaction of aminoguanidine with the α-oxoaldehydes glyoxal, methylglyoxal, and 3-deoxyglucosone under physiological conditions, *Biochem. Pharmacol.,* 2000, **60**, 55–65.
585 S. Agalou, N. Karachalias, A. B. Dawnay, and P. J. Thornalley, Reaction kinetics of the scavenging of α-oxoaldehydes by aminoguanidine and physiological conditions, in *G*, 2002, 513–515.

586 T. Taguchi, M. Sugiura, Y. Hamada, and I. Miwa, In vivo formation of a Schiff base of aminoguanidine with pyridoxal phosphate, *Biochem. Pharmacol.*, 1998, **55**, 1667–1671.
587 T. Taguchi, M. Sugiura, Y. Hamada, and I. Miwa, Inhibition of advanced protein glycation by a Schiff base between aminoguanidine and pyridoxal, *Eur. J. Pharmacol.*, 1999, **378**, 283–289.
588 T. Taguchi, H. Miyoshi, M. Sugiura, M. Takeuchi, K. Yanagisawa, Y. Watanabe, I. Miwa, and Z. Makita, A glycation inhibitor, aminoguanidine and pyridoxal adduct, suppresses the development of diabetic nephropathy, in *G*, 2002, 435–437.
589 P. Urios, A.-M. Borsos, J. Garaud, S. Feing-Kwong-Chan, and M. Sternberg, At low concentration, aminoguanidine markedly increases pentosidine formation in collagen incubated with glucose, whereas decreasing it at high level, in *G*, 2002, 413–414.
590 J. Liggins, N. Rodda, V. Burnage, J. Iley, and A. Furth, Effect of low concentrations of aminoguanidine on formation of advanced glycation endproducts *in vitro*, in *F*, 1998, 424.
591 S. Battah, N. Ahmed, and P. J. Thornalley, Kinetics and mechanism of the reaction of metformin with methylglyoxal, in *G*, 2002, 355–356.
592 K. Niigata, T. Kimura, S. Hayashibe, H. Shikama, T. Takasu, and E. Hirasaki, *N*-Amidinotriazole compounds as Maillard reaction inhibitors for therapeutic use, Japanese Patent, 1994, 06,192,089, 1–8, *via Chem. Abstr.*, 1994, **121**, 222017s.
593 K. Niigata, T. Maruyama, H. Shikama, T. Takasu, M. Umeda, and E. Hirasaki, Preparation of hydroxypyrazoles having inhibiting activity of Maillard reaction, Japanese Patent, 1994, 06,287,179, 1–5, *via Chem. Abstr.*, 1995, **122**, 105877g.
594 K. Niigata, T. Maruyama, S. Hayashibe, H. Shikama, T. Takasu, M. Umeda, and E. Hirasaki, Preparation of 1-amidinopyrazole derivatives for inhibiting Maillard reaction, Japanese Patent, 1994, 06,298,737, 1–17, *via Chem. Abstr.*, 1995, **122**, 133181u.
595 K. Niigata, T. Maruyama, H. Shikama, T. Takasu, M. Umeda, and E. Hirasaki, Preparation of 5-amino-1*H*-pyrazole-1-carboxamidine derivatives as Maillard reaction inhibitors, Japanese Patent, 1994, 06,298,738, 1–12, *via Chem. Abstr.*, 1995, **122**, 160633z.
596 K. Niigata, S. Hayashibe, H. Shikama, T. Takasu, M. Umeda, and E. Hirasaki, Preparation of amidinoindazole derivatives as Maillard reaction inhibitors, Japanese Patent, 1994, 06,287,180, 1–8, *via Chem. Abstr.*, 1995, **122**, 133180t.
597 J. W. Baynes, Pyridoxamine, a versatile inhibitor of advanced glycation and lipoxidation reactions, in *G*, 2002, 31–35.
598 S. R. Thorpe, N. L. Alderson, M. E. Chachich, A. Januszweski, N. N. Youssef, S. M. Jimenez, T. Gardiner, N. Frizzell, P. Canning, A. Lichanska, J. W. Baynes, and A. W. Stitt, Role of dyslipidemia and AGE/ALE formation in the progression of nephropathy and retinopathy in STZ-diabetic rats, in *G*, 2002, 169–173.
599 S. Nakamura, Z. Makita, S. Ishikawa, K. Yasumura, W. Fujii, K. Yanagisawa, T. Kawata, and T. Koike, Progression of nephropathy in spontaneous diabetic rats is prevented by OPB-9195, a novel inhibitor of advanced glycation, *Diabetes*, 1997, **46**, 895–899.
600 R. Wada, Y. Nishizawa, N. Yagihashi, M. Takeuchi, Y. Ishikawa, K. Yasumura, M. Nakano, and S. Yagihashi, Inhibition of the development of experimental diabetic neuropathy by suppression of AGE formation with a new antiglycation agent, in *G*, 2002, 101–105.
601 Y. Ueda, T. Miyata, Y. Izuhara, R. Inagi, K. Tatsumi, C. van Ypersele de Strihou, M. Nangaku, and K. Kurokawa, Mechanism of the inhibitory effect of 2-isopropylidenehydrazono-4-oxothiazolidin-5-ylacetanilide on advanced glycation endproduct and advanced lipoxidation endproduct formation, in *G*, 2002, 453–454.
602 S. Vasan, Zhang, Xin, Zhang, Xini, A. Kapurniotu, J. Bernhagen, S. Teichberg, J. Basgen, D. Wagle, D. Shih, I. Terlecky, R. Bucala, A. Cerami, J. Egan, and P. Ulrich, An agent cleaving glucose-derived protein cross-links *in vitro* and *in vivo*, *Nature*, 1996, **382**, 275–278.

603 R. Bucala, New horizons in AGE research, in *G*, 2002, 113–117.
604 B. H. R. Wolffenbuttel, C. M. Boulanger, F. R. L. Crijns, M. S. P. Huiberts, P. Poitevin, G. N. M. Swennen, S. Vasan, J. J. Egan, P. Ulrich, A. Cerami, and B. I. Levy, Breakers of advanced glycation end products restore large artery properties in experimental diabetes, *Proc. Natl. Acad. Sci. USA,* 1998, **95**, 4630–4634.
605 M. Asif, J. Egan, S. Vasan, G. N. Jyothirmayi, M. R. Masurekar, S. Lopez, C. Williams, R. L. Torres, D. Wagle, P. Ulrich, A. Cerami, M. Brines, and T. J. Regan, An advanced glycation endproduct cross-link breaker can reverse age-related increases in myocardial stiffness, *Proc. Natl. Acad. Sci. USA,* 2000, **97**, 2809–2813.
606 P. V. Vaitkevicius, M. Lane, H. Spurgeon, D. K. Ingram, G. S. Roth, J. J. Egan, S. Vasan, D. R. Wagle, P. Ulrich, M. Brines, J. P. Wuerth, A. Cerami, and E. G. Lakatta, A crosslink breaker has sustained effects on arterial and ventricular properties in older rhesus monkeys, *Proc. Natl. Acad. Sci. USA,* 2001, **98**, 1171–1175.
607 H. Shoda, S. Miyata, B. F. Liu, H. Yamada, T. Ohara, K. Suzuki, M. Oimomi, and M. Kasuga, Inhibitory effects of tenilsetam on the Mailard reaction, *Endocrinology,* 1997, **138**, 1886–1892.
608 H. Y. Kim and K. Kim, Effect of flavonoids on formation of advanced glycation endproducts in vitro, in *G*, 2002, 511.
609 P. R. Odetti, A. Borgoglio, A. de Pascale, R. Rolaandi, and L. Adezati, Prevention of diabetes — increased aging effect on rat collagen linked fluorescence by aminoguanidine and rutin, *Diabetes,* 1990, **39**, 796–801.
610 T. Nagasawa, N. Tabata, Y. Ito, and N. Nishizawa, Suppression of early and advanced glycation by dietary water-soluble rutin derivative in diabetic rats, in *G*, 2002, 403–405.
611 N. Matsuura, C. Sasaki, T. Aradate, M. Ubukata, H. Kojima, M. Ohara, and J. Hasegawa, Plantagoside as Maillard reaction inhibitor — its inhibitory mechanism and application, in *G*, 2002, 411–412.
612 J. Vertommen, M. van den Enden, L. Simoens, and I. de Leeuw, Flavonoid treatment reduces glycation and lipid peroxidation in experimental diabetic rats, *Phytother. Res.,* 1994, **8**, 430–432.
613 S. Battah, N. Ahmed, and P. J. Thornalley, Novel anti-glycation therapeutic agents: Glyoxalase I mimetics, in *G*, 2002, 107–111.
614 T. Horiuchi, T. Kurokawa, and N. Saito, Purification and properties of fructosyl-amino acid oxidase from *Corynebacterium* sp. 2-4-1, *Agric. Biol. Chem.,* 1989, **53**, 103–110.
615 T. Horiuchi and T. Kurokawa, Purification and properties of fructosylamine oxidase from *Aspergillus* sp. 1005, *Agric. Biol. Chem.,* 1991, **55**, 333–338.
616 N. Yoshida, Y. Sakai, A. Isogai, H. Fukuya, M. Yagi, Y. Tani, and N. Kato, Primary structures of fungal fructosyl amino acid oxidases and their application to the measurement of glycated proteins, *Eur. J. Biochem.,* 1996, **242**, 499–505.
617 M. Takahashi, M. Pischetsrieder, and V. M. Monnier, Isolation, purification, and characterization of Amadoriase isoenzymes (fructosyl amine-oxygen oxidoreductase EC 1.5.3) from *Aspergillus* sp, *J. Biol. Chem.,* 1997, **272**, 3437–3443.
618 M. Takahashi, M. Pischetsrieder, and V. M. Monnier, Molecular cloning and expression of Amadoriase isoenzyme (fructosyl amine:oxygen oxidoreductase, EC 1.5.3) from *Aspergillus fumigatus*, *J. Biol. Chem.,* 1997, **272**, 12505–12507.
619 C. Gerhardinger, M. S. Marion, A. Rovner, M. Glomb, and V. M. Monnier, Novel degradation pathway of glycated amino acids into *free* fructosamine by a *Pseudomonas* sp. soil strain extract, *J Biol. Chem.,* 1995, **270**, 218–224.
620 X. Wu, B. A. Palfey, V. V. Mossine, and V. M. Monnier, Kinetic studies, mechanism, and substrate specificity of Amadoriase I from *Aspergillus sp*, *Biochemistry,* 2001, **40**, 12886–12895.

621 Y. Al-Abed, T. Mitsuhashi, H. Li, J. A. Lawson, G. A. FitzGerald, H. Founds, T. Donnelly, A. Cerami, P. Ulrich, and R. Bucala, Inhibition of advanced glycation end product formation by acetaldehyde: Role in the cardioprotective effect of ethanol, *Proc. Natl. Acad. Sci. USA,* 1999, **96**, 2385–2390.

622 T. Yokozawa, H. Y. Kim, and E. J. Cho, Erythritol attenuates the diabetic oxidative stress through modulating glucose metabolism and lipid peroxidation in streptozotocin-induced diabetic rats, *J. Agric. Food Chem.,* 2002, **50**, 5485–5489.

623 K. Eichner, The influence of water content on non-enzymic browning reactions in dehydrated food and model systems and the inhibition of fat oxidation by browning intermediates, in *Water Relations of Foods,* R.B. Duckworth (ed), Academic Press, London, 1975, 417–434.

624 T.P. Labuza and M. Saltmarch, Kinetics of browning and protein quality loss in whey powders during steady state and nonsteady state storage conditions. *J. Food Sci.,* 1982, **47**, 92–96, 113.

625 T.P. Labuza, K. Bohnsack, and M.N. Kim, Kinetics of protein quality change in egg noodles stored under constant and fluctuating temperatures. *Cereal Chem.,* 1982, **59**, 142–147.

626 I.N. Shipanova, M.A. Glomb, and R.H. Nagaraj, Protein modification by methylglyoxal: chemical nature and synthetic mechanism of a major fluorescent adduct. *Arch. Biochem. Biophys.,* 1997, **344**, 29–36.

627 T.W.C. Lo, M.E. Westwood, A.C. McLellan, T. Selwood, and P.J. Thornalley, Binding and modification of proteins by methylglyoxal under physiological conditions: a kinetic and mechanistic study with N^α-acetylarginine, N^α-acetylcysteine, and N^α-acetyllysine, and bovine serum albumin. *J. Biol. Chem.,* 1994, **269**, 32299–32305.

628 H. Kato and F. Hayase, An approach to estimate the chemical structure of melanoidins, in *G,* 2002, 3–7.

629 A. Dawney, A.P. Wieslander, and D.J. Millar, Role of glucose degradation products in the generation of characteristic AGE fluorescence in peritoneal dialysis fluid? in *F,* 1998, 333–338.

630 D. Ruggiero-Lopez, M. Lecomte, N. Rellier, M. Lagarde, and N. Wiernsperger, Reaction of metformin with reducing sugars and dicarbonyl compounds, in *F,* 1998, 441.

631 R.H. Nagaraj and V.M. Monnier, Isolation and characterization of a blue fluorophore from human eye lens crystallins: in vitro formation from Maillard reaction with ascorbate and ribose, *Biochim. Biophys. Acta,* 1992, **1116**, 34–42.

632 K. Nakamura, T. Hasegawa, Y. Fukunaga, and K. Ienaga, Crosslines A and B as candidates for the fluorophores in age- and diabetes-related cross-linked proteins, and their diacetates produced by Maillard reaction of α-N-acetyl-L-lysine with D-glucose, *J. Chem. Soc., Chem. Commun.,* 1992, 992–994.

633 D.G. Dyer, J.A. Dunn, S.R. Thorpe, K.E. Bailie, T.J. Lyons, D.R. McCance, and J.W. Baynes, Accumulation of Maillard reaction products in skin collagen in diabetes and aging, *J. Clin. Invest.,* 1993, **91**, 2463–2469.

634 T.J. Lyons, G. Silvestri, J.A. Dunn, D.G. Dyer, and J.W. Baynes, Role of glycation in modification of lens crystallins in diabetic and non-diabetic senile cataracts, *Diabetes,* 1991, **40**, 1010–1015.

635 M.C. Wells-Knecht, T.J. Lyons, D.R. McCance, S.R. Thorpe, and J.W. Baynes, Age-dependent accumulation of *ortho*-tyrosine and methionine sulfoxide in human skin collagen is not increased in diabetes. Evidence against a generalized increase in oxidative stress in diabetes, *J. Clin. Invest.,* 1997, **100**, 839–846.

636 S. Rodrígues, M.E. Centurión, and E. Agulló, Chitosan-yeast interaction in cooked food: influence of the Maillard reaction, *J. Food Sci.,* 2002, **67**, 2576–2578.

637 J.W. Wong and T. Shibamoto, Genotoxicity of Maillard reaction products, in *The Maillard Reaction: Consequences for the Chemical and Life Sciences*, R. Ikan (ed), Wiley, Chichester, 1996, 129–159.
638 T. Hofmann, Application of site specific [^{13}C] enrichment and ^{13}C-NMR spectroscopy for the elucidation of the formation pathway leading to a red colored 1*H*-pyrrol-3(2*H*)-one during Maillard reaction of furan-2-carboxaldehyde and L-alanine. *J. Agric. Food Chem.*, 1998, **46**, 941–945.
639 C. Billaud and J. Adrian, Louis–Camille Maillard, 1878–1936, *Food Rev. Intern.*, 2003, **19**, 345–374.
640 G.P. Rizzi, Free radicals in the Maillard reaction, *Food Rev. Intern.*, 2003, **19**, 375–395.
641 D. Taeymans, J. Wood, P. Ashby, I. Blank, A. Studer, R.H. Stadler, P. Gondé, P. van Eijck, S. Lalljie, H. Lingnert, M. Lindblom, R. Matissek, D. Müller, D. Tallmadge, J. O'Brien, S. Thompson, D. Silvani, and T. Whitmore, A review of acrylamide: an industry perspective on research, analysis, formation, and control, *Crit. Rev. Food Sci. Nutr.*, 2004, **44**, 323–347.

Subject Index

Numbers written bold signify structures

Abhexone, 64, 65
ABTS trapping, 128, 137
Acetaldehyde, 83, 170
Acetic acid, 83
N-Acetonyl-4-aminobutanal, 69
N-Acetonyl-2-pyrroline, 70
N-Acetylcysteine, 158
5-Acetyl-2,3-dihydro-1H-pyrrolizine, 39, 66
5-Acetyl-2,3-dihydro-1,4-thiazine, 82, 83
2-Acetylfuran, 64, 66
2-Acetyl-3-D-glucopyranosylfuran, 10
5-Acetyl-4-hydroxy-1,3-dimethylpyrazole, 135
Acetyl-lysylpyrropyridine, 35
2-Acetyl-1-methylpyrrole, 26
2-Acetylpyrazine, 75
2-Acetylpyrido[3,4d]imidazole, 75, 76
2-Acetylpyrrole, 26, 65
2-Acetyl-1-pyrroline (ACPY), 40, 67–69, 84, 85
2-Acetyl-2-pyrroline, 68
2-Acetyltetrahydropyridine (ACTPY), 39, 67, 69, 70, 84, 85
2-Acetyl-4(5)-tetrahydroxybutylimidazole, 99
2-Acetylthiazole, 83
2-Acetylthiazolidine, 79, 80
2-Acetyl-2-thiazoline, 79, 80, 83
2-Acetylthiophen, 79
Acrolein, 78, 92
Acrylamide, 91

AFGP **66**, 106
AGE, 35, 117, 118
    formation *in vivo*, 112
AGE-LDL, 162
AGE-peptides, 162
Aging, 118
Alapyridaine, 88
Aldehyde-amine condensation, 20
Aldol condensation, 19, 36
ALE, 118
ALI **98**, 108, 109
Allergenicity, 98
Allysine, 19
ALT-711, 166
Alzheimer's disease, 98, 108, 167
Amadoriase, 168–170
Amadori compounds, 159, 160
    absorption/elimination, 103
    acetaldehyde derivative **107**, 168
    formation *in vivo*, 105
    rearrangement, 5
    reversibility, 13
Aminoacetone, 72, 73
Aminocarbolines, 93
Aminoguanidine (AG), 45, 114, 115, 118, 122, 123, 162–164, 166
4-Amino-5-hexenoic acid, 145
3-Amino-1,2,4-triazines **104–105**, 163
Anserine, 18
Antibodies 2A2, 120
    6D12 (CML), 115, 120, 122
    1F6 (fluorolink), 120
    KNH-30 (CEL), 120

Subject Index

mAb5A3 (THP), 111
mAb3C, mAb6B (MGO-Arg adducts), 111
Antioxidants, 96
Antioxidative activity, 128
  differences between amino acids, 130
  differences between sugars, 129
  mechanism, 137
  molecular mass, 134
  ratio of sugar to amino acid, 131
Antiradical efficiency, (AE), 138
AOXP, 128
Arg-hydroxy-triosidine **101**, 35, 108, 109
Arginine, 56
Argpyrimidine **75**, 35, 49, 106, 109, 111
Arrhenius plot, colour, 38
  volatiles, 39, 41
Ascorbic acid, 101, 146
  supplementation in diabetes, 149
Asparagine, 92
Aspartame, 61, 148
L-Aspartic acid, 119

Bakery products, 116, 140
Balenine, 19
Beer, 60, 152
Benzothiazole, 81
Bibliography, 172
Birnessite, 143
Biscuits, 95, 126, 132
Bis(2-methyl-3-furyl) disulfide, 83
Bitter compounds **33-44**, 86, 88, 148
Blocking agents, 55
Blood, 132
Blue-M1 **30**, 21, 22, 35, 54, 58
Blue-M2, 58
Bread, 60, 77, 81, 95, 134, 135, 140
Brewer's grains, 132
Butanedione, 83
Butter, 126

Caco-2 cells, 90
Cancer treatment, 123

Capillary electrophoresis, 50, 61
Caramelisation, 60, 150
Carbonate catalysis, 12
Carboxyethyllysine (CEL) **63**, 48, 106, 117–120
1-Carboxy-3-hydroxypropyllysine (CHPL), 48
Carboxymethylarginine **70**, 106, 109, 110
Carboxymethylethanolamine, 45
Carboxymethyllysine (CML) **62**, 45, 48, 49, 106, 114–120, 122, 166
N-(Carboxymethyl) phosphatidylserine, 45
Carcinogenicity, 90
Carnosine, 19, 44
α_{s1}-Casein, 50
β-Casein, 34, 50, 160
Cerium sulfate, 128
Ceroid, 47
Chitosan, 46
Cisplatin, 123
Cocoa, 14, 138
Coffee, 23, 30, 35, 71, 76, 77, 116, 126, 138, 141, 142
Cola, 61
Collagen, 118, 119
Colour, 52
Colour activity value (CAV), 52
Colour dilution factor (CD), 28, 52
Colour intensity, 60
Comparative taste dilution analysis (cTDA), 88
Continuous-flow reactor, 39
Cooling compounds, 89
Copper, 102, 103, 138, 144
Coronary heart disease, 99
Creatine, 93, 95, 112
Creatinine, 93–96
Crocin bleaching, 128, 133, 137
Crosslines **87**, 35, 107, 109
CROSSPY **84**, 107, 109
Crystallin, 109, 118, 121
Cyclopentane formation, 55, 82

Cyclopent(b)azepinones, 87
2-Cyclopentene-1,4-dione, 26
Cysteine, 110, 164

Dehydration, 153
Dehydroascorbic acid (DHAA), 60, 146, 147
6-Deoxy-6-fluoroascorbic acid (FAA), 148
1-Deoxyglucosone (1-DG), 163
3-Deoxyglucosone (3-DG), 6, 163
3-Deoxypentosone (3-DP), 50
Diacetylformoin, 56, 64, 65, 69, 70
N,N'-Dialkylpyrazine cation radical, 27, 29, 30
α-Dicarbonyl compounds, 15, 36
1,4-Dideoxy-5,6-hexosone, 36
2,3-Diethyl-5-methylpyrazine, 84
Diglucosyl-β-pyrone, 50
Dihydrodiacetylformoin, 68
2,3-Dihydro-5,6-dihydroxypyran-4-one, 130
2,3-Diketogulonic acid, 146
Dimethyldiguanide (metformin), 164, 165
2,4-Dimethyl-5-ethylthiazole, 79
4,5-Dimethyl-2-formylpyrrole, 26
Dimethylmaleimides, 110, 111
2,3-Dimethylpyrazine, 39, 73
2,5-Dimethylpyrazine, 72, 73
2,6-Dimethylpyrazine, 72, 73
2,6-Dimethylpyridine, 26
Dimethyl sulfide, 83
Dimethyl trisulfide, 83
Dipeptide, 33, 43
Dipyridoimidazole, 93
Disaccharides, 15
DMPD, 128, 138
DOGDIC **95**, 108, 109, 112, 113
DOLD **81**, 107, 109
DOPDIC **96**, 108, 109, 112, 113
DPPH trapping, 128, 134, 136–138

EAGLE, 118
Eggs, 10, 140, 153, 158
ELISA, 109

Embden-Meyerhof-Parnas glycolytic pathway, 15
Emulsifying power, 141
Enalapril, 166
Enhancers **45-47**, 88, 89
1,2-Enolisation, 7, 53
2,3-Enolisation, 7, 53
Enteral formula, 11
Enzymes, 168
Erythritol, 170
ESR, 27, 28, 30, 58–60, 130, 137, 143, 144
Essential amino acids, 101
Ethyldimethylpyrazine, 72, 73, 84
Ethylmethylpyrazine, 72

FAST (Fluorescence of AMP and Soluble Tryptophan) index, 11
FFI **82**, 107, 109
Field peas, 101
Fish, 93, 95, 96, 127, 154
Flavonoids, 167
Flavour, 62
Fluorescence, 35, 134, 167
Fluorophore LM-1 (see vesperlysine A), 35, 107, 109, 121
Foamability, foam stability, 141
Food dyes, reduction **102–103**, 139
Formaldehyde, 67, 117
Formyllysine (FoL), 48
2-Formyl-2-methylthiazolidine, 79
2-Formyl-5-methylthiophen, 79
2-Formylpyrrole, 26
2-Formylthiophen, 79
FRAP, 128, 132
Free radicals (see also ESR), 23, 26, 98, 99
Fructosamine-3-kinase, 114, 170
Fruit, 153, 158
Fulvic acid, 143
Fumonisin **3**, 46
Furaneol™ (see 4-hydroxy-2,5-dimethylfuran-3-one)
Furanones, 8
Furfural, 13, 147

Subject Index

2-Furfurylidene-4-hydroxy-5-methyl-3(2*H*)-furanone, 33
2-Furfurylidene-5-methyl-3(2*H*)-pyrrolinone, 33
2-Furoic acid, 147
Furosine, 8, 9, 10, 116, 127
2-Furylmethanethiol, 76–78, 83

GALA **64**, 106
Garlic, 6
GLARG **77**, 106, 109
Glass transition temperature, 25, 42
Glucosepan **97**, 108–110, 112, 113
Glucosylisomaltol (GIM), 11, 50, 51
Glucosyl-β-pyrone (GP), 50, 51
Glutaraldehyde, 117, 140
Glutathione, 117
Glyceraldehyde, 33, 117
Glycolaldehyde, 29, 121
Glycoxidation, 116
Glyoxal, 27, 49, 109
Glyoxalase, 168
GODIC **93**, 108, 109
GOLA **78**, 107, 109
GOLD **79**, 107, 109, 117, 118
Grape juice, 137, 142, 153
Gravy, 116

Haemoglobin, 105, 114, 120, 121, 132, 162, 170
Haemodialysis, 122
N^π-HAL, 110
Heterocyclic aromatic amines (HAAs) **48-61** , 93–95, 97
Hexanal formation, 138
  oxidation, 128, 139
High pressure, 33, 151
Homofuraneol, 19
Honey, 139
Hue index, 61
Human serum albumin (HSA), 49
Humic acids, 143
Humin, 143
Hydrochlorothiazide, 145
Hydrogen sulfide, 76, 83
α-Hydroxycarbonyl group, 15

4-Hydroxy-2,5-dimethylfuran-3-one (Furaneol™, HDMF), 19, 63–65, 67, 83, 85, 130
2-(1-Hydroxyethyl)-2-thiazoline, 80
5-Hydroxymaltol, 147
2-Hydroxymethylfuran, 64
Hydroxymethylfurfural (HMF), 6, 8, 57, 127, 142, 147, 148
3-Hydroxy-2-methylpyridine, 69
Hydroxymethylpyrroles, 21
3-Hydroxy-2-methylfuran-4-one (see norfuraneol)
4-Hydroxynonenal (HNE), 118
2-(1-Hydroxy-2-oxopropyl)pyrrolidine, 71
3-Hydroxy-2-pyrone (HP), 147

Imidazoles, 75
Imidazolone A, 48, 116, 120
Imidazolonylornithine **74**, 35, 106, 116
Indolizinium-6-olates **43-44** , 88
Indophenol, 128
Infant food, 10, 11
  formula, 9, 134
Infants, 116
Isoamyl acetate, 148

Jam, 10, 157

Kahweofuran **32**, 77
Kaolin, 144
Keratoconus, 122
Kinetics, Maillard reaction, 37

β-Lactoglobulin, 12, 50
Lactose, 34
Lactulose, 127
Lactulosyllysine, 10
Lard, 127
Lemon juice, 147, 153
Lens, 109, 121
Limit-peptide pigment (LPP), 58
Lipids, 46
Lipid peroxidation, 118
Lipofuscin, 47

Liqueurs, 60
Literature, 4
Lymphocytopenia, 99
Lysine, nutritionally blocked, 6, 9, 160
Lysozyme, 48
Lysylhydroxytriosidine **83**, 35, 107, 109
Lysylpyrrolaldehyde, 11
Lysylpyrropyridine **88**, 35, 107, 109, 111, 116

Macropeptide, 12
Maillard, Louis-Camille, 1
Maillard reaction scheme, 2, 3
  symptoms, 4
  inhibition, 152, 161
Maize, 69
Malondialdehyde (MDA), 118, 122
Malt, 67, 132
Maltol 14, 67, 83
Maltose, 50
Maltotetrose, 50
Maltotriose, 24, 50
Maltoxazine, 39, 66, 67, 84
Margarine, 126
Marinade, 96
Meal replacer, 116
Meat, 76, 77, 93, 95–97, 127, 138, 153, 154, 158
Medium, 56
Melanin, 97
Melanoidin, 8, 20–22, 25, 97, 135
  digestion, 103
  Standard, 24, 129, 136, 148
Melanoprotein, 59
2-Mercaptoethanal (methional), 78, 83
N-(2-Mercaptoethyl)-1,3-thiazolidine, 80, 81, 84
2-Mercapto-3-pentanone, 79
3-Mercapto-2-pentanone, 78, 79
Metals, interaction, 102, 125
Metformin (see dimethyldiguanide), 164, 165
Methanethiol, 83
Methionine sulfoxide, 117

3-Methylbutanal, 83
3-Methylcyclopen-2-en-2-olone (cyclotene), 39, 67, 82, 83
N-Methyl-2-formylpyrrole, 23
2-Methyl-3-furanthiol, 77-79, 83
5-Methylfurfural, 64
Methylglyoxal (MGO, see 2-oxopropanal)
N-Methyl-2-hydroxymethylpyrrole polymer, 35
4-Methylimidazole, 99
Methyl linoleate, 128, 136
2-Methylpyrazine, 33, 71, 73
N-Methylpyrrole, 23
5-Methylpyrrole-2-aldehyde, 66
2-Methyltetrahydro-3-furanone, 64
Microwave heating, 81, 96
Milk, 9, 116, 126, 127, 134, 140, 154
MODIC **94**, 108, 109
MOLD **80**, 107, 109, 117–119
Monier-Williams procedure, 156
Montmorillonite, 144
MRX, 35, 110
Mushrooms, 152
Mutagenicity, 91, 93

NADPH, 117
^{13}C-NMR, 21
Norfuraneol (3-hydroxy-2-methylfuran-4-one), 19, 31, 32, 33, 78, 83
Nucleic acid, glycation, 121

Odour quality, 84
Oils, 127
OPB-9195, 122, 166
ORAC, 128, 129
Orange juice, 11, 137, 147, 153
Ornithine, 67
2-Ornithyldihydroimidazolium-4-imide **99**, 108, 109
Oxalyllysine (OL), 48
Oxazoles, 75
Oxidative stress, 114, 115
  reactive oxygen species, 120
2-Oxopropanal (MGO, methylglyoxal, pyruvaldehyde), 28, 49, 109,

Oxygen, 2
  uptake, 128

Pancake, 95
Pasta, 10
Pectin, 141
Peeling-off pathway, 50, 51
Pentodilysine **86**, 35, 107, 109
Pentosidine **90**, 34, 35, 108–114, 116–121, 123, 140, 164, 166
C-Pentosidine **91**, 35, 108, 109, 112
Pentosinane **92**, 108, 109, 112, 113
Peptides, 43, 131
Peritoneal dialysis, 123
Peroxide value, 128
pH, 31, 124, 136, 153
Phase-I enzyme, 90
Phase II enzyme, 90
Phenacylthiazolium compounds, 166, 167
Phenylethanal, 83
Phosphate catalysis, 12, 13, 32, 43, 44
Phospholipids, 44, 47
Pigs, 101
Polarography, 128
Polyglycine, 26
Polyphenols, 121, 167
Potassium ferricyanide method, 130, 136
Potato, 95, 153, 154, 157, 158
Pronyl-N^α-acetyllysine methyl ester, 134
Pronyllysine **67**, 106
Pro-oxidant activity, 133
2-Propionylpyrido[3,4d]imidazole, 75
2-Propionyl-2-thiazoline, 84
Protective effects, 99
Protein glycation, 47
Protein hydrolysates, 131
Pyranopyrazine **89**, 107, 109
Pyrazines, 71, 142
Pyrazin-2-ones **2**, 43
Pyridine, 26, 136
Pyridines, 69
Pyridosine, 9
Pyridoxal, 164

Pyridoxamine, 118, 165, 166
Pyrolysis-GC-MS, 61, 64
Pyrraline **65**, 35, 50, 106, 111, 112, 116, 120, 123, 167
Pyrroles, 65
1-Pyrroline, 70
1-[H]-Pyrrolo-[2,1c]-1,4-thiazine, 81
Pyruvaldehyde (see 2-oxopropanal)

Quartz, 144
Quinones, 19

Radical scavenging, 128
Rancimat, 128, 130, 138
Redox potential, 125, 128
Reductone, 14
Refrigeration, 152
Retroaldolisation (dealdolisation), 15, 36
Ribonuclease, 16, 49
Rice, 68, 126
Roux, 140

S11 **72**, S12 **71**, S17 **73**, 106, 110
Serine, 114
Shellfish, 140
Skim milk powder (SMP), 12
Smell, 62
Soil, 143
Sotolone, 64, 65
Soybean trypsin inhibitor, 100
Soy sauce, 91, 116, 126, 132
Spaghetti, 12
Starch, hydrolysates, 132, 154
Strecker degradation, 18
Structures
  4–18, 53
  19–31, 54
  33–42, 86
  48–61, 94
  62–77, 106
  78–89, 107
  90–101, 108
Sucrose, heating, 61
Sugar-amine condensation, 5
Sugar fragmentation, 15–17, 55

Sulfite/sulfur dioxide, 55, 152, 154
  analysis, 155
Sun-tan, artificial, 144

Taste, 62
Taurine, 18
TBA value, 128
Tea, 97, 138
TEAC, 128, 132, 138
Tenilsetam **106**, 167, 168
Tetraethylpyrazine, 72
5,6,7,8-Tetrahydroindolizin-8-one, 66
Tetramethylpyrazine, 72
Textiles, 144
Thiamine, 79
Thiazoles, 47, 79
Thiazolium salt XTT, 127, 128
3-(2-Thienyl)-2-piperazinone (tenilsetam), 167, 168
Thiophens, 47
THP **76**, 49, 106, 111
Threshold, 62, 82
Tomato, 10, 11, 126, 160
Toxic products, 90
TRAP, 128

Trapping agents, 162
Triglucosyl-β-pyrone, 50
Trihydroxytriosidine **68**, 106
Trimethylpyrazine, 72, 73
Triosidinecarbaldehyde **69**, 106
Tripeptide, 33
o-Tyrosine, 117

VCEAC, 128
Vegetables, 153, 158, 159
Vesperlysine A **85** (Fluorophore LM-1), 35, 107, 109, 121
Vinegar, 60
Volatile compounds, 62
  classification, 62
  retention and release, 148
Water activity (a_w), 124, 136, 158
Wheat, 75, 126, 135
Whisky, 60
Williams-Landel-Ferry kinetics, 43
Wool, 144
Wound dressing, sphagnum, 144

Yeast, 67
Yellowing, 144, 145